Lecture Notes in Mathematics

Edited by A. Dold, F. Takens and B. Teissier

Editorial Policy
for the publication of monographs

1. Lecture Notes aim to report new developments in all areas of mathematics – quickly, informally and at a high level. Monograph manuscripts should be reasonably self-contained and rounded off. Thus they may, and often will, present not only results of the author but also related work by other people. They may be based on specialized lecture courses. Furthermore, the manuscripts should provide sufficient motivation, examples and applications. This clearly distinguishes Lecture Notes from journal articles or technical reports which normally are very concise. Articles intended for a journal but too long to be accepted by most journals, usually do not have this "lecture notes" character. For similar reasons it is unusual for doctoral theses to be accepted for the Lecture Notes series.

2. Manuscripts should be submitted (preferably in duplicate) either to one of the series editors or to Springer-Verlag, Heidelberg. In general, manuscripts will be sent out to 2 external referees for evaluation. If a decision cannot yet be reached on the basis of the first 2 reports, further referees may be contacted: the author will be informed of this. A final decision to publish can be made only on the basis of the complete manuscript, however a refereeing process leading to a preliminary decision can be based on a pre-final or incomplete manuscript. The strict minimum amount of material that will be considered should include a detailed outline describing the planned contents of each chapter, a bibliography and several sample chapters.
Authors should be aware that incomplete or insufficiently close to final manuscripts almost always result in longer refereeing times and nevertheless unclear referees' recommendations, making further refereeing of a final draft necessary.
Authors should also be aware that parallel submission of their manuscript to another publisher while under consideration for LNM will in general lead to immediate rejection.

3. Manuscripts should in general be submitted in English.
Final manuscripts should contain at least 100 pages of mathematical text and should include
– a table of contents;
– an informative introduction, with adequate motivation and perhaps some historical remarks: it should be accessible to a reader not intimately familiar with the topic treated;
– a subject index: as a rule this is genuinely helpful for the reader.

Continued on back inside cover

Lecture Notes in Mathematics

1749

Editors:
A. Dold, Heidelberg
F. Takens, Groningen
B. Teissier, Paris

Martin Fuchs   Gregory Seregin

Variational Methods for Problems from Plasticity Theory and for Generalized Newtonian Fluids

 Springer

Authors

Martin Fuchs
Universität des Saarlandes
Fachrichtung 6.1 Mathematik
Postfach 151150
66041 Saarbrücken, Germany

e-mail: fuchs@math.uni-sb.de

Gregory Seregin
V.A. Steklov Mathematical Institute
St. Petersburg Branch
Fontanka 27
191011 St. Petersburg, Russia

e-mail: seregin@pdmi.ras.ru

Cataloging-in-Publication Data applied for

Die Deutsche Bibliothek - CIP-Einheitsaufnahme

Fuchs, Martin:
Variational methods for problems from plasticity theory and for
generalized Newtonian fluids / Martin Fuchs ; Gregory Seregin. -
Berlin ; Heidelberg ; New York ; Barcelona ; Hong Kong ; London ;
Milan ; Paris ; Singapore ; Tokyo : Springer, 2000
 (Lecture notes in mathematics ; 1749)
 ISBN 3-540-41397-9

Mathematics Subject Classification (2000): 74, 74G40, 74G65, 76A05, 76M30,
49N15, 49N60, 35Q

ISSN 0075-8434
ISBN 3-540-41397-9 Springer-Verlag Berlin Heidelberg New York

Springer-Verlag Berlin Heidelberg New York
a member of BertelsmannSpringer Science+Business Media GmbH

© Springer-Verlag Berlin Heidelberg 2000
Printed in Germany

Typesetting: Camera-ready T_EX output by the author
SPIN: 10734279   41/3142-543210 - Printed on acid-free paper

Contents

Introduction   1

1   Weak solutions to boundary value problems in the deformation theory of perfect elastoplasticity   5

1.0   Preliminaries   5

1.1   The classical boundary value problem for the equilibrium state of a perfect elastoplastic body and its primary functional formulation   6

1.2   Relaxation of convex variational problems in non reflexive spaces. General construction   15

1.3   Weak solutions to variational problems of perfect elastoplasticity   27

2   Differentiability properties of weak solutions to boundary value problems in the deformation theory of plasticity   40

2.0   Preliminaries   40

2.1   Formulation of the main results   42

2.2   Approximation and proof of Lemma 2.1.1   52

2.3   Proof of Theorem 2.1.1 and a local estimate of Caccioppoli-type for the stress tensor   57

2.4   Estimates for solutions of certain systems of PDE's with constant coefficients   71

2.5   The main lemma and its iteration   76

2.6   Proof of Theorem 2.1.2   89

2.7   Open Problems   98

2.8   Remarks on the regularity of minimizers of variational functionals from the deformation theory of plasticity with power hardening .   100

Appendix A   **107**

    A.1   Density of smooth functions in spaces of tensor–valued functions   107

    A.2   Density of smooth functions in spaces of vector–valued functions   111

    A.3   Some properties of the space $BD(\Omega; \mathbb{R}^n)$ 116

    A.4   Jensen's inequality 126

3   Quasi-static fluids of generalized Newtonian type   **131**

    3.0   Preliminaries 131

    3.1   Partial C^1 regularity in the variational setting 143

    3.2   Local boundedness of the strain velocity 167

    3.3   The two–dimensional case 180

    3.4   The Bingham variational inequality in dimensions two and three   193

    3.5   Some open problems and comments concerning extensions 204

4   Fluids of Prandtl–Eyring type and plastic materials with logarithmic hardening law   **207**

    4.0   Preliminaries 207

    4.1   Some function spaces related to the Prandtl–Eyring fluid model 211

    4.2   Existence of higher order weak derivatives and a Caccioppoli–type inequality 216

    4.3   Blow–up: the proof of Theorem 4.1.1 for $n = 3$ 228

    4.4   The two–dimensional case 235

    4.5   Partial regularity for plastic materials with logarithmic hardening 237

    4.6   A general class of constitutive relations 248

Appendix B   **251**

    B.1   Density results 251

Notation and tools from functional analysis   **254**

Bibliography   **260**

Index   **268**

Introduction

In this monograph we develop a rigorous mathematical analysis of variational problems describing the equilibrium configuration for certain classes of solids and also for the stationary flow of some incompressible generalized Newtonian fluids. However, even if we restrict ourselves to the time–independent setting, it is not possible to include all aspects which might be of interest. So we made a selection which has been influenced by our scientific activities over the last years. In particular, we concentrate on variational problems from the deformation theory of plasticity and on fluid models whose stress–strain relation can be formulated in terms of a dissipative potential. The mathematical form of both problems is very close and may be reduced to the study of variational integrals with convex integrands depending only on the symmetric part of the gradient of the unknown vector–valued functions. In the applications these functions represent either displacement fields of a body or velocity fields of incompressible flows.

Such a reduction to variational problems sometimes might be not quite correct from the physical point of view. For instance, the flow theory of plasticity seems to be more adequate to reality than, say, the deformation theory. The point is that the current deformed state of a plastic body essentially depends on the history of loading, while in the deformation theory of plasticity, like in physically nonlinear elasticity, the current deformed state is determined only by the current values of the loads and does not depend on the history at all. However, for special kinds of loadings, excluding, of course, cycles and similar processes, the difference between these two theories is not so great. On the other hand, if the flow of an elasto–plastic body is quasi–static, i. e. inertial terms in the equations of motion can be neglected, then the usual time discretisation gives a sequence of variational problems having the same mathematical structure as in the case of the deformation theory of plasticity. For fluids, sometimes we do not take into account the so–called convective term. This again allows us to apply the powerful methods of the Calculus of Variations but one should remember that such mathematical models are reasonable only for slow flows. We do not pretend to investigate the mathematical properties for all reasonable models in plasticity theory and the theory of non–Newtonian fluids, and we restrict ourselves to the simplest ones at least in the sense of the mathematical formulation. But the reader will see that even these simple models are quite difficult from the mathematical point of view. We recommend to the reader to do what really we did ourselves: before starting with mathematical considerations to learn the basic physical foundations of the corresponding phenomena, by looking, for example at the monographs of [K], [Kl], [AM]. In fact, there are many books devoted to the correct mathematical formulation of problems from the mechanics of continua and fluids and which influenced us in various ways. Without being complete let us mention the monographs of G. Duvaut and J.L. Lions [DL], O.A. Ladyzhenskaya [L1], P.P. Mosolov and V.P. Mjasnikov [MM2], J. Nečas and I. Hlaváček [NH], E. Zeidler [Z], G. Astarita and G. Marrucci [AM], R. Bird, R. Armstrong and O. Hassager [BAH], and I. Ionescu and M. Sofonea [IS]. After presenting some physical background for the problems under consideration these authors introduce appropriate concepts of weak solutions and prove existence theorems in spaces of generalized functions. From the point of view of applications the most interesting problem concerns the smoothness properties of these generalized solutions which will be the main issue of our book. But let us first briefly comment the approach towards existence: by convexity, L^p–Korn's inequality and additional quite natural conditions concerning the data, the so–called direct method from the calculus of variations gives existence in some Sobolev space provided the integrand is of superlinear growth. We therefore obtain existence for plasticity with power hardening studied in chapter 2, section 8, and also for the generalized Newtonian fluids investigated in chapter 3. However, for integrands with linear growth as they occur in perfect plasticity, the problem of finding suitable classes of admissible deformations is not so trivial. The right space consists of all functions having bounded deformation, a definition and further discussion can be found in the works of G. Anzellotti and M. Giaquinta [AG1], H. Mathies, G. Strang and E. Christiansen [MSC], P. Suquet [Su] and R. Temam and G. Strang [ST2]. Existence of weak solutions in this space was proved by G. Anzellotti and M. Giaquinta [AG2], R. Kohn and R. Temam [KT], R. Hardt and D. Kinderlehrer [HK], R. Temam [T] and G. Seregin [Se1]. In the present book we use the approach of the second author outlined in [Se1] which is close to the one of R. Temam and G. Strang [ST1]. Roughly speaking, we concentrate on a dual variational problem for the stress tensor which in our opinion gives more chances for investigating differentiability properties and moreover it has a clear physical meaning. Chapter 1 contains a general scheme for the relaxation of convex variational problems being coercive on some non–reflexive spaces like L^1 or W_1^1 and as an application we will give a detailed description of our approach towards perfect plasticity. Besides plasticity with power hardening and perfect plasticity there is another model which in a certain sense lies in between these two cases and which is known as plasticity with logarithmic hardening. Here the integrand is not of power growth for some $p \geq 1$ which makes it necessary to introduce "new" function spaces in which solutions should be located. Integrands of logarithmic type also occur in the study of Prandtl–Eyring fluids. For this reason we decided to present a unified treatment of plasticity with logarithmic hardening and of Prandtl–Eyring fluids in chapter 4.

As already mentioned our main concern is the analysis of regularity of weak solutions. In chapter 2 we adress this question for perfect elastoplasticity including
the Hencky–Il'yushin model as a special case. According to our knowledge the
problem of regularity for integrands with linear growth and for vector–valued
functions is discussed only in the paper [AG3] of G. Anzellotti and M. Giaquinta
where they prove partial C^1–regularity, i.e. smoothness up to a relatively closed
set with vanishing Lebesgue measure, assuming in addition that the integrand
is strictly convex which is not the case for Hencky–Il'yushin plasticity. Our approach towards regularity is completely different and based on the idea that the
stress tensor is – from the physical point of view – the most important quantity
since for example it determines the elastic and plastic zones within the body.
We therefore study the dual variational problem for the stress tensor which has
a unique solution but the form of this problem differs from the setting in standard variational calculus: the functional does not involve any derivatives of the
unknown functions, the yield condition acts as a pointwise constraint and the
equilibrium equations for the stresses have to be incorporated in the admissible
class. Despite of these difficulties we obtain additional regularity for the stress
tensor and with the help of the duality relations which can be regarded as a
weak form of the constitutive equations we also establish some regularity for the
displacement field.

In chapter 3 we study the local regularity of solutions to variational problems
describing the stationary flow of generalized Newtonian fluids including also viscoplastic fluids of Bingham type. The Bingham case and its extension inherits
the main difficulties since the dissipative potential is not smooth at the origin.
Nevertheless we get partial regularity of the strain velocity on the complement
of the rigid zone. Other results for the Bingham model concern the boundedness
of the strain velocity and, for two–dimensional problems, just continuity of this
tensor. For other classes of generalized Newtonian fluids whose dissipative potential is obtained via a smooth lower order perturbation of a quadratic leading
part we prove (partial) regularity in the usual sense. During our investigations
we first assume that the velocity is also small in order to get a minimisation
problem. If we drop this assumption, then the variational problem has to be
replaced either by a variational inequality or by some elliptic system which differs from the time–independent Navier–Stokes system since its leading part is
nonlinear. By introducing some artificial new volume load we can reduce this
situation to the one studied before by the way obtaining the same regularity
results at least in dimensions two and three. It should be noted that all our
investigations are limited to incompressible flows which means that we have to
consider fields with vanishing divergence.

Chapter 4 presents the regularity theory for the logarithmic case working with
a natural notion of weak solutions. For plasticity with logarithmic hardening we
directly investigate the minimizing deformation field proving partial regularity
in three dimensions and global smoothness in the two–dimensional case. From

this and the constituent equations we deduce corresponding results for the stress tensor. As to the Prandtl–Eyring fluid model exactly the same smoothness theorems hold provided we assume smallness of the velocity field. It is an open problem if this restriction can be removed.

In the end, let us point out that chapters 2, 3 and 4, although devoted to different subjects, are strongly connected by the methods we use. For example, we apply suitable regularizations to get variational problems which are easier to handle, we prove weak differentiability of stresses or strain velocities, and we use blow–up arguments for establishing (partial) regularity. For keeping the exposition selfcontained we decided to include some background material concerning function spaces. The reader will find the necessary information at the end of chapter 2 and chapter 4. Throughout this book we will make use of various tools from functional analysis, and we expect the reader to be familiar with the notion of Lebesgue and Sobolev spaces as well as with the basic ideas of variational calculus. We tried to give references for this background material in each chapter, the nonstandard auxiliary results are presented in Appendix A and B. For the reader's convenience we added a seperate section concerning frequently used notation: here the non–specialist will find a list of definitions and also a collection of results from functional analysis which are fundamental for our purposes.

Last but not least we want to thank Mrs. C. Peters for doing an excellent typing job. The second author also acknowledges support from INTAS, grant No. 96–835.

Chapter 1

Weak solutions to boundary value problems in the deformation theory of perfect elastoplasticity

1.0 Preliminaries

Chapter 1 is organized as follows. In the first section we discuss the classical boundary value problem for the equilibrium configuration of a perfect elastoplastic body within the framework of the so–called deformation theory of plasticity and give some preliminary version of a functional formulation which takes the form of a minimax problem. At this stage we impose the condition that the strain tensor is a summable function. We then show that the minimax problem is equivalent to two variational problems for the displacement field and the stress tensor, respectively. But the minimisation problem for the displacement field is in general not solvable if we require summability of the strain tensor which is a consequence of considering some concrete examples. In section two we therefore describe an abstract procedure providing a suitable relaxation of variational problems being coercive on non–reflexive classes like L^1 or W_1^1. The main section of chapter 1 is section three in which we apply the abstract scheme to concrete problems of perfect elastoplasticity. As a result we obtain a relaxation of the above mentioned minimax problem with corresponding weak solutions, i.e.: the solutions of the relaxed minimax problem are, in some natural sense, generalized solutions of the original minimax problem. Moreover, this approach immediately leads to a relaxed version of the minimisation problem for the displacement field: the correct class for the admissible displacement fields turns out to be the space of functions of bounded deformation where the strain tensor is just a bounded measure.

1.1 The classical boundary value problem for the equilibrium state of a perfect elastoplastic body and its primary functional formulation

Let $\mathbb{M}^{n \times n}_{\text{sym}}$ be the space of all symmetric matrices of order n. We will use the following notation

$$u \cdot v = u_i v_i, \quad |u| = \sqrt{u \cdot u},$$

$$u \otimes v = (u_i v_j), \quad u \odot v = \tfrac{1}{2}(u \otimes v + v \otimes u) \text{ for } u = (u_i), \; v = (v_i) \in \mathbb{R}^n,$$

$$\varepsilon : \varkappa = \varepsilon_{ij} \varkappa_{ij}, \quad |\varepsilon| = \sqrt{\varepsilon : \varepsilon},$$

$$\sigma \nu = (\sigma_{ij} \nu_j) \in \mathbb{R}^n \text{ for } \varepsilon = (\varepsilon_{ij}), \quad \varkappa = (\varkappa_{ij}),$$

$$\sigma = (\sigma_{ij}) \in \mathbb{M}^{n \times n}_{\text{sym}} \quad \text{and} \quad \nu = (\nu_i) \in \mathbb{R}^n,$$

where the convention of summation over Latin indices running from 1 to n is adopted.

We consider an elastoplastic body whose undeformed state is represented by a set $\overline{\Omega} \subset \mathbb{R}^n$, $n = 2, 3$. It is assumed that Ω is a bounded Lipschitz domain with boundary consisting of two measurable parts $\partial_1 \Omega$ and $\partial_2 \Omega$ satisfying the conditions $\partial \Omega = \overline{\partial_1 \Omega} \cup \overline{\partial_2 \Omega}$ and $\partial_1 \Omega \cap \partial_2 \Omega = \varnothing$. Under the action of given forces the body is deformed. In mathematical terms this process is described by the displacement field $\overline{\Omega} \ni x \mapsto u(x) \in \mathbb{R}^n$ and the tensor of small deformations (strain tensor) which is the symmetric part of the gradient of the vector–valued field u, i.e.

$$\varepsilon(u) = \big(\varepsilon_{ij}(u)\big), \quad 2\varepsilon_{ij}(u) = u_{i,j} + u_{j,i}, \quad u_{i,j} = \partial u_i / \partial x_j,$$

$$i, j = 1, 2, \ldots, n.$$

The strain tensor contains all the important information about the geometry of the deformation whereas the stress state of the elastoplastic body is characterized by the so–called stress tensor $\sigma : \overline{\Omega} \to \mathbb{M}^{n \times n}_{\text{sym}}$.

Within the framework of the deformation theory of plasticity the classical boundary value problem modelling the static equilibrium configuration of the given elastoplastic body can be formulated as follows (see, for instance, [Kl]). We look for the displacement field $u : \overline{\Omega} \to \mathbb{R}^n$ and the stress tensor $\sigma : \overline{\Omega} \to \mathbb{M}^{n \times n}_{\text{sym}}$ which satisfy the following three groups of relations:

Equilibrium equations for the stresses

(1.1) $\text{div } \sigma(x) + f(x) = 0, \ x \in \Omega; \quad \sigma(x)\nu(x) = F(x), \ x \in \partial_2\Omega.$

Deformation relations

(1.2) $2\varepsilon(u) = (u_{i,j} + u_{j,i}) \text{ on } \Omega; \quad u = u_0 \text{ on } \partial_1\Omega.$

Constitutive equations

(1.3)
$$
\begin{cases}
\varepsilon\big(u(x)\big) = \varepsilon_e(x) + \varepsilon_p(x), \ x \in \Omega; \quad \mathcal{F}\big(\sigma(x)\big) \leq 0, \ x \in \Omega; \\[2mm]
\varepsilon_e(x) = \frac{1}{n^2 K_0}\mathbf{1}\text{tr } \sigma(x) + \frac{1}{2\mu}\sigma^D(x), \ x \in \Omega; \\[2mm]
\varepsilon_p(x) : \big(\tau - \sigma(x)\big) \leq 0 \text{ for any } \tau \in \mathbb{M}^{n \times n}_{\text{sym}} \text{ s.t. } \mathcal{F}(\tau) \leq 0 \\
\text{and for all } x \in \Omega.
\end{cases}
$$

Here div $\sigma = (\sigma_{ij,j})$ is the divergence of the tensor σ, $f : \Omega \to \mathbb{R}^n$, $F : \partial_2\Omega \to \mathbb{R}^n$ are given volume and surface forces, $\nu = (\nu_i)$ is the unit outward normal to $\partial\Omega$, ε_e and ε_p are the elastic and plastic parts of the strain tensor and K_0, μ denote positive constants characterizing the elastic properties of the deformed body. The symbol σ^D stands for the deviator of the tensor σ, i.e.

$$
\sigma^D = \sigma - \frac{1}{n} \text{ tr } \sigma\mathbf{1},
$$

where $\mathbf{1}$ is the identity and tr $\sigma = \sigma_{ii}$.

Finally, the convex function $\mathcal{F} : \mathbb{M}^{n \times n}_{\text{sym}} \to \mathbb{R}$ determines the yield surface in the space of stresses, i.e. the set of tensors σ s.t.

(1.4) $\mathcal{F}(\sigma) = 0.$

Equation (1.4) is called the yield condition, since a necessary condition for $|\varepsilon_p(x)| \neq 0$ is exactly the equality $\mathcal{F}\big(\sigma(x)\big) = 0$. We confine ourself to the von Mises yield condition by setting

(1.5) $\mathcal{F}(\tau) = |\tau^D| - \sqrt{2}k_*, \ \tau \in \mathbb{M}^{n \times n}_{\text{sym}},$

where k_* is a positive constant. The plasticity model described by the relations (1.1) – (1.3) and (1.5) is called Hencky–Il'yushin plasticity. Concerning the mathematical treatment of other plasticity models we refer to the papers [An], [RS], [Se8], [Fu5] and the references quoted therein.

The first serious problem we are faced with when studying the boundary value problem (1.1) – (1.3), (1.5) is connected with the correct functional formulation. Indeed, the primary version of a functional formulation based on the assumption of summability of the strain tensor has the form of the minimax problem. We seek a pair $(u, \sigma) \in (V_0 + u_0) \times K$ which is a saddle point of the Lagrangian $(v, \tau) \mapsto \ell(v, \tau)$ on the set $(V_0 + u_0) \times K$, i.e.

$$(1.6) \qquad \ell(u, \tau) \leq \ell(u, \sigma) \leq \ell(v, \sigma) \quad \forall v \in V_0 + u_0, \ \tau \in K,$$

where

$$\ell(v, \tau) = \int_\Omega \left(\varepsilon(v) : \tau - g^*(\tau) - f \cdot v \right) dx - \int_{\partial_2 \Omega} F \cdot v \, d\ell.$$

$$(1.7) \qquad V_0 + u_0 = \left\{ v \in D^{2,1}(\Omega) : v = 0 \text{ on } \partial_1 \Omega \right\} + u_0$$

denotes the set of kinematically admissible displacements,

$$K = \left\{ \tau \in \Sigma : \mathcal{F}(\tau(x)) \leq 0 \text{ for a.a. } x \in \Omega \right\}$$

is the set of admissible stress tensors, and we further let

$$\mathbb{M}^{n \times n}_{\text{sym}} \ni \tau \mapsto g^*(\tau) = \frac{1}{2n^2 K_0} \operatorname{tr}^2 \tau + g_0^*(|\tau^D|),$$

$$\mathbb{R} \ni t \mapsto g_0^*(t) = \begin{cases} \frac{1}{4\mu} t^2, & |t| \leq \sqrt{2} k_*, \\ +\infty, & |t| > \sqrt{2} k_*. \end{cases}$$

$$D^{p,q}(\Omega) = \left\{ v \in L^1(\Omega; \mathbb{R}^n) : \|v\|_{p,q} = \|\operatorname{div} v\|_{L^p(\Omega)} \right.$$

$$\left. + \||v| + |\varepsilon^D(v)|\|_{L^q(\Omega)} < +\infty \right\}, \quad p, q \geq 1, \quad \text{and}$$

$$\Sigma = \left\{ \tau = (\tau_{ij}) : \|\tau\|_\Sigma = \|\operatorname{tr} \tau\|_{L^2(\Omega)} + \|\tau^D\|_{L^\infty(\Omega)} < +\infty \right\}$$

denote the spaces of admissible fields of displacements and stresses, respectively, $L^p(\Omega; \mathbb{R}^m)$, $W_p^\ell(\Omega; \mathbb{R}^m)$ are the familiar Lebesgue and Sobolev spaces of vector-valued functions taking their values in $\mathbb{R}^m$ (compare [A] and [LU]). It is known (see [MM1], [MM2]) that the space $D^{1,1}(\Omega)$ is imbedded continuously in the spaces $L^{\frac{n}{n-1}}(\Omega; \mathbb{R}^n)$ and $L^1(\partial\Omega; \mathbb{R}^n)$ and compactly in the spaces $L^p(\Omega; \mathbb{R}^n)$ for $p \in [1, \frac{n}{n-1}[$. So the minimax problem (1.6) will be well-posed at least formally if we assume that

$$(1.8) \qquad f \in L^n(\Omega; \mathbb{R}^n), \quad F \in L^\infty(\partial_2\Omega; \mathbb{R}^n), \quad u_0 \in W_2^1(\Omega; \mathbb{R}^n).$$

Unfortunately, the minimax problem (1.6) even for smooth data and under some additional natural conditions which will be imposed below, has in general no solution for the following reasons. The Lagrangian ℓ generates two variational problems being in duality. In the first of them a displacement field $u \in V_0 + u_0$ is determined as a minimizer of the problem

$$(1.9) \qquad I(u) = \inf\{I(v) : v \in V_0 + u_0\},$$

where

$$(1.10) \qquad \begin{cases} V_0 + u_0 \ni v \mapsto I(v) = \sup\{\ell(v, \tau) : \tau \in K\} = \int_\Omega g(\varepsilon(v))\,dx - \overline{M}(v), \\[2mm] \overline{M}(v) = \int_\Omega f \cdot v\,dx + \int_{\partial_2\Omega} F \cdot v\,d\ell. \end{cases}$$

Here $g : \mathbb{M}_{\mathrm{sym}}^{n \times n} \to \mathbb{R}$ is equal to the conjugate function of g^*, i.e.

$$g(\varkappa) = \sup\{\varkappa : \tau - g^*(\tau) : \tau \in \mathbb{M}_{\mathrm{sym}}^{n \times n}\},$$

which can be expressed through the conjugate function of g_0^* in the following way:

$$\mathbb{M}_{\mathrm{sym}}^{n \times n} \ni \varkappa \mapsto g(\varkappa) = \frac{1}{2}K_0 \operatorname{tr}^2 \varkappa + g_0(|\varkappa^D|),$$

$$\mathbb{R} \ni t \mapsto g_0(t) = \sup\{st - g_0^*(s) : s \in \mathbb{R}\}.$$

In the case of Hencky–Il'yushin plasticity we have

$$(1.11) \qquad g_0(t) = \begin{cases} \mu t^2, & |t| \le t_0 = \frac{k_*}{\sqrt{2}\mu} \\[2mm] k_*(\sqrt{2}|t| - \frac{k_*}{2\mu}), & |t| > t_0. \end{cases}$$

In the second variational problem, called the Haar–Karman principle, we look for a stress tensor $\sigma \in Q_f \cap K$ such that

$$(1.12) \qquad R(\sigma) = \sup\{R(\tau) : \tau \in Q_f \cap K\},$$

where

(1.13) $Q_f \cap K \ni \tau \mapsto R(\tau) = \ell(u_0, \tau)$

is the functional of the problem and

$$Q_f = \{\tau \in \Sigma : \int_\Omega \tau : \varepsilon(v) dx = \overline{M}(v) \; \forall v \in V_0\}$$

is the set of stress tensors satisfying the equilibrium equations (1.1) in the weak sense.

As it will be shown below problem (1.12) has a unique solution σ under the condition $K \cap Q_f \neq \varnothing$. Moreover, if the so–called "safe load hypothesis" is imposed, i.e.

(1.14) $\exists \sigma^1 \in Q_f$ and $\exists \delta_1 > 0 : \mathcal{F}(\sigma^1) \leq -\delta_1$ a.e. in Ω,

then we have two important results:

(1.15) $\inf\{I(v) : v \in V_0 + u_0\} = R(\sigma)$,

(1.16)
$$\begin{cases} (u, \sigma) \in (V_0 + u_0) \times K \\ \text{is a saddle point of problem (1.6) if and only if} \\[2mm] u \in V_0 + u_0 \text{ is a minimizer of problem (1.9) and} \\[2mm] \sigma \in Q_f \cap K \text{ is a maximizer of problem (1.12).} \end{cases}$$

So, according to (1.16), the minimax problem (1.6) is solvable if and only if the variational problem (1.9) admits a solution. At this stage we should add a remark justifying the considerations from above. Suppose that u and σ are sufficiently regular solutions of (1.1)–(1.3) with $\mathcal{F}$ defined in (1.5). Then it is easy to see that u solves (1.9) whereas σ is R–maximizing. Conversely, any smooth saddle point (u, σ) of the Lagrangian ℓ provides solutions of the relations (1.1)–(1.3). From this point of view and also with respect to (1.16) it is natural to investigate the variational problems (1.9) and (1.12) and to prove some smoothness properties of suitable generalized solutions. However, although the convex functional I is continuous on $D^{2,1}(\Omega)$ and coercive on $V_0 + u_0$, i.e.

$$I(v) \to \infty \text{ if } \|v\|_{2,1} \to \infty \text{ and } v \in V_0 + u_0,$$

we can not prove the existence of a minimizer of problem (1.9) by applying the direct method of the Calculus of Variations. The point is that the nonreflexive space $D^{2,1}(\Omega)$ shows the same "unpleasant" properties as the classes L^1 and W_1^1.

In order to demonstrate that problems (1.6) and (1.9) in general have no solutions, let us consider as an example the plane deformation of a concentric ring whose interior contour is fixed whereas the exterior one is twisted. Let us introduce polar coordinates (ρ, θ) with origin at the center of the ring and let R_1 and R_2 be the interior and exterior radius of the ring. Next, let u_ρ, u_θ and $\sigma_{\rho\rho}$, $\sigma_{\rho\theta}$, $\sigma_{\theta\theta}$ be the physical components of the displacement field u and the stress tensor σ. Our boundary conditions take the form

$$(1.17) \quad \begin{cases} u = (u_\rho, u_\theta), \quad \partial_2\Omega = \varnothing; \quad u = 0 \text{ at } \rho = R_1, \\[2mm] u_\rho = 0, \quad u_\theta = U_0 = \text{constant} > 0 \text{ at } \rho = R_2. \end{cases}$$

We suppose that

$$(1.18) \quad f = 0, \quad U_0 > U_* = \frac{\alpha^2 - 1}{\alpha^2} \frac{k_*}{2\mu} R_2,$$

where $\alpha = R_2/R_1$.

Let us introduce the tensor σ by setting

$$(1.19) \quad \sigma = \begin{pmatrix} \sigma_{\rho\rho} & \sigma_{\rho\theta} \\[2mm] \sigma_{\rho\theta} & \sigma_{\theta\theta} \end{pmatrix}, \quad \sigma_{\rho\rho} = \sigma_{\theta\theta} = 0, \quad \sigma_{\rho\theta} = k_*\left(\frac{R_1}{\rho}\right)^2.$$

It is easy to check that $\sigma \in K$ and div $\sigma = 0$ on Ω and thus

$$(1.20) \quad \sigma \in Q_f \cap K.$$

We let

$$(1.21) \quad u_0 = (u_\rho^0, u_\theta^0), \quad u_\rho^0 = 0, \quad u_\theta^0 = \frac{U_0}{U_*} \frac{k_*}{2\mu}\left(\rho - \frac{R_1^2}{\rho}\right).$$

If we introduce the function

$$(1.22) \quad \tilde{u} = (\tilde{u}_\rho, \tilde{u}_\theta), \quad \tilde{u}_\rho = 0, \quad \tilde{u}_\theta = \frac{\rho}{R_2}U_0 + \frac{U_*}{\alpha^2 - 1}\left(\frac{\rho}{R_2} - \frac{R_2}{\rho}\right),$$

then elementary calculations show that

$$(1.23) \quad \begin{cases} \frac{1}{2\mu}\sigma_{\rho\theta} = \varepsilon_{\rho\theta}(u_0) = \varepsilon_{\rho\theta}(\tilde{u}), \\ \varepsilon_{\rho\rho}(u_0) = \varepsilon_{\theta\theta}(u_0) = \varepsilon_{\rho\rho}(\tilde{u}) = \varepsilon_{\theta\theta}(\tilde{u}) = 0, \\ \tilde{u}_\theta(R_1) = \frac{U_0 - U_*}{\alpha}, \quad u_\theta^0(R_1) = 0, \\ \tilde{u}_\theta(R_2) = u_\theta^0(R_2) = U_0. \end{cases}$$

Next, let τ be an arbitrary function from $\Sigma_2 \cap K$ where

$$(1.24) \quad \Sigma_s = \{\tau \in \Sigma : \operatorname{div} \tau \in L^s(\Omega; \mathbb{R}^2)\}, \ s \geq 1.$$

By Lemma A.1.5, $\tau \in L^2(\Omega; \mathbb{M}_{\mathrm{sym}}^{2\times2})$ and therefore, according to Lemma A.1.3, a sequence $\tau^m \in C^\infty(\overline{\Omega}; \mathbb{M}_{\mathrm{sym}}^{2\times2})$ exists such that

$$(1.25) \quad \begin{cases} \tau^m \to \tau \quad \text{in } L^2(\Omega; \mathbb{M}_{\mathrm{sym}}^{2\times2}), \\ \operatorname{div} \tau^m \to \operatorname{div} \tau \quad \text{in } L^2(\Omega; \mathbb{R}^2), \\ \|\tau^{mD}\|_{L^\infty(\Omega)} = \|\tau^D\|_{L^\infty(\Omega)}. \end{cases}$$

From the last identity in (1.25) we infer that

$$(1.26) \quad \tau^m \in \Sigma_2 \cap K.$$

Now, let us consider the integral

$$I_m = \int_\Omega [\varepsilon(u_0) : (\tau^m - \sigma) + (u_0 - \tilde{u}) \cdot \operatorname{div}(\tau^m - \sigma) - a(\sigma, \tau^m - \sigma)] \, dx$$

where

$$(1.27) \quad a(\tau, \sigma) = \frac{1}{4K_0} \operatorname{tr}\tau \operatorname{tr}\sigma + \frac{1}{2\mu}\tau^D : \sigma^D, \ \tau, \sigma \in \mathbb{M}_{\mathrm{sym}}^{2\times2}.$$

Since the integral contains only smooth functions we can integrate by parts. As a result we get

$$I_m = \int_{\partial\Omega} (\tau^m - \sigma)\nu \cdot (u_0 - \tilde{u}) \, d\ell + \int_\Omega \left[\varepsilon(\tilde{u}) : (\tau^m - \sigma) - a(\sigma, \tau^m - \sigma)\right] dx.$$

It follows from (1.23) that

$$\varepsilon(\tilde{u}) : (\tau^m - \sigma) - a(\sigma, \tau^m - \sigma) = 2\left(\varepsilon_{\rho\theta}(\tilde{u}) - \frac{1}{2\mu}\sigma_{\rho\theta}\right)(\tau^m_{\rho\theta} - \sigma_{\rho\theta}) \equiv 0,$$

and therefore

$$I_m = \int\limits_{\rho=R_1} (\tau^m - \sigma)\nu \cdot (u_0 - \tilde{u})d\ell = - \int\limits_0^{2\pi} (\tau^m_{\rho\theta} - \sigma_{\rho\theta})(u^0_\theta - \tilde{u}_\theta)\Big|_{\rho=R_1} R_1 d\theta$$

$$= - \int\limits_0^{2\pi} (\tau^m_{\rho\theta}(R_1) - k_*) \frac{U_* - U_0}{\alpha} R_1 d\theta.$$

For all ρ and θ we have (by (1.26)) $|\tau^m_{\rho\theta}| \leq k_*$ which gives $I_m \leq 0$ for any $m \in \mathbb{N}$. After passing to the limit $m \to \infty$ we get the variational inequality

$$(1.28) \qquad \int\limits_\Omega \left[\varepsilon(u_0) : (\tau - \sigma) + (u_0 - \tilde{u}) \cdot \mathrm{div}(\tau - \sigma) - a(\sigma, \tau - \sigma)\right] dx \leq 0$$

for all $\tau \in \Sigma_2 \cap K$. For $\tau \in Q_f \cap K$ inequality (1.28) reduces to

$$\int\limits_\Omega \left[\varepsilon(u_0) : (\tau - \sigma) - a(\sigma, \tau - \sigma)\right] dx \leq 0$$

which means that σ is the unique solution of the dual problem (1.12). Now, let us suppose that a pair $(u, \sigma) \in (V_0 + u_0) \times K$ is a saddle point of the minimax problem (1.6). By (1.16) the stress tensor σ is determined according to (1.19). Since $|\sigma^D| < \sqrt{2}k_*$ inside of Ω, the plastic part of the deformation is equal to zero and we have

$$\varepsilon(u) = \frac{1}{4K_0} 1\,\mathrm{tr}\,\sigma + \frac{1}{2\mu}\sigma^D = \varepsilon(\tilde{u}) \text{ in } \Omega.$$

Moreover, $u = \tilde{u}$ at $\rho = R_2$ implies $u \equiv \tilde{u}$ in Ω. Recall that $\tilde{u} = (0, \frac{U_0 - U_*}{\alpha})$ at $\rho = R_1$. On the other hand, $u \in V_0 + u_0$ and therefore $u = 0$ at $\rho = R_1$ which leads to a contradiction.

Thus the assumption of summability of the strain tensor is too strong and does not enable us to prove existence theorems for problem (1.6). Hence we need a suitable relaxed version of (1.6) whose dual problem is still given by (1.12) and which will allow us to construct physically reasonable generalized solutions. A detailed analysis of this construction will be presented in the next sections. It is worth to remark that in the present case minimizing sequences of problem

(1.9) converge to a function having a jump discontinuity on the internal contour ($\rho = R_1$) which is in accordance with the theory of perfect elastoplasticity. Moreover, this theory admits discontinuities of slip type which means that the normal component of the displacement field w.r.t. to a surface of discontinuity remains continuous.

Up to now we limited our discussion to the Hencky–Il'yushin model. At the end of chapter 2 the reader will find some comments concerning other models treated in plasticity with hardening. We also refer to chapter 4.

1.2 Relaxation of convex variational problems in non reflexive spaces. General construction

Let V, U and P be Banach spaces so that $V \subset U$, and let V_0 be a subspace of V. Next, let $A : V \to P$ denote a linear bounded operator, and suppose that $G : P \to \overline{\mathbb{R}}$ and $M : U \to \overline{\mathbb{R}}$ are convex, lower semicontinuous functionals (for general terminology we refer the reader to the book [ET]). We also assume that these functionals are proper in the sense that they are not equal to $+\infty$ identically and do not take the value $-\infty$.

We denote by P^* and U^* the dual spaces to P and U, by $\langle \cdot, \cdot \rangle$ and $(\cdot, \cdot)$ the duality relations between P and P^*, U and U^*, respectively, and by G^* the conjugate functional of G, i.e.

$$G^*(p^*) = \sup \left\{ \langle p^*, p \rangle - G(p) : p \in P \right\}, \ p^* \in P^*.$$

Let us consider the variational problem

(2.1) to find $u \in V_0 + u_0$ such that $I(u) = \inf \left\{ I(v) : v \in V_0 + u_0 \right\}$

where

$$I(v) = G(Av) + M(v)$$

and $u_o \in V$ is fixed. To formulate the problem dual to (2.1) we first introduce the Lagrangian ℓ by letting

$$\ell(v, q^*) = \langle q^*, Av \rangle - G^*(q^*) + M(v), \quad q^* \in P^*, \quad v \in V_0 + u_0.$$

Then the dual problem reads

(2.2) to find $p^* \in P^*$ such that $R(p^*) = \sup \left\{ R(q^*) : q^* \in P^* \right\}$

where

$$R(q^*) = \inf \left\{ \ell(v, q^*) : v \in V_0 + u_0 \right\}.$$

Concerning the solvability of problem (2.2) we have the following

THEOREM 1.2.1 ([ET]). *Suppose that the next two conditions hold*

(2.3) $C := \inf \left\{ I(v) : v \in V_0 + u_0 \right\} \in]-\infty, +\infty[;$

$$(2.4) \quad \begin{cases} \exists u_1 \in V_0 + u_0 : G(Au_1) < +\infty, M(u_1) < +\infty; \\ \\ \text{the function } p \mapsto G(Au_1 + p) \text{ is continuous at zero.} \end{cases}$$

Then problem (2.2) has at least one solution and the identity

$$(2.5) \quad C = \sup\left\{ R(q^*) : q^* \in P^* \right\}$$

is valid.

Proof. We consider the variational problem equivalent to (2.1)

$$(2.6) \quad \begin{cases} \text{to find } u \in V \text{ such that } \tilde{I}(u) = \inf\{\tilde{I}(v) : v \in V\} \text{ where} \\ \\ \tilde{I}(v) = G(Av) + \tilde{M}(v), \\ \\ \tilde{M}(v) = \begin{cases} M(v), & v \in V_0 + u_0 \\ +\infty, & v \notin V_0 + u_0, v \in V. \end{cases} \end{cases}$$

Now, let us define the perturbed functional

$$\Phi(v,p) = G(Av + p) + \tilde{M}(v), \quad v \in V, \quad p \in P.$$

Following [ET] we introduce the problem dual to (2.6), i.e.

$$(2.7) \quad \text{to find } p^* \in P^* \text{ such that } -\Phi^*(0,p^*) = \sup\left\{ -\Phi^*(0,q^*) : q^* \in P^* \right\}.$$

Here $\Phi^* : V^* \times P^* \to \overline{\mathbb{R}}$ is the conjugate function of Φ. In our case we have

$$-\Phi^*(0,q^*) = -\sup_{v \in V} \sup_{p \in P} \left\{ \langle q^*, p \rangle - G(Av + p) - \tilde{M}(v) \right\}$$

$$= -\sup_{v \in V} \sup_{q \in P} \left\{ -\langle q^*, Av \rangle + \langle q^*, q \rangle - G(q) - \tilde{M}(v) \right\}$$

$$= -\sup_{v \in V} \left\{ -\langle q^*, Av \rangle + G^*(q^*) - \tilde{M}(v) \right\} = \inf_{v \in V_0 + u_0} \ell(v, q^*) = R(q^*).$$

Thus problem (2.7) is equivalent to problem (2.2). Sufficient conditions under which problem (2.7) is solvable and we have the identity

$$(2.8) \quad \inf\left\{ \tilde{I}(v) : v \in V \right\} = \sup\{-\Phi^*(0,q^*) : q^* \in P^*\}$$

are for example (see [ET], Chapter III, Proposition 2.3)

$$(2.9) \quad \inf\left\{ \tilde{I}(v) : v \in V \right\} \in]-\infty, +\infty[,$$

(2.10)   the function Φ is convex,

(2.11)
$$\begin{cases} \exists \tilde{u}_1 \in V \text{ such that the function } p \mapsto \Phi(\tilde{u}_1, p) \\ \text{is finite and continuous at zero.} \end{cases}$$

Obviously (2.9) is equivalent to condition (2.3). Convexity of Φ is provided by convexity of G and $\tilde{M}$. Finally, condition (2.11) follows from (2.4) if we put $\tilde{u}_1 = u_1$. Since identities (2.8) and (2.5) are equivalent, Theorem 1.2.1 is proved.

□

Along with the problems (2.1) and (2.2) we consider the following minimax problem

(2.12)
$$\begin{cases} \text{to find a pair } (u, p^*) \in (V_0 + u_0) \times P^* \text{ such that} \\ \ell(u, q^*) \le \ell(u, p^*) \le \ell(v, p^*), \quad \forall v \in V_0 + u_0, \ q^* \in P^*. \end{cases}$$

(Such a pair is called a saddle point.)

Since $G : P \to \overline{\mathbb{R}}$ is a proper, convex, lower semicontinuous functional, we have

(2.13) $G(p) = \sup \left\{ \langle p^*, p \rangle - G^*(p^*) : q^* \in P^* \right\}$

and therefore

(2.14) $I(v) = \sup \left\{ \ell(v, q^*) : q^* \in P^* \right\}.$

Thus under conditions (2.3), (2.4) we have the identity

(2.15) $\displaystyle \inf_{v \in V_0 + u_0} \sup_{q^* \in P^*} \ell(v, q^*) = C = \sup_{q^* \in P^*} \inf_{v \in V_0 + u_0} \ell(v, q^*)$

and the general theory of duality provides the following statement:

(2.16)
$$\begin{cases} \text{a pair } (u, p^*) \in (V_0 + u_0) \times P^* \\ \text{is saddle point of the minimax problem (2.12) if and only if} \\ u \in V_0 + u_0 \text{ is a minimizer of problem (2.1) and} \\ p^* \in P^* \text{ is a maximizer of problem (2.2).} \end{cases}$$

So, by Theorem 1.2.1 and (2.16), solvability of problem (2.1) is equivalent to solvability of the minimax problem (2.12).

From now on we will impose the following additional conditions on the data:

(2.17)
$$\begin{cases} \text{the imbedding of } V \text{ into } U \text{ is continuous;} \\ V_0 \text{ is dense in } U; \\ U \text{ is a reflexive space.} \end{cases}$$

(2.18)
$$\begin{cases} \exists u_2 \in V_0 + u_0 : u_2 \in \text{int dom } M, \\ \text{dom } M := \{u \in U : M(u) < \infty\}. \end{cases}$$

(2.19) $\quad I(v) \to +\infty$ if $\|v\|_V \to +\infty$ and $v \in V_0 + u_0$.

Condition (2.19) means coercivity of I on the set $V_0 + u_0$.

Standard arguments now show that if the space V is reflexive, then the coercivity condition (2.19) together with convexity and lower semicontinuity of the functionals G and M provide existence of at least one solution of problem (2.1) and thus existence of at least one solution of the minimax problem (2.12). So it remains to discuss the case of a nonreflexive space V.

The example considered in the previous section shows that if the space V is nonreflexive, then in general problems (2.1) and (2.12) have no solutions. We therefore have to construct suitable relaxations which should satisfy at least two restrictions:

1. conservation of the greatest lower bound of problem (2.1),

2. conservation of the dual problem.

The first condition is clearly satisfied if we can guarantee that all minimizing sequences of problem (2.1) converge in some sense to a solution of the relaxed problem. As to the second restriction one should say that a solution of the dual variational problem exists and as a rule, it is unique and it has a clear physical or geometrical meaning. Thus there is no need to change the dual variational problem.

The construction of a suitable variational relaxation to problem (2.1) now follows the paper [Se1]. We first introduce an auxiliary operator A^* whose domain $D(A^*)$ is the set

$$P_0^* = \{p^* \in P^* : \exists u^* \in U^* \text{ such that } \langle p^*, Au \rangle = (u^*, u) \; \forall u \in V_0\}.$$

By condition (2.17), for each $p^* \in P_0^*$, there is only one element $u^* \in U^*$ satisfying the identity $\langle p^*, Au \rangle = (u^*, u)$ on V_0. So we can define the linear operator $A^* : P_0^* \to U^*$ just through the relation

$$\langle p^*, Au \rangle = (A^* p^*, u) \quad \forall p^* \in P_0^*, \ u \in V_0.$$

If u_0 is a fixed element from V, then we have the identity

(2.20) $\quad \langle p^*, Au \rangle = \mathcal{E}(u_0, p^*) + (A^* p^*, u) \quad \forall u \in V_0 + u_0, \ p^* \in P_0^*$

where

$$\mathcal{E}(u_0, p^*) = \langle p^*, Au_0 \rangle - (A^* p^*, u_0).$$

We enlarge the set $V_0 + u_0$ by letting

(2.21) $\quad V_+ = \{u \in U : \sup_{\|p^*\|_{P^*} \leq 1, p^* \in D(A^*)} |\mathcal{E}(u_0, p^*) + (A^* p^*, u)| < +\infty\},$

and introduce a relaxation Φ of the functional I with the help of the Lagrangian L through the relations

(2.22) $\quad \begin{cases} L(v, q^*) = \mathcal{E}(u_0, q^*) + (A^* q^*, v) - G^*(q^*) + M(v), \\[2mm] q^* \in D(A^*), \ v \in V_+; \\[2mm] \Phi(v) = \sup_{q^* \in D(A^*)} L(v, q^*), \quad \Phi : V_+ \to \overline{\mathbb{R}}. \end{cases}$

Let us remark some useful facts which follow from (2.20) and our definitions (2.21) and (2.22):

(2.23) $\quad V_0 + u_0 \subset V_+,$

(2.24) $\quad \Phi(v) \leq I(v) \quad \forall v \in V_0 + u_0.$

Indeed, if $v \in V_0 + u_0$, then we have from (2.20)

$$\sup_{\|p^*\|_{P^*} \leq 1, p^* \in D(A^*)} |\mathcal{E}(u_0, p^*) + (A^* p^*, v)|$$

$$= \sup_{\|p^*\|_{P^*} \leq 1, p^* \in D(A^*)} |\langle p^*, Av \rangle|$$

$$\leq \sup_{\|p^*\|_{P^*} \leq 1} |\langle p^*, Av \rangle| = \|Av\|_P,$$

and inclusion (2.23) follows. Next, by (2.20)

(2.25) $L(v, q^*) = \ell(v, q^*) \quad \forall v \in V_0 + u_0, \; q^* \in D(A^*).$

But then, for $v \in V_0 + u_0$, we have (recall (2.14))

$$\Phi(v) = \sup_{q^* \in D(A^*)} \ell(v, q^*) \leq \sup_{q^* \in P^*} \ell(v, q^*) = I(v).$$

LEMMA 1.2.1 *Suppose that for any $p \in \mathrm{dom}\, G^*$ there exists a sequence $p_m^* \in D(A^*)$ such that*

(2.26) $\begin{cases} p_m^* \stackrel{*}{\rightharpoonup} p^* \text{ in } P^*, \\[2mm] G^*(p_m^*) \to G^*(p^*). \end{cases}$

Then the identity

(2.27) $\Phi(v) = I(v) \quad \forall v \in V_0 + u_0$

is valid.

Proof.   Since G is proper, convex and lower semicontinuous, the function G^* is of the same type. Thus

$$G(p) = \sup\{\langle q^*, p \rangle - G^*(q^*) : q^* \in \mathrm{dom}\, G^*\}$$

and therefore

(2.28) $I(v) = \sup\{\ell(v, q^*) : q^* \in \mathrm{dom}\, G^*\} \quad \forall v \in V_0 + u_0.$

Suppose first that $I(v) = +\infty$. Then, for any $A > 0$, there is a $q^* \in \mathrm{dom}\, G^*$ such that $\ell(v, q^*) > A$. Using the conditions of Lemma 1.2.1 one can find a sequence $p_m^* \in D(A^*)$ having the properties

(2.29) $p_m^* \stackrel{*}{\rightharpoonup} q^* \text{ in } P^*$

and

(2.30) $G^*(p_m^*) \to G^*(q^*), \quad \ell(v, p_m^*) \to \ell(v, q^*).$

It follows from the definition of Φ and identity (2.25) that

$$\Phi(v) \geq L(v, p_m^*) = \ell(v, p_m^*) \to \ell(v, q^*) > A.$$

By the arbitrariness of A, we get $\Phi(v) = +\infty$.

If $I(v) < +\infty$, then according to (2.28) for any $\varepsilon > 0$ there exists $q^* \in$ dom G^* such that

$$I(v) < \ell(v, q^*) + \varepsilon.$$

Let a sequence $p_m^* \in D(A^*)$ be chosen such that (2.29) and (2.30) hold. Then

$$I(v) < \ell(v, q^*) + \varepsilon = \lim_{m \to \infty} \ell(v, p_m^*) + \varepsilon = \lim_{m \to \infty} L(v, p_m^*) + \varepsilon \leq \Phi(v) + \varepsilon.$$

From this we get $\Phi(v) \geq I(v)$ and Lemma 1.2.1 is proved.

$\square$

It is interesting to know under what conditions the following statement will be true

$$(2.31) \qquad V_+ = V_0 + u_0.$$

LEMMA 1.2.2 *Suppose that*

$(2.32) \qquad P$ *is a reflexive space,*

$(2.33) \qquad$ *the restriction of the operator A on V_0 has a closed range of values in P,*

$(2.34) \qquad \|w\| = \displaystyle\sup_{p^* \in D(A^*), \|p^*\|_{P^*} \leq 1} |(A^*p^*, w)| = 0 \Rightarrow w = 0.$

Then (2.31) holds.

Proof. We denote by $\tilde{A}$ the restriction of the operator A on V_0. Let us take an arbitrary element $u \in V_+$ and set $w = u - u_0$. For the linear functional $p^* \mapsto (A^*p^*, w)$ we have the estimate

$$|(A^*p^*, w)| \leq \|w\| \|p^*\|_{P^*} \quad \forall p^* \in D(A^*).$$

Since $u \in V_+$ we have $\|w\| < +\infty$. By condition (2.32) and the Hahn–Banach theorem we find an element $p \in P$ such that

$$(2.35) \qquad \langle p^*, p \rangle = (A^*p^*, w) \quad \forall p^* \in D(A^*).$$

Let us introduce the set

$$Q = \{p \in P : \langle p^*, p \rangle = 0 \text{ for any } p^* \in \ker A^* \}.$$

It is clear that $\tilde{A}(V_0) \subset Q$. Suppose that there is some $p_0 \in Q \setminus \tilde{A}(V_0)$. By condition (2.33) the set $\tilde{A}(V_0)$ is closed, hence there exists $p_0^* \in P^*$ such that

$$\langle p_0^*, p_0 \rangle = 1, \quad \langle p_0^*, p \rangle = 0 \quad \forall p \in \tilde{A}(V_0).$$

Thus $\langle p_0^*, Av \rangle = 0$ for all $v \in V_0$ which shows $p_0^* \in D(A^*)$ and $(A^* p_0^*, v) = 0$ for any $v \in V_0$. By condition (2.17) $p_0^* \in \ker A^*$. But then it follows from the definition of the set Q that $\langle p_0^*, p_0 \rangle = 0$. So we have proved that $Q = \tilde{A}(V_0)$. This means that for p satisfying identity (2.35), we can find $v \in V_0$ with the property $p = Av$ and $(A^* p^*, w - v) = 0$ for all $p^* \in D(A^*)$. From this and (2.34) it follows that $w = v$, hence $u \in V_0 + u_0$ and Lemma 1.2.2 is proved. $\qquad\square$

REMARK 1.2.1 We have shown that if conditions (2.17), (2.26), (2.32)–(2.34) hold, then

(2.36) $V_+ = V_0 + u_0$ and $\Phi(v) = I(v)$ $\forall v \in V_+$.

Before proving the main results concerning the relaxation of problems (2.1) and (2.12) we establish

LEMMA 1.2.3 *Consider a bounded sequence* $\{u_m\} \subset V_0 + u_0$ *converging to* u *weakly in* U. *Then*

(2.37) $u \in V_+$,

(2.38) $\liminf\limits_{m \to +\infty} I(u_m) \geq \Phi(u)$.

Proof.   We have

$$|\mathcal{E}(u_0, p^*) + (A^* p^*, u_m)| = |\langle p^*, Au_m \rangle| \leq \|A\| \sup_m \|u_m\|_V$$

for all $p^* \in D(A^*)$ such that $\|p^*\|_{P^*} \leq 1$. Passing to the limit and using the definition of the class V_+ we get (2.37). To prove (2.38) we take into account identity (2.25) in order to get

$$I(u_m) \geq \ell(u_m, q^*) = L(u_m, q^*) =$$

$$\mathcal{E}(u_0, q^*) + (A^* q^*, u_m) - G^*(q^*) + M(u_m),$$

$$q^* \in D(A^*),$$

and therefore

$$\liminf_{m \to \infty} I(u_m) \geq \mathcal{E}(u_0, q^*) + (A^* q^*, u) - G^*(q^*), \quad \text{i.e.}$$

$$\liminf_{m \to \infty} I(u_m) \geq L(u, q^*) \quad \forall q^* \in D(A^*).$$

From this and the definition of the functional Φ (see (2.22)) (2.38) follows. Lemma 1.2.3 is proved.

$\square$

We consider now the minimax problem

$$(2.39) \quad \begin{cases} \text{to find a pair } (u, p^*) \in V_+ \times D(A^*) \text{ such that} \\ \\ L(u, q^*) \leq L(u, p^*) \leq L(v, p^*) \quad \forall v \in V_+, \quad q^* \in D(A^*). \end{cases}$$

As usual this problem generates two variational problems being in duality:

$$(2.40) \quad \begin{cases} \text{to find } u \in V_+ \text{ such that} \\ \\ \Phi(u) = \inf \left\{ \Phi(v) : v \in V_+ \right\}, \end{cases}$$

where

$$\Phi(v) = \sup \left\{ L(v, q^*) : q^* \in D(A^*) \right\},$$

and

$$(2.41) \quad \begin{cases} \text{to find } p^* \in D(A^*) \text{ such that} \\ \\ \tilde{R}(p^*) = \sup \left\{ \tilde{R}(q^*) : q^* \in D(A^*) \right\}, \end{cases}$$

where

$$\tilde{R}(q^*) = \inf \left\{ L(v, q^*) : v \in V_+ \right\}.$$

The main statement of this section is contained in

MAIN THEOREM 1.2.2 *Suppose that the conditions* (2.3), (2.4), (2.17)–(2.19) *hold. Then we have*

(i) *Problems* (2.40) *and* (2.41) *are solvable. Moreover, if $u \in V_+$ is a solution to problem* (2.40) *and $p^* \in D(A^*)$ is a solution to problem* (2.41), *then the identity*

$$(2.42) \quad \Phi(u) = C = \tilde{R}(p^*)$$

is true.

(ii) *Problems (2.2) and (2.41) are equivalent, i.e. they have the same set of solutions.*

(iii) *A pair $(u, p^*) \in V_+ \times D(A^*)$ is a saddle point of the minimax problem (2.39) if and only if $u \in V_+$ is a minimizer of problem (2.40) and p^* in $D(A^*)$ is a maximizer of problem (2.41).*

(iv) *Any minimizing sequence of problem (2.1) contains a subsequence converging to some solution to problem (2.40) weakly in U.*

Proof.

(i) Let $\{u_m\} \subset V_0 + u_0$ be an arbitrary minimizing sequence of problem (2.1), i.e. $I(u_m) \to C = \inf\{I(v) : v \in V_0 + u_0\}$. Condition (2.19) implies boundedness of this sequence in V and, by (2.17), it contains a subsequence converging weakly in U. We denote by u its limit. Due to Lemma 1.2.3 we have $u \in V_+$ and $\Phi(u) \leq C$. Let p^* be any solution to problem (2.2). If we can show that u is a solution to problem (2.40), p^* is a solution to problem (2.41) and identity (2.42) holds, then statement (i) will be established. Moreover, by the arbitrariness of the minimizing sequence, statement (iv) and also the fact that any maximizer of problem (2.2) is a maximizer of problem (2.41) will be proved.

According to (2.5) we have

$$(2.43) \qquad \Phi(u) \leq C = R(p^*)$$

and therefore

$$(2.44) \qquad L(u, q^*) \leq C \leq \ell(v, p^*) \quad \forall q^* \in D(A^*), \quad v \in V_0 + u_0.$$

Let us show that $p^* \in D(A^*)$. To do this we rewrite the right hand side of inequality (2.44) in the form

$$-\langle p^*, Av \rangle \leq -C + \langle p^*, Au_2 \rangle - G^*(p^*) + M(u_2 + v) \quad \forall v \in V_0.$$

By condition (2.18) the functional M is continuous at the point u_2 and therefore it is bounded in some ball of the space U with center at u_2. So we have the estimate

$$-\langle p^*, Av \rangle \leq \text{constant } \|v\|_U.$$

It follows from the Hahn–Banach theorem that there exists $u^* \in U^*$ such that

$$\langle p^*, Av \rangle = (u^*, v) \quad \text{for all } v \in V_0.$$

This gives the claim $p^* \in D(A^*)$. Using identity (2.25), we may rewrite inequality (2.44) in the equivalent form

$$(2.45) \qquad L(u, q^*) \leq C \leq L(v, p^*) \quad \forall q \in D(A^*), \quad v \in V_0 + u_0.$$

Now we are going to prove that the right hand side of the last inequality holds for all $v \in V_+$. Obviously, it is enough to check this for all $v \in V_+ \cap \operatorname{dom} M$.

First let $w \in V_+ \cap \operatorname{int} \operatorname{dom} M$. Since $w - u_0 \in U$, condition (2.17) gives existence of a sequence $\{v_m\} \subset V_0$ such that $v_m \to w - u_0$ in U. The functional M is continuous on int dom M, hence $M(v_m + u_0) \to M(w)$. Since $v_m + u_0 \in V_0 + u_0$ for all $m \in \mathbb{N}$, we can insert $v_m + u_0$ into (2.45) and get after passing to the limit (2.45) for any $v \in V_+ \cap \operatorname{int} \operatorname{dom} M$. In case $w \in V_+ \cap \operatorname{dom} M$ we set $v(\lambda) = (1 - \lambda)w + \lambda u_2 \in \operatorname{int} \operatorname{dom} M$ for all $\lambda \in]0, 1]$. For $v = v(\lambda)$ inequality (2.45) is correct. The function $v \mapsto L(v, p^*)$ is convex and thus

$$C \leq (1 - \lambda)L(w, p^*) + \lambda L(u_2, p^*).$$

With λ going to zero we conclude that

$$(2.46) \qquad \begin{cases} u \in V_+, \quad p^* \in D(A^*) \\ \\ L(u, q^*) \leq C \leq L(v, p^*) \quad \forall v \in V_+, \quad q^* \in D(A^*). \end{cases}$$

From (2.46) we get

$$\inf_{v \in V_+} \sup_{q^* \in D(A^*)} L(v, q^*) = \inf_{v \in V_+} \Phi(v) \leq \Phi(u) \leq$$

$$\leq C \leq \tilde{R}(p^*) \leq \sup_{q^* \in D(A^*)} \tilde{R}(q^*) = \sup_{q^* \in D(A^*)} \inf_{v \in V_+} L(v, q^*)$$

and therefore

$$(2.47) \qquad \Phi(u) = \inf_{v \in V_+} \Phi(v) = C = \tilde{R}(p^*) = \sup_{q^* \in D(A^*)} \tilde{R}(q^*).$$

So we have shown that u is a minimizer of (2.40), p^* is a maximizer of (2.41) and identity (2.42) holds.

(ii) To prove statement (ii) it is enough to check that any maximizer p_0^* in $D(A^*)$ of problem (2.41) is a maximizer of problem (2.2). By (2.47), (2.25) and (2.5), we have

$$C = \tilde{R}(p_0^*) = \inf_{v \in V_+} L(v, p_0^*) \leq \inf_{v \in V_0 + u_0} L(v, p_0^*)$$

$$= \inf_{v \in V_0 + u_0} \ell(v, p_0^*) = R(p_0^*) \leq \sup_{q^* \in P^*} R(q^*) = C$$

and the claim follows.

(iii) This statement is a consequence of identity (2.42) and standard arguments from duality theory.

(iv) This statement has already been proved (see (i)). Theorem 1.2.2 is established.

□

Taking into account the statements of Theorem 1.2.2 it is quite natural to call all solutions to problem (2.40) weak solutions to the variational problem (2.1), and all solutions to problem (2.39) weak solutions to the minimax problem (2.12).

1.3 Weak solutions to variational problems of perfect elastoplasticity

We consider a model of a perfect elastoplastic body providing some generalization of the classical Hencky–Il'yushin plasticity model. Suppose that an even convex function $g_0 : \mathbb{R} \to \mathbb{R}$ of class C^1 is given satisfying the conditions

$$(3.1) \qquad \begin{cases} g_0(0) = 0, \quad g_0'(0) = 0, \quad \text{and} \\[2mm] g_0'(t) \to \sqrt{2}k_* \quad \text{as } t \to +\infty, \end{cases}$$

where k_* is a fixed positive constant. In the case of Hencky–Il'yushin plasticity g_0 has the form (see (1.11))

$$g_0(t) = \begin{cases} \mu t^2 & \text{if } |t| \le t_0 = \frac{k_*}{\sqrt{2}\mu} \\[2mm] k_*(\sqrt{2}|t| - \frac{k_*}{2\mu}) & \text{if } |t| > t_0, \end{cases}$$

where the positive constant μ is related to the elastic properties of the body under consideration. Another example of a function g_0 satisfying (3.1) is given by

$$(3.2) \qquad g_0(t) = \sqrt{2}k_*(\sqrt{1 + t^2} - 1).$$

It is easy to see that for the conjugate function g_0^* of any g_0 with property (3.1) we have

$$(3.3) \qquad g_0^*(s) = \sup\{st - g_0(t) : t \in \mathbb{R}\} = +\infty \text{ if } |s| > \sqrt{2}k_*.$$

Setting

$$(3.4) \qquad g(\varkappa) = \frac{1}{2}K_0 \mathrm{tr}^2 \varkappa + g_0(|\varkappa^D|), \quad \varkappa \in \mathbb{M}^{n \times n}_{\mathrm{sym}},$$

K_0 denoting a positive constant, we get

$$(3.5) \qquad \begin{aligned} g^*(\tau) &= \sup\{\varkappa : \tau - g(\varkappa) : \varkappa \in \mathbb{M}^{n \times n}_{\mathrm{sym}}\} \\[2mm] &= \frac{1}{2n^2 K_0}\mathrm{tr}^2 \tau + g_0^*(|\tau^D|) \ \forall \tau \in \mathbb{M}^{n \times n}_{\mathrm{sym}}. \end{aligned}$$

Now we may consider the minimax problem (1.6) for the Lagrangian

$$\begin{cases} \ell(v, \tau) = \int\limits_{\Omega} (\varepsilon(v) : \tau - g^*(\tau))\,dx - \overline{M}(v), \\[3mm] v \in V_0 + u_0, \ \tau \in K, \\[3mm] \text{where } \overline{M}(v) = \int\limits_{\Omega} f \cdot v\,dx + \int\limits_{\partial_2 \Omega} F \cdot v\,d\ell. \end{cases}$$

So we look for a pair $(u, \sigma) \in (V_0 + u_0) \times K$ such that

(3.6) $\ell(u, \tau) \leq \ell(u, \sigma) \leq \ell(v, \sigma)$ $\forall v \in V_0 + u_0,$ $\tau \in K.$

We recall the definitions of the sets V_0 and K which were given in the first section:

$$V_0 = \{v \in D^{2,1}(\Omega) : v = 0 \text{ on } \partial_1 \Omega\},$$

$$K = \{\tau \in \Sigma : \mathcal{F}(\tau(x)) = |\tau^D(x)| - \sqrt{2}k_* \leq 0 \text{ for a.a. } x \in \Omega\},$$

where

$$D^{p,q}(\Omega) = \{v \in L^1(\Omega; \mathbb{R}^n) :$$
$$\|v\|_{p,q} = \|\text{div } v\|_{L^p(\Omega)} + \||v| + |\varepsilon^D(v)|\|_{L^q(\Omega)} < +\infty\}, \, p, q \geq 1,$$
$$\Sigma = \{\tau = (\tau_{ij}) : \|\tau\|_\Sigma = \|\text{tr}\tau\|_{L^2(\Omega)} + \|\tau^D\|_{L^\infty(\Omega)} < +\infty\}.$$

For the space $D^{1,1}(\Omega)$ the following facts are known (see, for example, [MM1], [MM2]):

(3.7) $\begin{cases} \text{the space } D^{1,1}(\Omega) \text{ is imbedded continuously into the spaces} \\ L^{\frac{n}{n-1}}(\Omega; \mathbb{R}^n) \text{ and } L^1(\partial\Omega; \mathbb{R}^n), \text{ and compactly into the spaces} \\ L^p(\Omega; \mathbb{R}^n) \text{ for any } p \in [1, \frac{n}{n-1}[. \end{cases}$

It is therefore quite natural to impose the following conditions on the data of the problem:

(3.8) $f \in L^n(\Omega; \mathbb{R}^n),$ $F \in L^\infty(\partial_2\Omega; \mathbb{R}^n),$ $u_0 \in D^{2,1}(\Omega).$

As usual the minimax problem generates two variational problems being in duality:

(3.9)   to find $u \in V_0 + u_0$ such that $I(u) = \inf\{I(v) : v \in V_0 + u_0\},$

where

(3.10) $I(v) = \sup_{\tau \in K} \ell(v, \tau) = \int_\Omega g(\varepsilon(v)) \, dx - \overline{M}(v), \, v \in D^{2,1}(\Omega),$

and

(3.11)   to find $\sigma \in Q_f \cap K$ such that $R(\sigma) = \sup\{R(\tau) : \tau \in Q_f \cap K\},$

where

$$(3.12) \quad R(\tau) = \left\{ \begin{array}{ll} \ell(u_0, \tau) & \text{if } \tau \in Q_f \cap K \\ -\infty & \text{if } \tau \notin Q_f \cap K \end{array} \right\}, \tau \in K,$$

and the set Q_f is defined as

$$(3.13) \quad Q_f = \{\tau \in \Sigma : \int_\Omega \tau : \varepsilon(v) \, dx = \overline{M}(v) \quad \forall v \in V_0\}.$$

In order to handle the variational problem (3.9) with the help of the general scheme described in the previous section, we set

$$V = D^{2,1}(\Omega), \quad U = L^{\frac{n}{n-1}}(\Omega; \mathbb{R}^n), \quad U^* = L^n(\Omega; \mathbb{R}^n),$$

$$(u^*, u) = \int_\Omega u^* \cdot u \, dx,$$

$$(3.14) \quad \left\{ \begin{array}{l} P = \left\{ p = \{\tau, a\} : \|p\|_P^2 = \|\tau^D\|_{L^1(\Omega)}^2 + \frac{1}{n}\|\text{tr}\tau\|_{L^2(\Omega)}^2 \right. \\ \left. +\|a\|_{L^1(\partial_2\Omega)}^2 < +\infty \right\} \subset L^1(\Omega; \mathbb{M}_{\text{sym}}^{n \times n}) \times L^1(\partial_2\Omega; \mathbb{R}^n). \end{array} \right.$$

It is clear that

$$(3.15) \quad \left\{ \begin{array}{l} P^* = \left\{ p^* = \{\sigma, b\} : \sigma^D \in L^\infty(\Omega; \mathbb{M}_{\text{sym}}^{n \times n}), \text{tr}\sigma \in L^2(\Omega), \right. \\ \left. b \in L^\infty(\partial_2\Omega; \mathbb{R}^n) \right\} = \Sigma \times L^\infty(\partial_2\Omega; \mathbb{R}^n) \end{array} \right.$$

and also that $\left(p = \{\tau, a\} \in P, \ p^* = \{\sigma, b\} \in P^* \right)$

$$\langle p^*, p \rangle = \int_\Omega \sigma : \tau \, dx + \int_{\partial_2\Omega} a \cdot b \, d\ell.$$

The norm on the space P^* is defined in the usual way

$$\|p^*\|_{P^*} = \sup\{\langle p^*, p \rangle : \|p\|_P = 1\}.$$

This norm is equivalent to the following one

$$\left(\|\sigma^D\|_{L^\infty(\Omega)}^2 + \frac{1}{n}\|\text{tr}\sigma\|_{L^2(\Omega)}^2 + \|b\|_{L^\infty(\partial_2\Omega)}^2 \right)^{1/2}.$$

Obviously the spaces V, U, V_0 satisfy the conditions stated in (2.17).

Next, let us introduce the functionals $G : P \to \overline{\mathbb{R}}$ and $M : U \to \overline{\mathbb{R}}$

$$(3.16) \quad \begin{cases} G(p) = \int\limits_{\Omega} g(\tau) dx + \int\limits_{\partial_2 \Omega} F \cdot a \, d\ell, \quad p = \{\tau, a\} \in P, \\ M(v) = -\int\limits_{\Omega} f \cdot v \, dx, \quad v \in U. \end{cases}$$

These functionals are convex and finite and therefore continuous. It is also clear that condition (2.18) imposed on the functional M is satisfied. Moreover, we have

$$G^*(p^*) = \sup\{\langle p^*, p \rangle - G(p) : p \in P\}$$

$$(3.17) \quad = \begin{cases} \int\limits_{\Omega} g^*(\sigma) dx & \text{if } b = F \text{ on } \partial_2 \Omega \\ +\infty & \text{if } b \neq F \text{ on } \partial_2 \Omega \end{cases} , p^* = \{\sigma, b\}.$$

The linear operator $A : V \to P$ may be introduced as follows. We let

$$Av = \{\varepsilon(v), -v\big|_{\partial_2 \Omega}\}, \quad v \in V,$$

and get from (3.7)

$$\|Av\|_P = \left(\frac{1}{n} \|\operatorname{div} v\|_{L^2(\Omega)}^2 + \|\varepsilon^D(v)\|_{L^1(\Omega)}^2 + \|v\|_{L^1(\partial_2 \Omega)}^2 \right)^{1/2} \leq c(\Omega, n) \|v\|_{2,1},$$

hence the operator A is bounded. Taking into account all definitions introduced above, we may represent our functional $I : V \to \overline{\mathbb{R}}$ in the form

$$I(v) = \int\limits_{\Omega} g(\varepsilon(v)) dx - \int\limits_{\partial_2 \Omega} F \cdot v \, d\ell - \int\limits_{\Omega} f \cdot v \, dx = G(Av) + M(v).$$

Now we check if conditions (2.3), (2.4) and (2.19) hold. Since the functional G is convex and finite, i.e. dom $G = P$, the function $p \mapsto G(Au_1 + p)$ is continuous at zero for any $u_1 \in V_0 + u_0$. By finiteness of the functional M condition (2.4) is fulfilled.

The validity of conditions (2.3) and (2.19) is guaranteed by the so called "safe load condition", i.e. we assume that

$$(3.18) \quad \exists \sigma^1 \in Q_f : \mathcal{F}(\sigma^1) \leq -\sqrt{2} k_*(1 - \lambda) \text{ a.e. in } \Omega \text{ for some } \lambda \in]0, 1[.$$

Indeed, by convexity of the function g_0 and by condition (3.1) we may find a number $s_\lambda > 0$ such that $g_0'(s_\lambda) = \sqrt{2}k_* \frac{1+\lambda}{2}$. Then $g_0(t) \geq g_0(s_\lambda) + \sqrt{2}k_* \frac{1+\lambda}{2}(t - s_\lambda)$ for all t and therefore

$$I(v) \geq$$

$$\tfrac{1}{2}K_0 \int_\Omega \operatorname{div}^2 v \, dx + \int_\Omega \left[g_0(s_\lambda) + \sqrt{2}k_* \frac{1+\lambda}{2}(|\varepsilon^D(v)| - s_\lambda) \right] dx - \overline{M}(v)$$

$$= (g_0(s_\lambda) - \sqrt{2}k_* \tfrac{1+\lambda}{2} s_\lambda)|\Omega| + \tfrac{1}{2}K_0 \int_\Omega \operatorname{div}^2 v \, dx$$

(3.19)
$$+\sqrt{2}k_* \tfrac{1+\lambda}{2} \int_\Omega |\varepsilon^D(v)| dx - \overline{M}(v_0) + \int_\Omega \sigma^1 : \varepsilon(u_0) dx - \int_\Omega \sigma^1 : \varepsilon(v) dx$$

$$\geq \tfrac{1}{2}K_0 \int_\Omega \operatorname{div}^2 v \, dx + \sqrt{2}k_* \frac{1-\lambda}{2} \int_\Omega |\varepsilon^D(v)| dx$$

$$-\tfrac{1}{n}\int_\Omega |\mathrm{tr}\sigma^1| \, |\operatorname{div}v| dx + \int_\Omega \sigma^1 : \varepsilon(u_0) dx - \overline{M}(u_0)$$

$$+(g_0(s_\lambda) - \sqrt{2}k_* \tfrac{1+\lambda}{2} s_\lambda)|\Omega| \to +\infty \text{ as } \|v\|_{2,1} \to +\infty, \quad v \in V_0 + u_0.$$

So (2.19) follows from (3.18). Finally, condition (2.3) is provided by the estimate

$$C = \inf\{I(v) : v \in V_0 + u_0\} \geq R(\sigma^1) > -\infty.$$

There is a useful criterion which implies condition (3.18). Let us set

$$\lambda_* = \inf\{\sqrt{2}k_* \int_\Omega |\varepsilon(v)| dx : v \in V_0, \ \operatorname{div} v = 0 \text{ in } \Omega, \ \overline{M}(v) = 1\}.$$

As it was shown in [MM2] there exists a stress tensor $\sigma^* \in K$ such that

$$\int_\Omega \sigma^* : \varepsilon(v) dx = \lambda_* \overline{M}(v) \quad \forall v \in V_0.$$

Suppose that the loads f and F and the domain Ω are such that

(3.20) $\qquad \lambda_* > 1.$

Then, for $\lambda = \frac{1}{\lambda_*}$ and $\sigma^1 = \sigma_* \lambda$, we get $\sigma^1 \in Q_f$ and $|\sigma^{1D}| = \lambda|\sigma_*^D| \leq \lambda\sqrt{2}k_*$, i.e. condition (3.20) implies (3.18).

To sum up, we see that conditions (3.8) and (3.18) guarantee the validity of conditions (2.3), (2.4), (2.17)–(2.19), and we can state that problem (3.11) has at least one solution $\sigma \in Q_f \cap K$, that identity (1.15) holds and that statement (1.16) is valid. As it was demonstrated in the first section the variational problem (3.9) and the corresponding minimax problem (3.6) in general have no solutions. Their relaxations, providing existence of weak solutions, may now be constructed with the help of the abstract scheme of relaxation for convex variational problems in nonreflexive spaces which was discussed in the second section. To do this we need to define the auxiliary operator $A^* : D(A^*) \to U^*$. A pair $p^* = \{\sigma, b\}$ belongs to $D(A^*)$ if and only if there is an element $u^* \in U^* = L^n(\Omega; \mathbb{R}^n)$ such that

$$(3.21) \qquad \int_\Omega u^* \cdot v \, dx = \int_\Omega \sigma : \varepsilon(v) dx - \int_{\partial_2 \Omega} b \cdot v \, d\ell \quad \forall v \in V_0.$$

From (3.21) it follows that if $p^* = \{\sigma, b\} \in D(A^*)$, then $u^* = A^* p^* = -\operatorname{div} \sigma \in L^n(\Omega; \mathbb{R}^n)$. So we have

$$D(A^*) = \{p^* = \{\sigma, b\} \in P^* : \operatorname{div} \sigma \in L^n(\Omega; \mathbb{R}^n),$$

$$\int_{\partial_2 \Omega} b \cdot v \, d\ell = \int_\Omega (\sigma : \varepsilon(v) + v \cdot \operatorname{div} \sigma) dx \quad \forall v \in V_0\}.$$

In the first section (see (1.24)) we introduced the space

$$\Sigma_s = \{\tau \in \Sigma : \operatorname{div} \tau \in L^s(\Omega; \mathbb{R}^n)\}, \quad s \geq 1.$$

DEFINITION 1.3.1 *Let ν be the unit outward normal to the boundary $\partial\Omega$. Let $\tau \in \Sigma_n$ and $b \in L^\infty(\partial_2\Omega; \mathbb{R}^n)$. We say that $\tau\nu = b$ on $\partial_2\Omega$ if and only if $\{\tau, b\} \in D(A^*)$.*

This definition of course is in accordance with the usual pointwise identity $\tau\nu = b$ on $\partial_2\Omega$ provided the functions τ and b are smooth enough. So

$$(3.22) \qquad \tau\nu = b \text{ on } \partial_2\Omega, \ \tau \in \Sigma_n, \ b \in L^\infty(\partial_2; \mathbb{R}^n) \Leftrightarrow p^* = \{\tau, b\} \in D(A^*)$$

is a suitable extension of the classical notion.

Now we can describe the extension V_+ of the set $V_0 + u_0$

$$(3.23) \qquad \begin{cases} V_+ = \{v \in L^{\frac{n}{n-1}}(\Omega; \mathbb{R}^n) : \quad \sup_{\|p^*\|_{P^*} \leq 1, p^* = \{\sigma, b\} \in D(A^*)} \\ \left\langle -\int_{\partial_2 \Omega} b \cdot u_0 \, d\ell + \int_\Omega (\sigma : \varepsilon(u_0) + (u_0 - v) \cdot \operatorname{div} \sigma) \, dx \right\rangle < +\infty\}. \end{cases}$$

From the definition of the class V_+ it follows directly that $V_+ \subset BD(\Omega; \mathbb{R}^n)$ where $BD(\Omega; \mathbb{R}^n)$ is the space of functions of bounded deformation (see [MSC], [Su], [ST2]). Functions from this set are summable and the symmetric part of their gradient is a bounded measure. The norm in $BD(\Omega; \mathbb{R}^n)$ can be introduced for example as follows

$$\|v\|_{BD(\Omega;\mathbb{R}^n)} = \int_\Omega |v|dx + \int_\Omega |\varepsilon(v)|,$$

where

$$\int_\Omega |\varepsilon(v)| = \sup\left\{ -\int_\Omega v \cdot \operatorname{div} \tau \, dx : \tau \in C_0^\infty(\Omega; \mathbb{M}_{\text{sym}}^{n \times n}), \ |\tau| \le 1 \text{ in } \Omega \right\}.$$

By conditions (3.3), (3.17) $G^*(q^*) = G^*(\{\tau, b\}) = +\infty$ if $b \ne F$ on $\partial_2\Omega$ or $\tau \notin K$. For this reason we consider the relaxed Lagrangian of the form

$$(3.24) \quad \begin{cases} L(v, q^*) = \mathcal{E}(u_0, q^*) + (A^*q^*, v) - G^*(q^*) + M(v) \\[2mm] = -\int_{\partial_2\Omega} F \cdot u_0 d\ell + \int_\Omega \Big[\varepsilon(u_0) : \tau + (u_0 - v) \cdot \operatorname{div} \tau - g^*(\tau) - f \cdot v \Big] dx \end{cases}$$

for all $v \in V_+$ and $q^* = \{\tau, F\} \in D(A^*)$ such that $\tau \in K$. It is convenient to introduce the set

$$(3.25) \quad Q = \{\tau \in \Sigma_n : \tau\nu = F \text{ on } \partial_2\Omega\} = \{\tau \in \Sigma : \{\tau, F\} \in D(A^*)\}$$

and the new Lagrangian on $V_+ \times (Q \cap K)$

$$(3.26) \quad \tilde{L}(v, \tau) = L(v, q^*)$$

where

$$q^* = \{\tau, F\} \in D(A^*), \quad \tau \in K.$$

Now, instead of the minimax problem (3.6), we consider its relaxation

$$(3.27) \quad \begin{cases} \text{to find a pair } (u, \sigma) \in V_+ \times (Q \cap K) \text{ such that} \\[2mm] \tilde{L}(u, \tau) \le \tilde{L}(u, \sigma) \le \tilde{L}(v, \sigma) \quad \forall v \in V_+, \ \tau \in Q \cap K. \end{cases}$$

For the functional $\Phi : V_+ \to \mathbb{R}$ we have the formula

$$(3.28) \qquad \Phi(v) = \sup_{q^* \in D(A^*)} L(v, q^*) = \sup_{\substack{q^* = \{\tau, F\} \in D(A^*) \\ \tau \in K}} L(v, q^*) = \sup_{\tau \in Q \cap K} \tilde{L}(v, \tau),$$

and the relaxation of the variational problem (3.9) reads

$$(3.29) \qquad \text{to find } u \in V_+ \text{ such that } \Phi(u) = \inf_{v \in V_+} \Phi(v).$$

We next give a sufficient condition which implies that $\Phi = I$ on the set $V_0 + u_0$.

LEMMA 1.3.1 *Suppose that*

$$(3.30) \qquad \int_0^{+\infty} \left(\sqrt{2}k_* - g_0'(t)\right) dt < +\infty.$$

Then we have

$$(3.31) \qquad \Phi(v) = I(v) \quad \forall v \in V_0 + u_0.$$

Proof.   Recall that

$$g_0^*(\sqrt{2}k_*) = \sup\{\sqrt{2}k_* t - g_0(t) : t \in \mathbb{R}\}.$$

By convexity of g_0 and condition (3.1) we have

$$g_0^*(\sqrt{2}k_*) = \lim_{t \to +\infty} \int_0^t \left(\sqrt{2}k_* - g_0'(\theta)\right) d\theta - g_0(0)$$

and (3.30) implies

$$(3.32) \qquad |g_0^*(\sqrt{2}k_*)| < +\infty.$$

Since the function g_0^* is convex and lower semicontinuous, (3.32) gives

$$(3.33) \qquad g_0^* \in C^0\left([-\sqrt{2}k_*, \sqrt{2}k_*]\right)$$

so that

$$(3.34) \qquad \exists c > 0 : |g_0^*(t)| \le c \quad \forall t \in [-\sqrt{2}k_*, \sqrt{2}k_*].$$

Obviously in our case

$$p^* = \{\sigma, b\} \in \text{dom } G^* \Leftrightarrow \sigma \in K, \ b = F \text{ on } \partial_2 \Omega.$$

By Lemma A.1.4 for $t = 2$ a sequence $\tilde{\sigma}^m \in C^\infty(\overline{\Omega}; \mathbb{M}_{\text{sym}}^{n \times n})$ exists such that

(3.35)
$$\begin{cases} \tilde{\sigma}^m \to \sigma \text{ in } L^2(\Omega; \mathbb{M}_{\text{sym}}^{n \times n}) \\[2mm] \tilde{\sigma}^{mD} \overset{*}{\rightharpoonup} \sigma^D \text{ in } L^\infty(\Omega; \mathbb{M}_{\text{sym}}^{n \times n}) \\[2mm] \|\tilde{\sigma}^{mD}\|_{L^\infty(\Omega)} = \|\sigma^D\|_{L^\infty(\Omega)} (\Rightarrow \tilde{\sigma}^m \in K). \end{cases}$$

Let us take a sequence $\varphi_m \in C_0^\infty(\Omega)$ having the properties

$$0 \leq \varphi_m \leq 1 \text{ on } \Omega, \quad \varphi_m \to 1 \text{ a.e. in } \Omega$$

and let $\sigma^m = \varphi_m \tilde{\sigma}^m + (1 - \varphi)\sigma^1$ where σ^1 is the function from the "safe load condition" (3.18). It is clear that $\sigma^m \in K$. On the other hand, since div $\sigma^1 = -f$ a.e. in Ω and $f \in L^n(\Omega, \mathbb{R}^n)$, we have $\sigma^1 \in \Sigma_n$. But according to Lemma A.1.5 we have $\sigma^1 \in L^n(\Omega; \mathbb{M}_{\text{sym}}^{n \times n})$. So we can deduce

$$\sigma^m \in \Sigma_n \text{ and div } \sigma^m = (1 - \varphi_m)\text{div } \sigma^1 + \varphi_m \text{div } \tilde{\sigma}^m + (\tilde{\sigma}^m - \sigma^1)\nabla\varphi_m$$

for all m. Let us show that $p^{*m} = \{\sigma^m, F\} \in D(A^*)$, i.e. $\sigma^m \in Q$. In fact, by condition (3.18), we have for all $v \in V_0$

$$\int_\Omega \sigma^m : \varepsilon(v) dx = \int_\Omega (\varphi_m \tilde{\sigma}^m + (1 - \varphi_m)\sigma^1) : \varepsilon(v) dx$$

$$= \int_\Omega \sigma^1 : \varepsilon(v) dx + \int_\Omega \varphi_m(\tilde{\sigma}^m - \sigma^1) : \varepsilon(v) dx$$

$$= \int_\Omega f \cdot v \, dx + \int_{\partial_2 \Omega} F \cdot v \, d\ell - \int_\Omega \varphi_m \text{div}(\tilde{\sigma}^m - \sigma^1) \cdot v \, dx$$

$$- \int_\Omega (\tilde{\sigma}^m - \sigma^1) : (v \odot \nabla\varphi_m) dx$$

$$= \int_{\partial_2 \Omega} F \cdot v \, d\ell - \int_\Omega ((1 - \varphi_m)\text{div}\sigma^1 + \varphi_m \text{ div } \sigma^m) \cdot v \, dx$$

$$- \int_\Omega \sigma^1 : (v \odot \nabla(1 - \varphi_m)) dx - \int_\Omega \tilde{\sigma}^m : (v \odot \nabla\varphi_m) dx$$

$$= \int_{\partial_2 \Omega} F \cdot v \, d\ell - \int_\Omega v \cdot \text{div } \sigma^m dx.$$

Taking into account the estimate

$$\left(\int_\Omega |\sigma^m - \sigma|^2 dx\right)^{1/2} \leq \left(\int_\Omega \varphi_m^2 |\tilde\sigma^m - \sigma|^2 dx\right)^{1/2}$$
$$+ \left(\int_\Omega (1-\varphi_m)^2 |\sigma^1 - \sigma|^2 dx\right)^{1/2},$$

we obtain

(3.36)
$$\begin{cases}
\sigma^m \to \sigma \text{ in } L^2(\Omega; M_{sym}^{n\times n}), \\
\\
\sigma^m \to \sigma \text{ a.e. in } \Omega, \\
\\
\sigma^{mD} \stackrel{*}{\rightharpoonup} \sigma^D \text{ in } L^\infty(\Omega; M_{sym}^{n\times n}), \\
\\
\sigma^m \in K, \ \{\sigma^m, F\} \in D(A^*).
\end{cases}$$

Now from (3.33), (3.34), (3.36) and Lebesgue's theorem on dominated convergence it follows that

$$G^*(p^{*m}) = \int_\Omega \left(\frac{1}{2n^2 K_0} tr^2 \sigma^m + g_0^*(|\sigma^{mD}|)\right) dx$$

$$\to \int_\Omega \left(\frac{1}{2n^2 K_0} tr^2 \sigma + g_0^*(|\sigma^D|)\right) dx = G^*(p^*) \text{ as } m \to \infty.$$

But since $D(A^*) \ni p^{*m} \stackrel{*}{\rightharpoonup} p^* = \{\sigma, F\}$ in P^*, we get that all conditions of Lemma 1.2.1 hold and therefore identity (3.31) is valid. Lemma 1.3.1 is proved.
□

It remains to reformulate the Main Theorem 1.2.2 of the second section using the new definitions (see (3.24), (3.25), (3.26), (3.28)).

THEOREM 1.3.1 *Suppose that the conditions* (3.1), (3.8), (3.18) *hold. Then there exists at least one pair* $(u, \sigma) \in V_+ \times (Q \cap K)$ *being a solution to the minimax problem* (3.27). *Moreover,* σ *is a solution to the dual variational problem* (3.11), u *is a solution to the relaxed variational problem* (3.29) *and the identity*

$$\Phi(u) = \inf\{I(v) : v \in V_0 + u_0\} = \tilde L(u, \sigma) = R(\sigma)$$

holds.

A pair $(u, \sigma) \in V_+ \times (Q \cap K)$ is a saddle point of minimax problem (3.27) if and only if $u \in V_+$ is a minimizer of problem (3.29) and $\sigma \in Q_f \cap K$ is a maximizer of problem (3.11).

For any $v \in V_0 + u_0$, we have

$$\Phi(v) \leq I(v),$$

and if for example we assume that

$$\int\limits_0^{+\infty} (\sqrt{2}k_* - g_0'(t)) dt < +\infty,$$

then

$$\Phi(v) = I(v) \quad \forall v \in V_0 + u_0.$$

Finally, any minimizing sequence of problem (3.9) converges strongly in $L^1(\Omega; \mathbb{R}^n)$ and weakly in $L^{\frac{n}{n-1}}(\Omega; \mathbb{R}^n)$ to some solution to problem (3.29).

The statements of Theorem 1.3.1 allow us to call the pair $(u, \sigma) \in V_+ \times (Q \cap K)$ a weak solution to the minimax problem (3.6), and the function $u \in V_+$ a weak solution to problem (3.9).

Let us set

$$(3.37) \qquad t_0 = \sup\{t > 0 : g_0'(t) < \sqrt{2}k_*\}.$$

Note that in the case of Hencky–Il'yushin plasticity $t_0 = \frac{\sqrt{2}k_*}{2\mu} < +\infty$, whereas $t_0 = +\infty$ in the case of example (3.2).

Suppose that

$$(3.38) \qquad g_0 \text{ is of class } C^2 \text{ on }]-t_0, t_0[\text{ and } g_0'' > 0 \text{ on }]-t_0, t_0[.$$

LEMMA 1.3.2 *Let the conditions (3.1), (3.8), (3.18) and (3.38) be satisfied. Then the variational problem (3.11) has a unique solution.*

Proof. It is enough to prove that the function g_0^* is strictly convex. In fact, under condition (3.38), the function g_0^* is the Legendre transformation of g_0 in int dom g_0^*, i.e.

$$t = g_0'(s), \ g_0^*(t) = sg_0'(s) - g_0(s), \ g_0'(]-t_0, t_0[) =]-\sqrt{2}k_*, \sqrt{2}k_*[.$$

Since $(g_0^*)''(t) = 1/g_0''(s)$ for $t = g_0'(s)$ and any $s \in]-t_0, t_0[$, the function g_0^* is strictly convex on $]-\sqrt{2}k_*, \sqrt{2}k_*[$.

If $g_0^*(\pm\sqrt{2}k_*) = +\infty$, then g_0^* is strictly convex on $[-\sqrt{2}k_*, \sqrt{2}k_*]$. Let $g_0^*(\sqrt{2}k_*) < +\infty$. We assume that for some $t \in [0, \sqrt{2}k_*[$ there is a number $\lambda \in]0, 1[$ such that

$$\lambda g_0^*(\sqrt{2}k_*) + (1 - \lambda)g_0^*(t) = g_0^*(t_\lambda),$$

where $t_\lambda = \sqrt{2}k_*\lambda + (1 - \lambda)t$. Then, by the Lagrange theorem, we can find numbers t' and t'' such that

$$t < t' < t'' < \sqrt{2}k_*, \quad (g_0^*)'(t') = (g_0^*)'(t'') = \frac{g_0^*(\sqrt{2}k_*) - g_0^*(t)}{\sqrt{2}k_* - t}.$$

But this contradicts the monotonicity of $(g_0^*)'$ on $]0, \sqrt{2}k_*[$. Since the function g_0^* is even, we have shown that g_0^* is strictly convex on $[-\sqrt{2}k_*, \sqrt{2}k_*]$. Thus the function g^* defined by identity (3.5) is strictly convex. Lemma 1.3.2 is proved. □

Let us return to the example considered in the first section. We observe first that if $\partial_2\Omega = \varnothing$, then $Q \equiv \Sigma_n$. Thus the variational inequality (1.28) is equivalent to the following one

$$\tilde{L}(\tilde{u}, \tau) \leq \tilde{L}(\tilde{u}, \sigma) \leq \tilde{L}(v, \sigma) \quad \forall \tau \in Q \cap K, \ v \in V_+,$$

with functions $\tilde{u}$ and σ being determined by the relations (1.19) and (1.22). If we show that $\tilde{u} \in V_+$, then since $\sigma \in Q \cap K$, the pair $(\tilde{u}, \sigma)$ will be a weak solution to the minimax problem (3.6). It is clear that in this case all conditions of Lemma 1.3.2 hold and therefore problem (3.11) has the unique solution defined by relations (1.19). We know that problem (3.27) admits at least one solution $(\tilde{\tilde{u}}, \sigma) \in V_+ \times (Q \cap K)$, where σ is the function from (1.19), and that the inequality

$$\tilde{L}(\tilde{\tilde{u}}, \tau) \leq \tilde{L}(\tilde{\tilde{u}}, \sigma) \quad \forall \tau \in \Sigma_n \cap K$$

holds which is equivalent to the relation

$$\int_\Omega \left[\varepsilon(u_0) : (\tau - \sigma) + (u_0 - \tilde{\tilde{u}}) \cdot \operatorname{div}(\tau - \sigma) - a(\sigma, \tau - \sigma)\right] dx \leq 0$$

$$\forall \tau \in \Sigma_n \cap K.$$

Since $|\sigma^D| < \sqrt{2}k_*$ in Ω, we get

$$\begin{cases} \varepsilon(\tilde{\tilde{u}}) = \frac{1}{4K_0}\mathbf{1}\operatorname{tr}\sigma + \frac{1}{2\mu}\sigma^D = \varepsilon(\tilde{u}) \quad \text{in } \Omega; \\ \tilde{\tilde{u}} = \tilde{u} \quad \text{at } \rho = R_2. \end{cases}$$

This implies that $\overset{\approx}{u} = \tilde{u}$ in Ω. But then $\tilde{u} \in V_+$ and thus the pair $(\tilde{u}, \sigma)$ is the weak solution to the minimax problem (3.6), and this solution is known to be unique.

We finish this chapter by remarking that for Hencky–Il'yushin plasticity the question concerning the relaxation of the variational problem (3.9) (in this case the function g_0 is given by (1.11)) was studied by G. Anzellotti and M. Giaquinta [AG1], [AG2], R. Temam and G. Strang [ST1], R. Kohn and R. Temam [KT], R. Hardt and D. Kinderlehrer [HK] and by R. Temam [T]. For the case of bending perfect elastoplastic thin plates we refer the reader to [Se9], [CLT].

Chapter 2

Differentiability properties of weak solutions to boundary value problems in the deformation theory of plasticity

2.0   Preliminaries

In the first section of this chapter we briefly recall some of the concepts which have been introduced in chapter 1, in particular, we review how to formulate an appropriate relaxed version of the minimax problem describing the equilibrium configuration of a perfect elastoplastic body. We then collect some results concerning the structure of the corresponding weak solutions. The first theorem states that the stress tensor is weakly differentiable with derivatives in the space L^2_{loc}. For the Hencky–Il'yushin plasticity model we get, roughly speaking, that the elastic zone is an open set. As a byproduct we obtain a partial regularity theorem for strictly convex variational integrals with linear growth. The second section introduces some regularizations of our problem and we prove convergence of these approximations to the original problem. Section three is devoted to the proof of weak differentiablity of the stress tensor, moreover, we discuss some estimate of Caccioppoli type which is an essential tool in our discussion of regularity. The fourth section contains a version of a Campanato type estimate being valid for solutions to systems of partial differential equations occuring in linear elasticity. In the next section we prove certain decay estimates for quantities involving the mean oscillation of the stress tensor and also of the displacement field calculated with respect to balls in the domain of definition. Here we argue by contradiction using a blow–up procedure. In section six we apply the excess decay lemma by the way getting the regularity results stated in section one. A list of open problems concerning perfect plasticity is given in section seven, some comments on plasticity with power hardening can be found in a final section.

2.1   Formulation of the main results

For the reader's convenience we recall the functional formulation of the classical boundary value problem describing the equilibrium of a perfect elastoplastic body within the framework of the deformation theory of plasticity and briefly sketch the ideas leading to the notion of weak solutions. So (see Chapter 1, section 3 for details) our problem is

(1.1)
$$\begin{cases} \text{to find a pair } (u, \sigma) \in (V_0 + u_0) \times K \text{ such that} \\[2mm] \ell(u, \tau) \leq \ell(u, \sigma) \leq \ell(v, \sigma) \quad \forall v \in V_0 + u_0, \quad \tau \in K, \end{cases}$$

where

$$\ell(v, \tau) = \int_\Omega \left(\varepsilon(v) : \tau - g^*(\tau) \right) dx - \overline{M}(v)$$

is the Lagrangian of the minimax problem (1.1). We let

$$V_0 = \{ v \in V : v = 0 \text{ on } \partial_1 \Omega \},$$

$$V = D^{2,1}(\Omega) = \Big\{ v \in L^1(\Omega; \mathbb{R}^n) : \|v\|_{2,1} = \|\operatorname{div} v\|_{L^2(\Omega)}$$

$$+ \|v\|_{L^1(\Omega)} + \|\varepsilon^D(v)\|_{L^1(\Omega)} < +\infty \Big\},$$

$$K = \Big\{ \tau \in \Sigma : \mathcal{F}(\tau) \equiv |\tau^D| - \sqrt{2} k_* \leq 0 \text{ a.e. in } \Omega \Big\},$$

$$\Sigma = \Big\{ \tau = (\tau_{ij}) : \|\tau\|_\Sigma = \|\operatorname{tr} \tau\|_{L^2(\Omega)} + \|\tau^D\|_{L^\infty(\Omega)} < +\infty \Big\},$$

$$\overline{M}(v) = \int_\Omega f \cdot v \, dx + \int_{\partial_2 \Omega} F \cdot v \, d\ell.$$

$g^*(\tau) = \frac{1}{2n^2 K_0} tr^2 \tau + g_0^*(|\tau^D|) = \sup \Big\{ \tau : \varkappa - g(\varkappa) : \varkappa \in \mathbb{M}^{n \times n}_{\text{sym}} \Big\}$ is the conjugate function of g: $\mathbb{M}^{n \times n}_{\text{sym}} \to \overline{\mathbb{R}}$, $g(\varkappa) = \frac{1}{2} K_0 tr^2 \varkappa + g_0(|\varkappa^D|)$, $\varkappa \in \mathbb{M}^{n \times n}_{\text{sym}}$, and $g_0^*(s) = \sup\{st - g_0(t) : t \in \mathbb{R}\}$ denotes the conjugate function of $g_0 : \mathbb{R} \to \overline{\mathbb{R}}$.

We suppose that the given functions f, F, u_0 and g_0 satisfy the following conditions:

(1.2) $f \in L^n(\Omega; \mathbb{R}^n)$, $F \in L^\infty(\partial_2 \Omega; \mathbb{R}^n)$, $u_0 \in V$,

$$\begin{cases} g_0 \text{ is an even convex function of class } C^1, \\[2mm] g_0(0) = 0, \ g_0'(0) = 0, \\[2mm] g_0'(t) \longrightarrow \sqrt{2}k_* \quad \text{as } t \to +\infty. \end{cases}$$

(1.3)

In addition, it is assumed that Ω is a bounded Lipschitz domain in $\mathbb{R}^n$ whose boundary $\partial\Omega$ consists of two measurable parts satisfying

$$\partial_1\Omega \cap \partial_2\Omega = \varnothing, \quad \overline{\partial_1\Omega} \cup \overline{\partial_2\Omega} = \partial\Omega.$$

The minimax problem (1.1) generates two variational problems being in duality:

(1.4)
$$\begin{cases} \text{to find } u \in V_0 + u_0 \text{ such that} \\[2mm] I(u) = \inf\{I(v) : v \in V_0 + u_0\}, \end{cases}$$

where

$$I(v) = \sup_{\tau \in K} \ell(v, \tau) = \int_\Omega g\big(\varepsilon(v)\big)\,dx - \overline{M}(v) \quad \forall v \in V,$$

and

(1.5)
$$\begin{cases} \text{to find } \sigma \in Q_f \cap K \text{ such that} \\[2mm] R(\sigma) = \sup\{R(\tau) : \tau \in Q_f \cap K\}, \end{cases}$$

where

$$R(\tau) = \left\{ \begin{array}{ll} \ell(u_0, \tau), & \tau \in Q_f \cap K \\[2mm] -\infty, & \tau \notin Q_f \cap K \end{array} \right\}, \tau \in K,$$

$$Q_f = \{\tau \in \Sigma : \int_\Omega \tau : \varepsilon(v)\,dx = \overline{M}(v) \ \forall v \in V_0\}.$$

We remark that according to Definition 1.3.1 we have

$$\tau \in Q_f \Leftrightarrow \operatorname{div}\tau = -f \text{ in } \Omega, \quad \tau\nu = F \text{ on } \partial_2\Omega,$$

i.e. the stress tensor τ satisfies the equilibrium equations of stresses in a suitable weak sense.

We also assume that the "safe load condition" holds, i.e.

(1.6)
$$\begin{cases} \exists \sigma^1 \in Q_f \text{ and } \lambda \in]0,1[\text{ such that} \\[2mm] \mathcal{F}(\sigma^1) \leq -\sqrt{2}k_*(1-\lambda) \text{ a.e. in } \Omega. \end{cases}$$

As it was shown in Chapter 1, section 3, (1.6) provides coercivity of the functional I on the set $V_0 + u_0$, i.e.

(1.7) $I(v) \to +\infty$ if $\|v\|_{2,1} \to +\infty$ and $v \in V_0 + u_0$.

Under conditions (1.2), (1.3), (1.6) problem (1.5) has at least one solution while problems (1.1) and (1.4) in general are not solvable. In Chapter 1, section 3, we presented suitable relaxations of problems (1.1) and (1.4) which have solutions and these solutions were called weak solutions of problem (1.1) and problem (1.4), respectively. Instead of problem (1.1) we consider the relaxed minimax problem

(1.8)
$$\begin{cases} \text{to find a pair } (u,\sigma) \in V_+ \times (Q \cap K) \text{ such that} \\[2mm] L(u,\tau) \leq L(u,\sigma) \leq L(v,\tau) \quad \forall v \in V_+, \ \tau \in Q \cap K, \end{cases}$$

where

$$L(v,\tau) = -\int_{\partial_2\Omega} F \cdot u_0 d\ell + \int_{\Omega} \left[\varepsilon(u_0):\tau + (u_0-v)\cdot \operatorname{div}\tau - g^*(\tau) - f\cdot v\right]dx$$

is the Lagrangian of the relaxed problem,

$$V_+ = \left\{ v \in L^{\frac{n}{n-1}}(\Omega;\mathbb{R}^n) : \sup_{\{\tau,b\}\in D(A^*),\|\tau\|_\Sigma+\|b\|_{L^\infty(\partial_2\Omega)}\leq 1} -\int_{\partial_2\Omega} b \cdot u_0 d\ell \right.$$
$$\left. + \int_{\Omega}\left[\tau:\varepsilon(u_0)+(u_0-v)\cdot\operatorname{div}\tau\right]dx < +\infty \right\}$$

is the extension of the set $V_0 + u_0$ of admissible displacement fields, and

$$D(A^*) = \left\{\{\tau,b\} \in \Sigma \times L^\infty(\partial_2\Omega;\mathbb{R}^n) : \operatorname{div}\tau \in L^n(\Omega;\mathbb{R}^n),\right.$$
$$\int_{\Omega}\left(\tau:\varepsilon(v)+v\cdot\operatorname{div}\tau\right)dx = \int_{\partial_2\Omega} b\cdot v\, d\ell \quad \forall v \in V_0\right\},$$
$$Q = \left\{\tau \in \Sigma : \{\tau,F\} \in D(A^*)\right\}.$$

As it was remarked above we have the inclusion

$$(1.9) \quad \begin{cases} V_+ \subset BD(\Omega; \mathbb{R}^n) = \{v \in L^1(\Omega; \mathbb{R}^n) : \\ \\ \|v\|_{BD(\Omega;\mathbb{R}^n)} = \int_\Omega |v| dx + \int_\Omega |\varepsilon(v)| < +\infty\}. \end{cases}$$

According to Theorem 1.3.1 problem (1.8) has at least one solution. Moreover, a pair $(u, \sigma) \in V_+ \times (Q \cap K)$ is a saddle point of the minimax problem (1.8) if and only if $\sigma \in Q_f \cap K$ is a maximizer of problem (1.5) and $u \in V_+$ is a minimizer of the following problem

$$(1.10) \quad \Phi(u) = \inf\{\Phi(v) : v \in V_+\},$$

where $\Phi(v) = \sup\{L(v, \tau) : \tau \in Q \cap K\}$.
We also remark that $\Phi(v) \leq I(v)$ for any $v \in V_0 + u_0$ and that

$$(1.11) \quad \int_0^{+\infty} (\sqrt{2}k_* - g_0'(t)) dt < +\infty$$

implies the validity of

$$I(v) = \Phi(v) \quad \forall v \in V_0 + u_0.$$

We call all solutions to (1.8) weak solutions of the minimax problem (1.1) and all solutions to problem (1.10) weak solutions to the variational problem (1.4).

Setting $t_0 = \sup\{t > 0 : g_0'(t) < \sqrt{2}k_*\}$ we assume

$$(1.12) \quad \begin{cases} g_0 \text{ is of class } C^2 \text{ on}] - t_0, t_0[, \\ \\ g_0'' > 0 \text{ on}] - t_0, t_0[. \end{cases}$$

In this case the solution of problem (1.5) which has the clear physical meaning of the stress tensor is unique. In particular, the decomposition of our elastoplastic body into elastic $(\mathcal{F}(\sigma) < 0)$ and plastic $(\mathcal{F}(\sigma) = 0)$ zones is also unique.

In what follows we will assume that the conditions (1.2), (1.3), (1.6) providing the existence of weak solutions to the minimax problem (1.1) and also the conditions (1.12) providing unique solvability of problem (1.5) with respect to the stress tensor hold.

The main purpose of this chapter is the investigation of the differentiability properties of weak solutions to problem (1.1) in the interior of Ω. For this reason we need a local variant of the inequality

(1.13) $L(u, \tau) \leq L(u, \sigma)$ for all $\tau \in Q \cap K$.

Let $\varphi \in C_0^1(\Omega)$ be an arbitrary cut–off function satisfying $0 \leq \varphi \leq 1$ in Ω. Next, let $\varkappa$ be an arbitrary tensor–valued map from the set $\Sigma_n \cap K$ where

(1.14) $\Sigma_s = \{\tau \in \Sigma : \text{div } \tau \in L^s(\Omega; \mathbb{R}^n)\}, s \geq 1.$

We set $\tau = \sigma + \varphi(\varkappa - \sigma)$ and show that $\tau \in Q \cap K$. It is clear that since $0 \leq \varphi \leq 1$ we have $\tau \in K$. Taking into account that $\varkappa$ and $\sigma \in \Sigma_n$ (div $\sigma = -f \in L^n(\Omega; \mathbb{R}^n)$) we get from Lemma A.1.5

$$\varkappa, \sigma \in L^n(\Omega; \mathbb{M}_{\text{sym}}^{n \times n}) \big(\Rightarrow \tau \in L^n(\Omega; \mathbb{M}_{\text{sym}}^{n \times n}) \big)$$

and therefore

$$\text{div } \tau = \text{div } \sigma + \varphi \, \text{div}(\varkappa - \sigma) + (\varkappa - \sigma)\nabla\varphi \in L^n(\Omega; \mathbb{R}^n).$$

Next, we consider the expression

(1.15)
$$\begin{cases} A = \displaystyle\int_\Omega \Big(\tau : \varepsilon(v) + v \cdot \text{div } \tau \Big) dx = \int_\Omega \Big[\sigma : \varepsilon(v) + v \cdot \text{div } \sigma \\[2mm] \quad + \varphi\varepsilon(v) : (\varkappa - \sigma) + \varphi v \cdot \text{div}(\varkappa - \sigma) + v \cdot \big((\varkappa - \sigma)\nabla\varphi\big)\Big] dx, v \in V_0. \end{cases}$$

Since $\sigma \in Q$ we may write (keeping v fixed)

(1.16)
$$\begin{cases} A = \displaystyle\int_{\partial_2\Omega} F \cdot v \, d\ell + \int_\Omega \Big[\varphi\varepsilon(v) : (\varkappa - \sigma) + \varphi v \cdot \text{div}(\varkappa - \sigma) \\[2mm] \quad + v \cdot \big((\varkappa - \sigma)\nabla\varphi\big)\Big] dx, \ v \in V_0. \end{cases}$$

According to Lemma A.1.3 (applied to $\varkappa - \sigma \in \Sigma_n$) there exists a sequence $\sigma^m \in C^\infty(\overline{\Omega}; \mathbb{M}_{\text{sym}}^{n \times n})$ such that

$$\begin{cases} \sigma^m \to \varkappa - \sigma \text{ in } L^n(\Omega; \mathbb{M}_{\text{sym}}^{n \times n}), \\[2mm] \text{div } \sigma^m \to \text{div}(\varkappa - \sigma) \text{ in } L^n(\Omega; \mathbb{R}^n), \\[2mm] \sigma^{mD} \overset{*}{\to} (\varkappa - \sigma)^D \text{ in } L^\infty(\Omega; \mathbb{M}_{\text{sym}}^{n \times n}), \end{cases}$$

and thus (recall that $v \in D^{2,1}(\Omega) \subset L^{\frac{n}{n-1}}(\Omega; \mathbb{R}^n)$)

$$A_m = \int_{\partial_2\Omega} F \cdot v \, d\ell + \int_\Omega \left[\varphi\varepsilon(v) : \sigma^m + \varphi v \cdot \operatorname{div} \sigma^m + v \cdot (\sigma^m \nabla\varphi) \right] dx \to A$$

for any $v \in V_0$.

On the other hand we may use integration by parts to see

$$A_m = \int_\Omega \left[\varphi\varepsilon(v)\sigma^m + v \cdot \operatorname{div}(\varphi\sigma^m) \right] dx + \int_{\partial_2\Omega} F \cdot v \, d\ell = \int_{\partial_2\Omega} F \cdot v \, d\ell \; \forall v \in V_0.$$

So $A = \int_{\partial_2\Omega} F \cdot v \, d\ell$ and (1.15) means that $\tau \in Q$. Inserting τ into inequality (1.13) and using convexity of g^*, we get the appropriate local variant of (1.13)

$$(1.17) \quad -\int_\Omega \left[\varphi u \cdot \operatorname{div}(\varkappa - \sigma) + (u \odot \nabla\varphi) : (\varkappa - \sigma) + \varphi\big(g^*(\varkappa) - g^*(\sigma)\big) \right] dx \leq 0$$

which is valid for any $\varkappa \in \Sigma_n \cap K$ and $\varphi \in C_0^1(\Omega)$ such that $0 \leq \varphi \leq 1$ in Ω. Here the pair (u, σ) is an arbitrary weak solution to problem (1.1).

Our approach to the local regularity of weak solutions is now based on the observations that the stress tensor is uniquely determined and that the most important physical characteristics are expressed in terms of this tensor, for example the distribution of elastic and plastic zones. For this reason it seemed to us that we should expect better regularity for stresses than for strains which turned out to be true. So we first will try to get regularity for the stresses and then, using the variational inequality (1.17), we discuss regularity of the strains. This is the crucial point of our approach. It is worth to remark that all results on regularity for weak solutions to variational problems in perfect elastoplasticity which are known to the authors have been established in this way (see [EK], [Se3] – [Se10], [BF]). We do not touch the regularity problem for twisting elastoplastic bars being treated for example in [BS], [CR], [F] with the help of different methods.

Let us briefly describe the main results of this chapter. To avoid nonessential technical difficulties we assume in addition that

$$(1.18) \quad u_0 \in W_2^1(\Omega; \mathbb{R}^n).$$

Next, we suppose that the sets $\partial_1\Omega$ and $\partial_2\Omega$ are chosen in such a way that the following density condition holds, namely

$$(1.19) \quad \begin{cases} V_0 \cap W_2^1(\Omega; \mathbb{R}^n) \text{ is dense in } V_0 \\ \\ \text{with respect to the norm of the space } D^{2,1}(\Omega). \end{cases}$$

As it is shown in the Appendix (see A.2) the density condition (1.19) is satisfied in particular if $\partial_1\Omega = \partial\Omega$ or $\partial_2\Omega = \partial\Omega$.

In what follows we will always assume that the conditions (1.2), (1.3), (1.6), (1.12), (1.18) and (1.19) hold and we will not mention this explicitly in the next statements.

Our first result on the regularity of solutions to problem (1.5) is

THEOREM 2.1.1 *Suppose in addition to the above hypothesis that*

$$(1.20) \qquad |f| + |\varepsilon(f)| \in L^\infty_{loc}(\Omega)$$

and let the function g_0 satisfy the condition

$$(1.21) \qquad \exists c_* > 0 : g_0''(t) \le c_* g_0'(t)/t \ \text{ for all } 0 < t < t_0.$$

Then we have weak differentiability of the unique solution σ of (1.5), i.e.

$$(1.22) \qquad \sigma \in W^1_{2,loc}(\Omega; \mathbb{M}^{n\times n}_{sym}).$$

REMARK 2.1.1 Letting t tend to zero in inequality (1.21) and taking into account conditions (1.12), we get $c_* \ge 1$.

REMARK 2.1.2 It is very natural for plasticity theory to assume that the function $t \mapsto g_0'(t)$ is concave for $t > 0$. In this case condition (1.21) is satisfied with $c_* = 1$.

In the case of strict convexity of the function g_0, i.e. for $t_0 = +\infty$, we have

LEMMA 2.1.1 *Suppose that the following two conditions hold:*

$$(1.23) \qquad \text{the function } t \mapsto g_0'(t) \text{ is concave for } t \ge t_* > 0$$

and

$$(1.24) \qquad t_0 = +\infty.$$

Then

$$(1.25) \qquad \left|\{x \in \Omega : \mathcal{F}(\sigma(x)) = 0\}\right| = 0.$$

Now we can formulate the main result of this chapter.

THEOREM 2.1.2 *Suppose that the following conditions are satisfied:*

(1.26) $f \in W^2_{\bar{n},\text{loc}}(\Omega; \mathbb{R}^n)$ *for some* $\bar{n} > n$

and

(1.27)
$$
\begin{cases}
\text{for any } \lambda \in [0, \sqrt{2}k_*[\text{ there is a constant } H_\lambda > 0 \text{ such that} \\[2mm]
\left(\frac{g_0'(|\tau|)}{|\tau|} \tau - \frac{g_0'(|\varkappa|)}{|\varkappa|} \varkappa \right) : (\tau - \varkappa) \geq H_\lambda \max \left\{ g_0''(|\tau|), \frac{g_0'(|\tau|)}{|\tau|} \right\} |\tau - \varkappa|^2 \\[2mm]
\text{for all } \tau \in \mathbb{M}^{n \times n}_{\text{sym}} \text{ and all } \varkappa \in \mathbb{M}^{n \times n}_{\text{sym}} \text{ such that } g_0'(|\varkappa|) \leq \lambda.
\end{cases}
$$

Let σ denote the unique solution of (1.5). Then there exists an open subset $\Omega_0 \subset \Omega$ having the properties

(1.28) $\sigma \in C^\nu(\Omega_0; \mathbb{M}^{n \times n}_{\text{sym}})$ *for any $\nu \in [0, 1[$*

and, moreover,

(1.29) $\mathcal{F}(\sigma(x)) < 0$ *for all $x \in \Omega_0$,*

(1.30) $\mathcal{F}(\sigma) = 0$ *a.e. in $\Omega \setminus \Omega_0$.*

We note two important corollaries of Theorem 2.1.2. For any set $\Omega_1 \subset\subset \Omega_0$ we put $d = \sup\{|\sigma^D(x)| : x \in \overline{\Omega}_1\}$. By Theorem 2.1.2, $d < \sqrt{2}k_*$. Let us use the variational inequality (1.17) choosing a cut–off function φ so that $\varphi \equiv 0$ outside of Ω_0 and $\varphi \equiv 1$ on Ω_1. Then we put $\varkappa = \sigma + \gamma\tau$ where $\tau \in C^1_0(\Omega; \mathbb{M}^{n \times n}_{\text{sym}})$, $|\tau| = 0$ outside of Ω_1, $|\tau| \leq 1$ in Ω, and the number γ satisfies the condition $0 < \gamma < (\sqrt{2}k_* - d)/2$. It is clear that $\varkappa \in \Sigma_n \cap K$ and so inequality (1.17) implies the estimate

(1.31) $-\displaystyle\int_{\Omega_1} u \cdot \text{div}\tau \, dx \leq \frac{1}{\gamma} \int_{\Omega_1} \left[g^*(\sigma + \gamma\tau) - g^*(\sigma) \right] dx.$

On the sets $\{\varkappa \in \mathbb{M}^{n \times n}_{\text{sym}} : |\varkappa^D| < t_0\}$ and $]-t_0, t_0[$ the conjugate functions of g and g_0 coincide with their Legendre transformations and therefore we have

(1.32)
$$
\begin{cases}
g^*(\tau) = \varkappa : \tau - g(\varkappa), \quad \tau = \frac{\partial g}{\partial \varkappa}(\varkappa), \quad |\varkappa^D| < t_0, \\[3mm]
\text{or} \\[3mm]
g(\varkappa) = \varkappa : \tau - g^*(\tau), \quad \varkappa = \frac{\partial g^*}{\partial \tau}(\tau), \quad \mathcal{F}(\tau) < 0, \\[3mm]
\text{and} \\[3mm]
g_0^*(s) = st - g_0(t), \quad s = g_0'(t), \quad |t| < t_0, \\[3mm]
\text{or} \\[3mm]
g_0(t) = st - g_0^*(s), \quad t = (g_0^*)'(s), \quad |s| < \sqrt{2}k_*.
\end{cases}
$$

Thus letting $\gamma \to 0$ in (1.31), we obtain

$$\varepsilon(u) = \frac{\partial g^*}{\partial \tau}(\sigma) \quad \text{or} \quad \sigma = \frac{\partial g}{\partial \varkappa}(\varepsilon(u)) \quad \text{in} \quad \Omega_1.$$

We summarize the above discussion in the following assertion.

COROLLARY 2.1.1 *Let $u \in V_+$ be a weak solution to the variational problem (1.4), i.e. the pair $(u, \sigma) \in V_+ \times (Q \cap K)$ is a saddle point of the Lagrangian on the set $V_+ \times (Q \cap K)$. Then the strain tensor $\varepsilon(u)$ is Hölder continuous on Ω_0, and moreover the displacement vector u satisfies in Ω_0 the following system of differential equations of elliptic type:*

$$div\left(\frac{\partial g}{\partial \varkappa}(\varepsilon(u))\right) + f = 0.$$

Thus the higher–order regularity of u on Ω_0 can be established using a well–known scheme (see [Gi]), and it is determined only by the properties of g_0 and f.

COROLLARY 2.1.2 *Assume that the hypotheses of Theorem 2.1.2 and Lemma 2.1.1 hold. Then $|\Omega \setminus \Omega_0| = 0$, and we have partial regularity of the weak solution, that is regularity on an open set of complete measure.*

At the end of this section we demonstrate a few simple criteria for checking condition (1.27).

LEMMA 2.1.2 *Condition (1.27) is a consequence of (1.21) and the inequality*

$$(1.33) \quad \left| \frac{t}{g_0'(t)} - \frac{s}{g_0'(s)} \right| \leq \frac{|t - s|}{\sqrt{2k_*}}, \quad t, s > 0.$$

Proof. We have for all $\lambda \in [0, \sqrt{2k_*}[$ and any $\tau, \varkappa \in \mathbb{M}_{\text{sym}}^{n \times n}$ such that $g_0'(|\varkappa|) \leq \lambda$

$$\left(\frac{g_0'(|\tau|)}{|\tau|}\tau - \frac{g_0'(|\varkappa|)}{|\varkappa|}\varkappa\right) : (\tau - \varkappa) = \frac{g_0'(|\tau|)}{|\tau|}|\tau - \varkappa|^2$$

$$+ \left(\frac{g_0'(|\tau|)}{|\tau|} - \frac{g_0'(|\varkappa|)}{|\varkappa|}\right)\varkappa : (\tau - \varkappa) \geq \frac{g_0'(|\tau|)}{|\tau|}|\tau - \varkappa|^2$$

$$- \frac{g_0'(|\tau|)}{|\tau|}\frac{g_0'(|\varkappa|)}{|\varkappa|}|\varkappa||\tau - \varkappa|\left|\frac{|\varkappa|}{g_0'(|\varkappa|)} - \frac{|\tau|}{g_0'(|\tau|)}\right|$$

$$\geq \frac{g_0'(|\tau|)}{|\tau|}\left\{|\tau - \varkappa|^2 - \frac{\lambda|\tau - \varkappa|^2}{\sqrt{2k_*}}\right\} = \left(1 - \frac{\lambda}{\sqrt{2k_*}}\right)$$

$$\times \frac{g_0'(|\tau|)}{|\tau|}|\tau - \varkappa|^2 \geq \left(1 - \frac{\lambda}{\sqrt{2k_*}}\right)\frac{1}{c_*}\max\left\{g_0''(|\tau|), \frac{g_0'(|\tau|)}{|\tau|}\right\}|\tau - \varkappa|^2.$$

Thus we can take

$$H_\lambda = \frac{1}{c_*}\left(1 - \frac{\lambda}{\sqrt{2}k_*}\right)$$

and Lemma 2.1.2 is proved.

□

It is easy to check that conditions (1.21) and (1.33) hold for the integrand

$$g_0(t) = \begin{cases} \mu t^2, & |t| \le t_0 = \frac{k_*}{\sqrt{2}\mu} \\ \\ k_*\left(\sqrt{2}|t| - \frac{k_*}{2\mu}\right), & |t| > t_0, \end{cases}$$

describing Hencky–Il'yushin plasticity and therefore all hypothesis of Theorem 2.1.2 are satisfied. In this model it is quite natural to call the domain Ω_0 the elastic zone since in this domain the plastic part of the deformation is equal to zero and, by Corollary 2.1.1, the displacement field u is governed by the system of PDE's from the theory of linear elasticity.

We note also that for $0 \le t < t_0$ condition (1.33) is equivalent to the following one

(1.34) $\quad \left|\left(\frac{t}{g_0'(t)}\right)'\right| \le \frac{1}{\sqrt{2}k_*}, \quad 0 < t < t_0.$

In particular, for the integrand

$$g_0(t) = \sqrt{2}k_*(\sqrt{1+t^2} - 1)$$

we have $t_0 = +\infty$, the function

$$t \mapsto g_0'(t) = \frac{\sqrt{2}k_* t}{\sqrt{1+t^2}}$$

is concave for $t > 0$ and moreover

$$g_0''(t) \le \frac{\sqrt{2}k_*}{(1+t^2)^{3/2}} \le \frac{g_0'(t)}{t}, \quad 0 \le \left(\frac{t}{g_0'(t)}\right)' = \frac{1}{\sqrt{2}k_*}\frac{t}{\sqrt{1+t^2}} \le \frac{1}{\sqrt{2}k_*}$$

for $t > 0$, and so conditions (1.34), (1.21) for $c_* = 1$, (1.23) for any $t_* > 0$ hold. Thus we can apply Theorem 2.1.2 and Corollary 2.1.2 which provide partial regularity of any weak solution.

2.2   Approximation and proof of Lemma 2.1.1

We will study the case $\partial_1 \Omega \neq \varnothing$. The case $\partial_1 \Omega = \varnothing$ requires some non–essential modifications and we recommend the reader to carry out the details. We consider a family of variational problems depending on a parameter $\delta \in]0,1]$

(2.1)
$$\begin{cases} \text{to find } u^\delta \in V_* + u_0 \text{ such that} \\ \\ I_\delta(u^\delta) = \inf\{I_\delta(v) : v \in V_* + u_0\}, \end{cases}$$

where

$$V_* = V_0 \cap W_2^1(\Omega; \mathbb{R}^n), \quad I_\delta(v) = \frac{\delta}{2} \int\limits_\Omega |\varepsilon^D(v)|^2 dx + I(v).$$

Problem (2.1) has a unique minimizer $u^\delta \in V_* + u_0$ which satisfies a nonlinear system of PDE's of elliptic type

(2.2)
$$\begin{cases} \sigma^\delta = \delta \varepsilon^D(u^\delta) + \frac{\partial g}{\partial x}\left(\varepsilon(u^\delta)\right) \\ \\ = \delta \varepsilon^D(u^\delta) + K_0 \operatorname{div} u^\delta \mathbf{1} + g_0'(|\varepsilon^D(u^\delta)|) \frac{\varepsilon^D(u^\delta)}{|\varepsilon^D(u^\delta)|}, \end{cases}$$

(2.3)
$$\int\limits_\Omega \sigma^\delta : \varepsilon(v) dx = \overline{M}(v) \equiv \int\limits_\Omega f \cdot v \, dx + \int\limits_{\partial_2 \Omega} F \cdot v \, d\ell \quad \forall v \in V_*,$$

and therefore

(2.4) $\operatorname{div} \sigma^\delta + f = 0$ a.e. in Ω.

We have

LEMMA 2.2.1 *For any $\delta \in]0,1]$ the following estimate is true*

(2.5)
$$\sqrt{\delta}\|\varepsilon^D(u^\delta)\|_{L^2(\Omega)} + \|\operatorname{div} u^\delta\|_{L^2(\Omega)}$$
$$+ \|\varepsilon^D(u^\delta)\|_{L^1(\Omega)} + \|u^\delta\|_{L^{\frac{n}{n-1}}(\Omega)} \leq C$$

where the positive constant C depends only on

$$\|f\|_{L^n(\Omega)}, \quad \|F\|_{L^\infty(\partial_2\Omega)}, \quad \|u_0\|_{W_2^1(\Omega;\mathbb{R}^n)}$$

and λ being defined in (1.6).

Moreover, there exists a subsequence of the sequence $(u^\delta, \delta^\delta)$, which will be denoted in the same way, such that

(2.6_1) $u^\delta \rightharpoonup u$ in $L^{\frac{n}{n-1}}(\Omega; \mathbb{R}^n)$,

(2.6_2) $u^\delta \to u$ in $L^r(\Omega; \mathbb{R}^n)$ for $r \in [1, \frac{n}{n-1}[$,

(2.6_3) $\displaystyle\int_\Omega \tau : \varepsilon(u^\delta)dx \to \int_\Omega \tau : \varepsilon(u)dx \ \forall \tau \in C_0^\infty(\Omega; \mathbb{M}_{\mathrm{sym}}^{n \times n})$,

(2.6_4) $\mathrm{div}\, \sigma^\delta \rightharpoonup \mathrm{div}\, \sigma$ in $L^2(\Omega)$,

(2.6_5) $\delta \displaystyle\int_\Omega |\varepsilon^D(u^\delta)|^2 dx \to 0$,

(2.6_6) $\sigma^\delta \rightharpoonup \sigma$ in $L^2(\Omega; \mathbb{M}_{\mathrm{sym}}^{n \times n})$,

(2.6_7) $\sigma^{\delta D} - \delta\varepsilon^D(u^\delta) \overset{*}{\rightharpoonup} \sigma^D$ in $L^\infty(\Omega; \mathbb{M}_{\mathrm{sym}}^{n \times n})$,

where σ is the unique solution to problem (1.5), and u is a solution to problem (1.10).

Proof.   Let $s_\lambda > 0$ denote the solution of the equation

(2.7) $g_0'(s_\lambda) = \sqrt{2}k_* \dfrac{1 + \lambda}{2}$.

Using estimate (3.19) of the previous chapter, we get

$$+\infty > I_1(u_0) \geq I_\delta(u_0) \geq I_\delta(u^\delta) \geq \frac{\delta}{2} \int_\Omega |\varepsilon^D(u^\delta)|^2 dx$$

$$+ \frac{1}{2}K_0 \int_\Omega \mathrm{div}^2 u^\delta dx + \sqrt{2}k_* \frac{1 - \lambda}{2} \int_\Omega |\varepsilon^D(u^\delta)|dx$$

$$- \frac{1}{n} \int_\Omega |tr\sigma^1| |\mathrm{div}\, u^\delta|dx + \int_\Omega \sigma^1 : \varepsilon(u_0)dx - \overline{M}(u_0)$$

$$+ \left(g_0(s_\lambda) - \sqrt{2}k_* \tfrac{1+\lambda}{2} s_\lambda \right) |\Omega|.$$

From this and also from the imbedding theorem A.3.1 we deduce estimate (2.5).

It follows from (2.2) that the sequences $\{\sigma^\delta\}$ and $\{\sigma^{\delta D} - \delta\varepsilon^D(u^\delta)\}$ are bounded in $L^2(\Omega; M_{\text{sym}}^{n\times n})$ and $L^\infty(\Omega; M_{\text{sym}}^{n\times n})$, respectively. Therefore, we can state, in view of (2.5), that the claims $(2.6_1) - (2.6_4)$, (2.6_6) and (2.6_7) have been established. It remains to prove that u and σ are solutions to problems (1.10) and (1.5), respectively, and that (2.6_5) holds.

Since $\tau^\delta = \sigma^\delta - \delta\varepsilon^D(u^\delta) \in K$ and since the set K is weakly closed in $L^2(\Omega; M_{\text{sym}}^{n\times n})$, it follows that $\sigma \in K$. Now passing to the limit in (2.3) and using the density condition (1.19), we find that $\sigma \in Q_f$.

On the other hand, the duality relations imply that

$$\tau^\delta : \varepsilon(u^\delta) - g\big(\varepsilon(u^\delta)\big) - g^*(\tau^\delta) = 0 \quad \text{a.e. in } \Omega.$$

But then, by (2.2) and (2.3), we have

$$I_\delta(u^\delta) = \frac{\delta}{2}\int_\Omega |\varepsilon^D(u^\delta)|^2 dx + \int_\Omega \Big(\tau^\delta : \varepsilon(u^\delta) - g^*(\tau^\delta)\Big) dx - \overline{M}(u^\delta)$$

$$= -\frac{\delta}{2}\int_\Omega |\varepsilon^D(u^\delta)|^2 dx + \int_\Omega \sigma^\delta : \varepsilon(u^\delta) dx - \int_\Omega g^*(\tau^\delta) dx - \overline{M}(u^\delta)$$

$$= -\frac{\delta}{2}\int_\Omega |\varepsilon^D(u^\delta)|^2 dx - \int_\Omega g^*(\tau^\delta) dx + \int_\Omega \sigma^\delta : \varepsilon(u_0) dx - \overline{M}(u_0).$$

Taking into account the statements of Theorem 1.3.1 of the previous chapter, we get

$$\sup\{R(\tau) : \tau \in Q_f \cap K\} = \inf\{I(v) : v \in V_0 + u_0\} \leq I(u^\delta)$$

$$\leq I_\delta(u^\delta) = -\frac{\delta}{2}\int_\Omega |\varepsilon^D(u^\delta)|^2 dx + \int_\Omega (\sigma^\delta : \varepsilon(u_0) - g^*(\tau^\delta)) dx - \overline{M}(u_0).$$

Passing to the limit in this inequality, using the upper semicontinuity of the functional

$$\tau \mapsto -\int_\Omega g^*(\tau) dx$$

with respect to the weak–$*$ topology of Σ and the fact that $\sigma \in Q_f \cap K$, we obtain

$$(2.8) \qquad \sup\{R(\tau) : \tau \in Q_f \cap K\} \leq R(\sigma) + \limsup_{\delta\to 0}\left(-\frac{\delta}{2}\int_\Omega |\varepsilon^D(u^\delta)|^2 dx\right).$$

Inequality (2.8) implies that σ is a maximizer of problem (1.5) and also that

$$\limsup_{\delta \to 0} \left[-\frac{\delta}{2} \int_\Omega |\varepsilon^D(u^\delta)|^2 dx \right] \geq 0.$$

By passing to a subsequence (if necessary) we deduce claim (2.6_5) together with the identity

$$\lim_{\delta \to +\infty} I(u^\delta) = \inf\{I(v) : v \in V_0 + u_0\}$$

which shows that u^δ is a minimizing sequence of problem (1.4), and therefore it converges weakly in $L^{\frac{n}{n-1}}(\Omega; \mathbb{R}^n)$ to a solution of problem (1.10). Lemma 2.2.1 is proved.

$\square$

Proof of Lemma 2.1.1.   Let $\psi^\delta = \max\{t_*, |\varepsilon^D(u^\delta)|\}$. Estimate (2.5) implies that there exists a bounded positive Radon measure μ such that

$$\int_\Omega \varphi \psi^\delta dx \to \int_\Omega \varphi d\mu \quad \forall \varphi \in C_0^0(\Omega).$$

Let Q be a closed cube in Ω. Then

$$(2.9) \qquad \limsup_{\delta \to 0} \int_Q \psi^\delta dx \leq \mu(Q).$$

Further we have (recall $\tau^\delta = \sigma^\delta - \delta \varepsilon^D(u^\delta)$ and equation (2.2))

$$\int_Q |\tau^{\delta D}| dx \leq \int_Q g_0'(|\varepsilon^D(u^\delta)|) dx \leq \int_Q g_0'(\psi^\delta) dx.$$

From condition (1.23) and Jensen's inequality we infer the estimate

$$(2.10) \qquad \fint_Q |\tau^{\delta D}| dx \leq g_0'\left(\fint_Q \psi^\delta dx \right).$$

Taking into account (2.9), (2.10) and the results of Lemma 2.2.1 we obtain

$$(2.11) \qquad \begin{cases} \fint_Q |\sigma^D| dx \leq \liminf_{\delta \to 0} \fint_Q |\tau^{\delta D}| dx \\[2mm] \leq \liminf_{\delta \to 0} g_0'\left(\fint_Q \psi^\delta dx \right) \leq g_0'\left(\limsup_{\delta \to 0} \fint_Q \psi^\delta dx \right) \\[2mm] \leq g_0'\left(\frac{\mu(Q)}{|Q|} \right). \end{cases}$$

Let $x \in \Omega$ be the center of the cube Q. The function

$$x \mapsto \frac{d\mu}{dx}(x) = \lim_{|Q| \to 0} \frac{\mu(Q)}{|Q|}$$

is summable with respect to Lebesgue's measure in $\mathbb{R}^n$, see [DS], and is therefore finite almost everywhere with respect to this measure. But (2.11) implies that

$$|\sigma^D(x)| \leq g_0'\left(\frac{d\mu}{dx}(x)\right) \text{ for a.a. } x \in \Omega,$$

and by (1.24) $|\sigma^D| < \sqrt{2}k_*$ a.e. in Ω. Lemma 2.1.1 is proved.

$\square$

2.3 Proof of Theorem 2.1.1 and a local estimate of Caccioppoli-type for the stress tensor

We denote by $V_\times$ the finite–dimensional space of rigid displacements, i.e.

$$V_\times = \{v_\times = a + \omega x : a \in \mathbb{R}^n,\ \omega \text{ is a skew symmetric tensor}\}.$$

We continue to consider the regularized problem (2.1). Since $f \in L^n(\Omega; \mathbb{R}^n)$, it is easy to show, using the method of finite differences, that

$$(3.1) \qquad \varepsilon(u^\delta),\ \sigma^\delta \in W^1_{2,\text{loc}}(\Omega; \mathbb{M}^{n \times n}_{\text{sym}}),$$

and therefore we have

$$(3.2) \qquad \int_\Omega \sigma^\delta_{,k} : \varepsilon(v) dx = - \int_\Omega f \cdot v_{,k} dx \quad \forall v \in C^\infty_0(\Omega; \mathbb{R}^n),\ k = 1, 2, \ldots, n,$$

so that

$$(3.3) \qquad \sigma^\delta_{ij,j} = -f_i \text{ a.e. in } \Omega.$$

Let us introduce the bilinear forms $E(\tau; \cdot, \cdot)$ and $E^\delta(\tau; \cdot, \cdot)$ on the space $\mathbb{M}^{n \times n}_{\text{sym}}$ by defining

$$(3.4) \qquad \begin{cases} E(\tau; \varepsilon, \varkappa) = \left(\frac{\partial^2 g_0}{\partial \tau^2}(|\tau|)\varepsilon \right) : \varkappa \\[2mm] = \frac{g'_0(|\tau|)}{|\tau|} \varepsilon : \varkappa + \left(g''_0(|\tau|) - \frac{g'_0(|\tau|)}{|\tau|} \right) \frac{\tau : \varepsilon\ \tau : \varkappa}{|\tau|^2}, \\[2mm] E^\delta(\tau; \varepsilon, \varkappa) = \delta \varepsilon^D : \varkappa^D + K_0 \text{tr}\varepsilon\ \text{tr}\varkappa + E(\tau; \varepsilon^D, \varkappa^D), \end{cases}$$

where $\tau \in \mathbb{M}^{n \times n}_{\text{sym}}$ is a tensor parameter. Condition (1.21) provides the estimates

$$(3.5) \qquad h_-(|\tau|)|\varepsilon|^2 \le E(\tau; \varepsilon, \varepsilon) \le h_+(|\tau|)|\varepsilon|^2 \le c_* \frac{g'_0(|\tau|)}{|\tau|}|\varepsilon|^2,$$

in which

$$h_-(t) = \min\{g''_0(t), \frac{g'_0(t)}{t}\},\quad h_+(t) = \max\{g''_0(t), \frac{g'_0(t)}{t}\},\ t > 0.$$

Moreover, since

$$(3.6) \qquad \sigma^\delta_{,k} = \delta\varepsilon^D(u^\delta_{,k}) + K_0\mathrm{div}\, u^\delta_{,k} + \frac{\partial^2 g_0}{\partial\tau^2}(|\varepsilon^D(u^\delta)|)\varepsilon^D(u^\delta_{,k}),$$

it follows that

$$(3.7) \qquad \sigma^\delta_{,k} : \varkappa = E^\delta(\varepsilon^D(u^\delta); \varepsilon^D(u^\delta_{,k}), \varkappa) \quad \forall \varkappa \in \mathbb{M}^{n\times n}_{\mathrm{sym}}.$$

Now we are going to derive some inequalities for $\nabla\sigma^\delta$. We have (using summation w.r.t. k)

$$(3.8) \quad \begin{cases} |\nabla\sigma^\delta|^2 = \sigma^\delta_{,k} : \sigma^\delta_{,k} = E^\delta\big(\varepsilon^D(u^\delta); \varepsilon(u^\delta_{,k}), \sigma_{,k}\big) \\[2ex] \leq \Big(E^\delta\big(\varepsilon^D(u^\delta); \varepsilon(u^\delta_{,k}), \varepsilon(u^\delta_{,k})\big)\Big)^{1/2}\Big(E^\delta\big(\varepsilon^D(u^\delta); \sigma^\delta_{,k}, \sigma^\delta_{,k}\big)\Big)^{1/2} \\[2ex] \leq \big(\sigma^\delta_{,k} : \varepsilon(u^\delta_{,k})\big)^{1/2}\Big[\delta|\nabla\sigma^{\delta D}|^2 + K_0|\nabla\mathrm{tr}\sigma^\delta|^2 \\[2ex] +c_*\dfrac{g'_0(|\varepsilon^D(u^\delta)|)}{|\varepsilon^D(u^\delta)|}|\nabla\sigma^{\delta D}|^2\Big]^{1/2}. \end{cases}$$

We put

$$(3.9) \qquad \nu_2 = \sup\{g'_0(t)/t : t > 0\}$$

and show that $\nu_2 < +\infty$. Indeed, let us take and fix some $t_1 \in\,]0, t_0[$. Then, by condition (1.12), we have

$$\nu_2 = \max\Big\{\sup\{h(t) : t \in [0, t_1]\},\ \sup\{\frac{g'_0(t)}{t} : t > t_1\}\Big\},$$

where $h \in C^0([0, t_1])$ is defined according to

$$h(t) = \begin{cases} g'_0(t)/t & \text{if } t > 0 \\[1ex] g''_0(0) & \text{if } t = 0. \end{cases}$$

Thus

$$\nu_2 \leq \max\Big\{\sup\{h(t) : t \in [0, t_1]\},\ \frac{\sqrt{2}k_*}{t_1}\Big\} < +\infty.$$

Since $\delta \in]0, 1]$, it follows from (3.8) that

$$|\nabla \sigma^\delta|^2 \le \left(\sigma^\delta_{,k} : \varepsilon^D(u^\delta_{,k})\right)^{1/2} \left[(1 + \nu_2 c_*)|\nabla \sigma^{\delta D}|^2 + K_0|\nabla tr \sigma^\delta|^2\right]^{1/2}$$

$$\le c_1(c_*, \nu_2, K_0)\left(\sigma^\delta_{,k} : \varepsilon(u^\delta_{,k})\right)^{1/2}|\nabla \sigma^\delta|.$$

So we have established the estimate

(3.10) $|\nabla \sigma^\delta|^2 \le c_1^2 \sigma^\delta_{,k} : \varepsilon(u^\delta_{,k}).$

Note that in order to derive estimate (3.10) we made essential use of condition (1.21) of Theorem 2.1.1.

Let us introduce some additional tensor–valued and vector–valued fields as follows

(3.11) $\bar{\sigma}^\delta = \sigma^\delta - \sigma^0, \quad \bar{u}^\delta = u^\delta - \varkappa^0(x - x_0) - v_\times,$

where σ^0 and $\varkappa^0$ are arbitrary matrices from $\mathbb{M}^{n \times n}_{sym}$, and $v_\times$ is an arbitrary rigid displacement from $V_\times$. Next, let $\varphi \in C^3_0(\Omega)$ be an arbitrary cut–off function. By (3.1) we can apply (3.2) to the function

$$v = \varphi^6 \bar{u}^\delta_{,k}$$

and get

(3.12)

$$\begin{cases} \displaystyle\int_\Omega \sigma^\delta_{,k} : \varepsilon(\varphi^6 \bar{u}^\delta_{,k})dx = \int_\Omega f_{,k}(\varphi^6 \bar{u}^\delta_{,k})dx \\[4mm] \displaystyle = -\int_\Omega f \cdot \varphi^6 \Delta u^\delta dx - \int_\Omega f \cdot \varphi^6_{,k} \bar{u}^\delta_{,k} dx. \end{cases}$$

Since

$$\frac{1}{2}\Delta u^\delta = \operatorname{div} \varepsilon(u^\delta) - \frac{1}{2}\nabla \operatorname{div} u^\delta,$$

it follows that

$$
\begin{cases}
\quad -\int_\Omega f \cdot \varphi^6 \Delta u^\delta dx - \int_\Omega f \cdot \varphi^6_{,k} \overline{u}^\delta_{,k} dx \\[2mm]
= \; -2\int_\Omega f \cdot \varphi^6 \mathrm{div}\, \varepsilon(u^\delta) dx + \int_\Omega f \cdot \varphi^6 \nabla \mathrm{div}\, u^\delta dx - \int_\Omega f \cdot \varphi^6_{,k} \overline{u}^\delta_{,k} dx \\[2mm]
= \; 2\int_\Omega \varepsilon(f) : \varphi^6 \varepsilon(\overline{u}^\delta) dx + 2\int_\Omega (f \odot \nabla \varphi^6) : \varepsilon(\overline{u}^\delta) dx \\[2mm]
\quad + \int_\Omega \varphi^6 f \cdot \nabla \mathrm{div}\, u^\delta dx - \int_\Omega f \cdot \varphi^6_{,k} \overline{u}^\delta_{,k} dx \\[2mm]
= \; 2\int_\Omega \varphi^6 \varepsilon(f) : \varepsilon(\overline{u}^\delta) dx + \int_\Omega \varphi^6 f \cdot \nabla \mathrm{div}\, u^\delta dx + \int_\Omega f_i \varphi^6_{,j} \overline{u}^\delta_{j,i} dx \\[2mm]
= \; 2\int_\Omega \varphi^6 \varepsilon(f) : \varepsilon(\overline{u}^\delta) dx + \int_\Omega \varphi^6 f \cdot \nabla \mathrm{div}\, u^\delta dx \\[2mm]
\quad - \int_\Omega \nabla \varphi^6 \cdot \overline{u}^\delta \mathrm{div}\, f \, dx - \int_\Omega (f \odot \overline{u}^\delta) : \nabla^2 \varphi^6 dx.
\end{cases}
\tag{3.13}
$$

So (3.12) and (3.13) give us the decomposition

$$
J \equiv \int_\Omega \varphi^6 \sigma^\delta_{,k} : \varepsilon(u^\delta_{,k}) dx = J_1 + J_2 + J_3,
\tag{3.14}
$$

where

$$
\begin{cases}
J_1 \equiv \; -2\int_\Omega \sigma^\delta_{ij,k} \varphi^6_{,i} \varepsilon_{jk}(\overline{u}^\delta) dx, \\[3mm]
J_2 \equiv \; \int_\Omega \sigma^\delta_{ij,k} \varphi^6_{,i} \overline{u}^\delta_{k,j} dx, \\[3mm]
J_3 \equiv \; \int_\Omega \Big[2\varphi^6 \varepsilon(f) : \varepsilon(\overline{u}^\delta) + \varphi^6 f \cdot \nabla \mathrm{div}\, u^\delta \\[3mm]
\qquad\qquad - \nabla \varphi^6 \cdot \overline{u}^\delta \mathrm{div}\, f - (f \odot \overline{u}^\delta) : \nabla^2 \varphi^6 \Big] dx.
\end{cases}
\tag{3.15}
$$

Next we use the formulas

$$\varepsilon(\overline{u}^\delta) = \varepsilon^D(\overline{u}^\delta) + \frac{1}{n}\operatorname{div}\overline{u}^\delta\mathbf{1}, \quad \overline{\sigma}^\delta = \overline{\sigma}^{\delta D} + \frac{1}{n}\operatorname{tr}\overline{\sigma}^\delta\mathbf{1}$$

and identity (3.3). As a result we have the relation

$$(3.16) \quad \begin{cases} J_1 &= -2\int\limits_\Omega \sigma^\delta_{ij,k}\varphi^6_{,i}\varepsilon^D_{jk}(\overline{u}^\delta)dx - \frac{2}{n}\int\limits_\Omega \sigma^\delta_{ij,j}\varphi^6_{,i}\operatorname{div}\overline{u}^\delta dx \\[2mm]
&= -\frac{2}{n}\int\limits_\Omega \operatorname{tr}\sigma^\delta_{,k}\varphi^6_{,i}\varepsilon^D_{ik}(\overline{u}^\delta)dx - 2\int\limits_\Omega \sigma^{\delta D}_{ij,k}\varphi^6_{,i}\varepsilon^D_{jk}(\overline{u}^\delta)dx \\[2mm]
&\quad +\frac{2}{n}\int\limits_\Omega f\cdot\nabla\varphi^6\operatorname{div}\overline{u}^\delta dx = 2\int\limits_\Omega \left(f_k + \sigma^{\delta D}_{ks,s}\right)\varphi^6_{,i}\varepsilon^D_{ik}(\overline{u}^\delta)dx \\[2mm]
&\quad -2\int\limits_\Omega \sigma^{\delta D}_{ij,k}\varphi^6_{,i}\varepsilon^D_{jk}(\overline{u}^\delta)dx + \frac{2}{n}\int\limits_\Omega f\cdot\nabla\varphi^6\operatorname{div}\overline{u}^\delta dx \\[2mm]
&= 2\int\limits_\Omega (f\odot\nabla\varphi^6):\varepsilon^D(\overline{u}^\delta)dx + \frac{2}{n}\int\limits_\Omega f\cdot\nabla\varphi^6\operatorname{div}\overline{u}^\delta dx \\[2mm]
&\quad +12\int\limits_\Omega \varphi^5\sigma^{\delta D}_{ij,k}\left(-\varphi_{,i}\varepsilon^D_{jk}(\overline{u}^\delta) + \delta_{ik}\varphi_{,s}\varepsilon^D_{js}(\overline{u}^\delta)\right)dx \\[2mm]
&= 2\int\limits_\Omega (f\odot\nabla\varphi^6):\varepsilon(\overline{u}^\delta)dx + 12\int\limits_\Omega \varphi^5\sigma^{\delta D}_{,k}:S^{(k)}dx \end{cases}$$

where the matrices $S^{(k)}$, $k = 1, 2, \ldots, n$, are given by

$$S^{(k)} = \left(\delta_{ik}\varphi_{,s}\varepsilon^D_{js}(\overline{u}^\delta) - \varphi_{,i}\varepsilon^D_{jk}(\overline{u}^\delta)\right), \quad \delta_{ij} = \begin{cases} 1, & i = j \\ 0, & i \neq j. \end{cases}$$

We note the following relations

$$S^{(k)} : S^{(k)} \leq c_2(n)|\nabla\varphi|^2|\varepsilon^D(\overline{u}^\delta)|^2,$$

$$\sigma^{\delta D}_{,k} : S^{(k)} = \tfrac{1}{2}\sigma^{\delta D}_{,k} : \left(S^{(k)} + (S^{(k)})^T\right)$$

$$= \tfrac{\delta}{4}|S^{(k)} + (S^k)^T|^2 + E\left(\varepsilon^D(u^\delta); \varepsilon^D(u^\delta_{,k}), S^{(k)}\right)$$

which together with (3.4), (3.5), (3.6) imply

$$\sigma^{\delta D}_{,k} : S^{(k)} \leq |\nabla\varphi|\sqrt{c_2}\Big[\delta + c_* \frac{g'_0(|\varepsilon^D(u^\delta)|)}{|\varepsilon^D(u^\delta)|}\Big]^{1/2} |\varepsilon^D(\overline{u}^\delta)| \Big[\sigma^{\delta D}_{,k} : \varepsilon^D(u^\delta_{,k})\Big]^{1/2}.$$

Thus (3.16) and the last inequality give the bound

(3.17)
$$\begin{cases} J_1 \leq 2 \displaystyle\int_\Omega (f \odot \nabla\varphi^6) : \varepsilon(\overline{u}^\delta)dx \\[4mm] +12\sqrt{c_2}J^{1/2}\Big(\displaystyle\int_\Omega \varphi^4|\nabla\varphi|^2\Big[\delta + c_* \frac{g'_0(|\varepsilon^D(u^\delta)|)}{|\varepsilon^D(u^\delta)|}\Big]|\varepsilon^D(\overline{u}^\delta)|^2 dx\Big)^{1/2}. \end{cases}$$

Let us transform J_2, using integration by parts, in the following way

$$\begin{aligned} J_2 &= -\int_\Omega [\overline{\sigma}^\delta_{ij}\varphi^6_{,ik}\overline{u}^\delta_{k,j} + \overline{\sigma}^\delta_{ij}\varphi^6_{,j}\mathrm{div}\,\overline{u}^\delta_{,i}]dx \\[3mm] &= \int_\Omega \sigma^\delta_{ij,j}\varphi^6_{,ik}\overline{u}^\delta_k dx + \int_\Omega \overline{\sigma}^\delta_{ij}\varphi^6_{,ijk}\overline{u}^\delta_k dx \\[3mm] &\quad + \int_\Omega \sigma^\delta_{ij,j}\varphi^6_{,i}\mathrm{div}\,\overline{u}^\delta dx + \int_\Omega \overline{\sigma}^\delta_{ij}\varphi^6_{,ij}\mathrm{div}\,\overline{u}^\delta dx \\[3mm] &= -\int_\Omega (f \odot \overline{u}^\delta) : \nabla^2\varphi^6 dx - \int_\Omega f \cdot \nabla\varphi^6 \mathrm{div}\,\overline{u}^\delta dx \\[3mm] &\quad + \int_\Omega \overline{\sigma}^\delta_{ij}\varphi^6_{,ijk}\overline{u}^\delta_k dx + \int_\Omega \overline{\sigma}^\delta : \nabla^2\varphi^6\mathrm{div}\,\overline{u}^\delta \, dx. \end{aligned}$$

Combining this with (3.15), (3.17) gives:

(3.18)
$$\begin{cases} J \leq 12\sqrt{c_2}J^{1/2}\Big(\displaystyle\int_\Omega \varphi^4|\nabla\varphi|^2\Big[\delta + c_* \frac{g'_0(|\varepsilon^D(u^\delta)|)}{|\varepsilon^D(u^\delta)|}\Big]|\varepsilon^D(\overline{u}^\delta)|^2 dx\Big)^{1/2} \\[4mm] +\displaystyle\int_\Omega \overline{\sigma}^\delta_{ij}\varphi^6_{,ijk}\overline{u}^\delta_k dx + \int_\Omega \overline{\sigma}^\delta : \nabla\varphi^6\mathrm{div}\,\overline{u}^\delta dx + J_0, \end{cases}$$

where

$$
(3.19) \quad
\begin{cases}
J_0 = 2\int_\Omega \Big[(f \odot \nabla\varphi^6) : \varepsilon(\overline{u}^\delta) + \varphi^6\varepsilon(f) : \varepsilon(\overline{u}^\delta) \\[2mm]
\quad - f \cdot \nabla\varphi^6 \operatorname{div}\overline{u}^\delta - (f \odot \overline{u}^\delta) : \nabla^2\varphi^6\Big]dx \\[2mm]
\quad - \int_\Omega (\varphi^6 \operatorname{div} f \operatorname{div}\overline{u}^\delta + \nabla\varphi^6 \cdot \overline{u}^\delta \operatorname{div} f)dx.
\end{cases}
$$

After application of Young's inequality we finally get

$$
(3.20) \quad
\begin{cases}
J \le 144c_2 \int_\Omega \varphi^4 |\nabla\varphi|^2 \Big[\delta + c_* \dfrac{g_0'(|\varepsilon^D(u^\delta)|)}{|\varepsilon^D(u^\delta)|}\Big] |\varepsilon^D(\overline{u}^\delta)|^2 dx \\[4mm]
\quad + 2\int_\Omega \overline{\sigma}_{ij}^\delta \varphi_{,ijk}^6 \overline{u}_k^\delta dx + 2\int_\Omega \overline{\sigma}^\delta : \nabla^2\varphi^6 \operatorname{div}\overline{u}^\delta dx + 2J_0.
\end{cases}
$$

After these preparations we are ready to give the

Proof of Theorem 2.1.1.   We put $\sigma^0 = \varkappa^0 = 0$, $v_\times = 0$ in (3.11). In view of (2.2) we have

$$
(3.21) \quad \operatorname{div} u^\delta = \frac{1}{nK_0}\operatorname{tr}\sigma^\delta \quad \text{in } \Omega,
$$

hence it follows from (3.10), (3.20) and (1.3) that

$$
(3.22) \quad
\begin{cases}
\dfrac{1}{c_1^2}\int_\Omega \varphi^6 |\nabla\sigma^\delta|^2 dx \le 144c_2 \int_\Omega \varphi^4 |\nabla\varphi|^2 \Big[\delta|\varepsilon^D(u^\delta)|^2 + c_*\sqrt{2}k_* |\varepsilon^D(u^\delta)|\Big]dx \\[4mm]
\quad + 2J_0 + \dfrac{2}{nK_0}\int_\Omega \operatorname{tr}\sigma^\delta \, \sigma^\delta : \nabla^2\varphi^6 dx + 2\int_\Omega \sigma_{ij}^\delta \varphi_{,ijk}^6 u_k^\delta dx.
\end{cases}
$$

Now let us estimate the integral

$$
J_4 \equiv 2\int_\Omega \sigma_{ij}^\delta \varphi_{,ijk}^6 u_k^\delta dx = 2\int_\Omega \varphi^3 \sigma_{ij}^\delta A_{ijk} u_k^\delta dx
$$

where

$$
A_{ijk} = 6\Big\{5\big[4\varphi_{,i}\varphi_{,j}\varphi_{,k} + \varphi(\varphi_{,i}\varphi_{,jk} + \varphi_{,j}\varphi_{,ik} + \varphi_{,k}\varphi_{,ij})\big] + \varphi^2\varphi_{,ijk}\Big\}.
$$

During the following calculation we will use the restriction $n = 2$ or 3. By this and the imbedding theorem we have the inequality

$$\left(\int_\Omega |\varphi^3 \sigma^\delta|^n dx\right)^{1/n} \le c_3(n) d \left(\int_\Omega |\nabla(\varphi^3 \sigma^\delta)|^2 dx\right)^{1/2}$$

$$\le \sqrt{2} c_3 d \left[\left(\int_\Omega \varphi^6 |\nabla \sigma^\delta|^2 dx\right)^{1/2} + 3\left(\int_\Omega \varphi^4 |\nabla\varphi|^2 |\sigma^\delta|^2 dx\right)^{1/2}\right],$$

$$d = |\mathrm{spt}\varphi|^{\frac{2}{n} - \frac{1}{2}},$$

and thus

$$J_4 \le c_4(n) \left(\int_\Omega |\varphi^3 \sigma^\delta|^n dx\right)^{1/n} \left(\int_\Omega (A_{ijk} u_k^\delta A_{ijm} u_m^\delta)^{\frac{n}{2(n-1)}} dx\right)^{\frac{n-1}{n}}$$

$$\le c_5(n) d \left[\left(\int_\Omega \varphi^6 |\nabla \sigma^\delta|^2 dx\right)^{1/2} + 3\left(\int_\Omega \varphi^4 |\nabla\varphi|^2 |\sigma^\delta|^2 dx\right)^{1/2}\right]$$

$$\times \left(\int_\Omega (A_{ijk} u_k^\delta A_{ijm} u_m^\delta)^{\frac{n}{2(n-1)}} dx\right)^{\frac{n-1}{n}}.$$

So (3.22) can be rewritten in the form

$$\frac{1}{c_1^2} \int_\Omega \varphi^6 |\nabla \sigma^\delta|^2 dx \le 144 c_2 \int_\Omega \varphi^4 |\nabla\varphi|^2 \left(\delta |\varepsilon^D(u^\delta)|^2 + c_* \sqrt{2} k_* |\varepsilon^D(u^\delta)|\right) dx$$

$$+ 2 J_0 + \frac{2}{nK_0} \int_\Omega \mathrm{tr}\sigma^\delta \sigma^\delta : \nabla^2 \varphi^6 dx + c_5 d \left[\left(\int_\Omega \varphi^6 |\nabla \sigma^\delta|^2 dx\right)^{1/2}\right.$$

$$+ 3\left(\int_\Omega \varphi^4 |\nabla\varphi|^2 |\sigma^\delta|^2 dx\right)^{1/2}\right]\left(\int_\Omega (A_{ijk} u_k^\delta A_{ijm} u_m^\delta)^{\frac{n}{2(n-1)}} dx\right)^{\frac{n-1}{n}}.$$

It remains to apply Young's inequality in order to get the final estimate

$$\int_\Omega \varphi^6 |\nabla \sigma^\delta|^2 dx \leq c_6(n, K_0, k_*, c_*, c_1)$$

$$\left[\int_\Omega \langle \varphi^4 |\nabla \varphi|^2 (\delta |\varepsilon^D(u^\delta)|^2 + |\sigma^\delta|^2 + |\varepsilon^D(u^\delta)|) \right.$$

$$\left. + |\nabla^2 \varphi^6| |\sigma^\delta|^2 \rangle dx + |J_0| + d^2 \left(\int_\Omega (A_{ijk} u_k^\delta A_{ijm} u_m^\delta)^{\frac{n}{2(n-1)}} dx \right)^{\frac{2(n-1)}{n}} \right]$$

$$\leq c_7(n, K_0, k_*, c_*, c_1, \varphi) \left[\int_\Omega (\delta |\varepsilon^D(u^\delta)|^2 + |\varepsilon^D(u^\delta)| + |\sigma^\delta|^2) dx \right.$$

$$+ \left(\int_\Omega |u^\delta|^{\frac{n}{n-1}} dx \right)^{\frac{2(n-1)}{n}} + \left(\|f\|_{L^\infty(\mathrm{spt}\varphi)} + \|\varepsilon(f)\|_{L^\infty(\mathrm{spt}\varphi)} \right) \int_\Omega |\varepsilon(u^\delta)| dx$$

$$+ \left(\|f\|_{L^2(\Omega)} + \|\mathrm{div}\, f\|_{L^2(\mathrm{spt}\varphi)} \right) \|\mathrm{tr}\sigma^\delta\|_{L^2(\Omega)}$$

$$\left. + \left(\|f\|_{L^n(\Omega)} + \|\mathrm{div}\, f\|_{L^n(\mathrm{spt}\varphi)} \right) \|u^\delta\|_{L^{\frac{n}{n-1}}(\Omega)} \right].$$

Taking into account the hypotheses of Theorem 2.1.1 (see (1.20)), estimate (2.5) of Lemma 2.2.1 and relations (2.2), we derive from the last estimate that

(3.23) $\quad \|\nabla \sigma^\delta\|_{L^2(\Omega_0)} \leq c_8(\Omega_0) \quad \forall \Omega_0 \Subset \Omega.$

Letting $\delta \to 0$ and using Lemma 2.2.1, we have established (1.22). Theorem 2.1.1 is proved.

$\square$

LEMMA 2.3.1 *(Local estimate of Caccioppoli-type).* *Let all conditions of Theorem 2.1.2 hold. Then there is at least one pair $(u, \sigma) \in V_+ \times (Q \cap K)$ which is a weak solution to the minimax problem (1.1) and which satisfies the estimate*

(3.24)
$$\begin{cases} \dfrac{1}{R^{n-2}} \displaystyle\int_{B_{sR}(x_0)} |\nabla \sigma|^2 dx \leq \dfrac{c_9(n, K_0, k_*, \Omega_0, \nu_2, \|f\|_{W_{\frac{2}{n}}^2(\Omega_0; \mathbb{R}^n)})}{(1-s)^3 H_{\lambda_0}} \\[4mm] \times \left\{ R^2 + \left[\left(\displaystyle\fint_{B_R(x_0)} |\overline{\sigma}|^r dx \right)^{1/r} + \dfrac{1}{R} \left(\displaystyle\fint_{B_R(x_0)} |\overline{u}|^{r^*} dx \right)^{1/r^*} \right]^2 \right\} \end{cases}$$

being valid for all $B_R(x_0) \Subset \Omega_0 \Subset \Omega$, $0 < R < 1$ *and* $s \in]0, 1[$. *Here* $r \in]n, \frac{2n}{n-2}[$, $r \leq \bar{n}$, $r^* = \frac{r}{r-1}$, $\bar{\sigma} = \sigma - \sigma^0$, $\bar{u} = u - \varkappa^0(x - x_0) - v_\times$, *where* σ^0 *and* $\varkappa^0$ *are arbitrary tensors from* $\mathbb{M}_{sym}^{n \times n}$ *satisfying the relations*

$$(3.25) \qquad \left\{ \begin{array}{l} tr\sigma^0 = nK_0 tr\varkappa^0 \\[2mm] \sigma^{0D} = \frac{g_0'(|\varkappa^{0D}|)}{|\varkappa^{0D}|}\varkappa^{0D} \end{array} \right\} \Leftrightarrow \sigma^0 = \frac{\partial g}{\partial \varkappa}(\varkappa^0) \Leftrightarrow \varkappa^0 = \frac{\partial g^*}{\partial \tau}(\sigma^0),$$

$$(3.26) \qquad |\sigma^{0D}| \leq \lambda_0 < \sqrt{2}k_*$$

and $v_\times$ *denotes some rigid displacement.*

Proof. We first note that (1.27) implies (1.21). Indeed, let $t > 0$ and let us put $\varkappa = 0$, $\lambda = k_*$, $|\tau| = t$. Then we have

$$tg_0'(t) \geq H_{k_*} \max\{g_0''(t), \frac{g_0'(t)}{t}\}t^2$$

and therefore

$$(3.27) \qquad \frac{g_0'(t)}{t} \geq H_{k_*}.g_0''(t) \quad \forall t > 0 \; (\Rightarrow c_* = \frac{1}{H_{k_*}}).$$

Thus we can state that all conditions of Theorem 2.1.1 hold, and we have estimate (3.23) since condition (1.26) implies $f \in W^1_{\infty,loc}(\Omega; \mathbb{R}^n)$. Let us go back to estimate (3.20). The first identity in (3.25) allows us to write

$$(3.28) \qquad \operatorname{div} \bar{u}^\delta = \frac{1}{nK_0}tr\bar{\sigma}^\delta$$

where

$$\bar{\sigma}^\delta = \sigma^\delta - \sigma^0, \quad \bar{u}^\delta = u^\delta - \varkappa^0(x - x_0) - v_\times,$$

and therefore

$$(3.29) \qquad 2\int_\Omega \bar{\sigma}^\delta : \nabla^2 \varphi^6 \operatorname{div} \bar{u}^\delta dx \leq c_{10}(n, K_0) \int_\Omega |\nabla^2 \varphi^6| \, |\bar{\sigma}^\delta|^2 dx.$$

We integrate by parts in the first two terms of (3.19) and use identity (3.28) to get

$$(3.30) \quad \begin{cases} J_0 = -2 \int_\Omega \overline{u}^\delta \cdot \operatorname{div}\left(\varphi^6 \varepsilon(f) + f \odot \nabla \varphi^6\right) dx \\[2mm] \qquad - \frac{1}{nK_0} \int_\Omega \operatorname{tr} \overline{\sigma}^\delta [2f \cdot \nabla \varphi^6 + \varphi^6 \operatorname{div} f] dx \\[2mm] \qquad - \int_\Omega [2(f \odot \overline{u}^\delta) : \nabla^2 \varphi^6 + \nabla \varphi^6 \cdot \overline{u}^\delta \operatorname{div} f] dx \\[2mm] \qquad \leq c_{11}(n, K_0) \int_\Omega \left\{ |\overline{u}^\delta| [\varphi^6 |\operatorname{div} \varepsilon(f)| + |\nabla \varphi^6| |\nabla f| + |f| |\nabla^2 \varphi^6|] \right. \\[2mm] \qquad \left. + |\overline{\sigma}^\delta| [\varphi^6 |\nabla f| + |\nabla \varphi^6| |f|] \right\} dx. \end{cases}$$

It follows from (3.20), (3.29) and (3.30) that

$$(3.31) \quad \begin{cases} J \leq 144 c_2 \int_\Omega \varphi^4 |\nabla \varphi|^2 \left(\delta + \frac{1}{H_{k_*}} \frac{g_0'(|\varepsilon^D(u^\delta)|)}{|\varepsilon^D(u^\delta)|}\right) |\varepsilon^D(\overline{u}^\delta)|^2 dx \\[2mm] \qquad + c_{10}(n, K_0) \int_\Omega |\nabla^2 \varphi^6| |\overline{\sigma}^\delta|^2 dx + c_{12}(n) \int_\Omega |\overline{\sigma}^\delta| |\nabla^3 \varphi^6| |\overline{u}^\delta| dx \\[2mm] \qquad + 2c_{11}(n, K_0) \int_\Omega \left\{ |\overline{u}^\delta| [\varphi^6 |\operatorname{div} \varepsilon(f)| + |\nabla \varphi^6| |\nabla f| + |f| |\nabla^2 \varphi^6|] \right. \\[2mm] \qquad \left. + |\overline{\sigma}^\delta| [\varphi^6 |\nabla f| + |\nabla \varphi^6| |f|] \right\} dx. \end{cases}$$

Let

$$J_* \equiv \frac{1}{H_{k_*}} \int_\Omega \frac{g_0'(|\varepsilon^D(u^\delta)|)}{|\varepsilon^D(u^\delta)|} \varphi^4 |\nabla \varphi|^2 |\varepsilon^D(\overline{u}^\delta)|^2 dx$$

$$\leq \frac{1}{H_{k_*}} \int_\Omega h_+(|\varepsilon^D(u^\delta)|) \varphi^4 |\nabla \varphi|^2 |\varepsilon^D(\overline{u}^\delta)|^2 dx$$

where

$$h_+(t) = \max\left\{ g_0''(t), \frac{g_0'(t)}{t} \right\}, \quad t \geq 0.$$

We now make use of the variational identity (2.3) by setting $v = \Psi \overline{u}^\delta$ with a cut-off function $\Psi \in C_0^1(\Omega)$. As a result we get the equation

$$(3.32) \qquad \int_\Omega \Psi \overline{\sigma}^\delta : \varepsilon(\overline{u}^\delta) dx = \int_\Omega \Psi f \cdot \overline{u}^\delta dx - \int_\Omega \overline{\sigma}^\delta : (\overline{u}^\delta \odot \nabla \Psi) dx.$$

Taking into account the identities (3.25), (3.21) we can write

$$\overline{\sigma}^\delta : \varepsilon(\overline{u}^\delta) = K_0 \mathrm{div}^2 \overline{u}^\delta + \delta |\varepsilon^D(\overline{u}^\delta)|^2 + \delta \varkappa^{0D} : \varepsilon^D(\overline{u}^\delta)$$

$$+ \left(\frac{g_0'(|\varepsilon^D(u^\delta)|)}{|\varepsilon^D(u^\delta)|} \varepsilon^D(u^\delta) - \frac{g_0'(|\varkappa^{0D}|)}{|\varkappa^{0D}|} \varkappa^{0D} \right) : (\varepsilon^D(u^\delta) - \varkappa^{0D})$$

and after application of condition (1.27) we arrive at the estimate

$$\overline{\sigma}^\delta : \varepsilon(\overline{u}^\delta) \geq \delta \varkappa^{0D} : \varepsilon^D(\overline{u}^\delta) + H_{\lambda_0} h_+ (|\varepsilon^D(u^\delta)|) |\varepsilon^D(\overline{u}^\delta)|^2.$$

Inserting $\Psi = \varphi^4 |\nabla \varphi|^2$ into (3.32) and using the last estimate, we get

$$J_* \leq \frac{1}{H_{k_*} H_{\lambda_0}} \left\{ \int_\Omega \varphi^4 |\nabla \varphi|^2 f \cdot \overline{u}^\delta dx - \int_\Omega \sigma^\delta : (\nabla(\varphi^4 |\nabla \varphi|^2) \odot \overline{u}^\delta) dx \right.$$

$$\left. - \delta \int_\Omega \varphi^4 |\nabla \varphi|^2 \varkappa^{0D} : \varepsilon^D(\overline{u}^\delta) dx \right\}.$$

From the latter inequality and also from (3.31) and (3.10) it follows that

$$(3.33) \qquad \left\{ \begin{array}{l} \frac{1}{c_1^2} \int_\Omega \varphi^6 |\nabla \sigma^\delta|^2 dx \leq 144 c_2 \int_\Omega \varphi^4 |\nabla \varphi|^2 \delta |\varepsilon^D(\overline{u}^\delta)|^2 dx \\[3mm] + c_{10}(n, K_0) \int_\Omega |\nabla^2 \varphi^6| |\overline{\sigma}^\delta|^2 dx \\[3mm] + \int_\Omega |\overline{\sigma}^\delta| |\overline{u}^\delta| \left(\frac{144 c_2}{H_{k_*} H_{\lambda_0}} |\nabla(\varphi^4 |\nabla \varphi|^2)| + c_{12} |\nabla^3 \varphi^6| \right) dx \\[3mm] + \frac{144 c_2}{H_{k_*} H_{\lambda_0}} \delta \int_\Omega \varphi^4 |\nabla \varphi|^2 |\varkappa^{0D}| |\varepsilon^D(\overline{u}^\delta)| dx \\[3mm] + \frac{144 c_2}{H_{k_*} H_{\lambda_0}} \int_\Omega \varphi^4 |\nabla \varphi|^2 |f| |\overline{u}^\delta| dx \\[3mm] + 2 c_{11}(n, K_0) \int_\Omega \left\{ |\overline{u}^\delta| [\varphi^6 |\mathrm{div}\, \varepsilon(f)| + |\nabla \varphi^6| |\nabla f| \right. \\[3mm] \left. + |f| |\nabla^2 \varphi^6|] + |\overline{\sigma}^\delta| [\varphi^6 |\nabla f| + |\nabla \varphi^6| |f|] \right\} dx. \end{array} \right.$$

As remarked earlier condition (1.26) implies $f \in W^1_{\infty, \mathrm{loc}}(\Omega; \mathbb{R}^n)$. Taking into account estimate (3.27), we see that all hypotheses of Theorem 2.1.1 hold, and thus we have estimate (3.23). By the imbedding theorem and Lemma 2.2.1 we further have

$$(3.34) \quad \begin{cases} \overline{\sigma}^\delta \to \overline{\sigma} & \text{in} \quad L^r(\Omega_0; \mathbb{M}_{\text{sym}}^{n \times n}) \\ \\ \overline{u}^\delta \to \overline{u} & \text{in} \quad L^{r*}(\Omega; \mathbb{R}^n) \end{cases}$$

where the pair $(u, \sigma) \in V_+ \times (Q \cap K)$ is a weak solution to the minimax problem (1.1). Passing to the limit in inequality (3.33) and using the statements of Lemma 2.2.1 we get

$$(3.35) \quad \begin{cases} \displaystyle\int_\Omega \varphi^6 |\nabla \sigma|^2 dx \le \frac{c_{13}(u, K_0, k_*, c_1)}{H_{\lambda_0}} \Big\{ \int_\Omega |\nabla^2 \varphi^6| |\overline{\sigma}|^2 dx \\ \\ + \displaystyle\int_\Omega |\overline{\sigma}| |\overline{u}| \big(|\nabla(\varphi^4 |\nabla \varphi|^2)| + |\nabla^3 \varphi^6| \big) dx \\ \\ + \displaystyle\int_\Omega |\overline{u}| \big[\varphi^6 |\text{div} \, \varepsilon(f)| + |\nabla \varphi^6| |\nabla f| + |f| (\varphi^4 |\nabla \varphi|^2 + |\nabla^2 \varphi^6|) \big] dx \\ \\ + \displaystyle\int_\Omega |\overline{\sigma}| \big[\varphi^6 |\nabla f| + |\nabla \varphi^6| |f| \big] dx \Big\}. \end{cases}$$

Let the cut–off function φ satisfy the conditions

$$0 \le \varphi \le 1 \text{ in } \Omega, \; \text{spt} \, \varphi \subset B_R(x_0) \Subset \Omega_0 \Subset \Omega, \; \varphi \equiv 1 \text{ in } B_{sR}(x_0), \; 0 < s < 1,$$

$$|\nabla^k \varphi| \le \frac{\text{const}}{R^k(1-s)^k}, \; k = 1, 2, 3.$$

Then estimate (3.35) implies that

$$\frac{1}{R^{n-2}} \int_{B_{sR}(x_0)} |\nabla \sigma|^2 dx \le R^2 \frac{c_{14}(n, K_0, k_*, \nu_2)}{H_{\lambda_0}} \Big\{ \frac{1}{R^2(1-s)^2} \fint_{B_R(x_0)} |\overline{\sigma}|^2 dx$$

$$+ \frac{1}{R^3(1-s)^3} \fint_{B_R(x_0)} |\overline{\sigma}| |\overline{u}| dx$$

$$+ \fint_{B_R(x_0)} |\overline{u}| \Big[|\text{div} \, \varepsilon(f)| + \frac{1}{R(1-s)} |\nabla f| + |f| \frac{1}{R^2(1-s)^2} \Big] dx$$

$$+ \fint_{B_R(x_0)} |\overline{\sigma}| \big[|\nabla f| + \frac{1}{R(1-s)} |f| \big] dx \Big\} \le \frac{c_{15}(n, K_0, k_*, \nu_2)}{H_{\lambda_0}(1-s)^3} \Big\{ \fint_{B_R(x_0)} |\overline{\sigma}|^2 dx +$$

$$+\frac{1}{R}\Big(\fint_{B_R(x_0)}|\overline{\sigma}|^r dx\Big)^{1/r}\Big(\fint_{B_R(x_0)}|\overline{u}|^{r^*} dx\Big)^{1/r^*}$$

$$+R\|f\|_{W^1_\infty(\Omega_0;\mathbf{R}^n)}\Big[\fint_{B_R(x_0)}|\overline{\sigma}|dx+\frac{1}{R}\fint_{B_R(x_0)}|\overline{u}|dx\Big]$$

$$+R^2\Big(\fint_{B_R(x_0)}|\overline{u}|^{r^*}dx\Big)^{1/r^*}\Big(\fint_{B_R(x_0)}|\operatorname{div}\varepsilon(f)|^r dx\Big)^{1/r}\Big\}$$

$$\leq\frac{c_{15}}{H_{\lambda_0}(1-s)^3}\Big\{\Big(\fint_{B_R(x_0)}|\overline{\sigma}|^r dx\Big)^{2/r}+\frac{1}{2}\Big(\fint_{B_R(x_0)}|\overline{\sigma}|^r dx\Big)^{2/r}$$

$$+\frac{1}{2R^2}\Big(\fint_{B_R(x_0)}|\overline{u}|^{r^*}dx\Big)^{2/r^*}$$

$$+R\|f\|_{W^1_\infty(\Omega_0;\mathbf{R}^n)}\Big[\Big(\fint_{B_R(x_0)}|\overline{\sigma}|^r dx\Big)^{1/r}+\frac{1}{R}\Big(\fint_{B_R(x_0)}|\overline{u}|^{r^*}dx\Big)^{1/r^*}\Big]$$

$$+\frac{1}{R}\Big(\fint_{B_R(x_0)}|\overline{u}|^{r^*}dx\Big)^{1/r^*}\Big(\fint_{B_R(x_0)}|\operatorname{div}\varepsilon(f)|^{\overline{n}}dx\Big)^{\frac{1}{\overline{n}}}R^3\Big\}$$

$$\leq\frac{c'_{15}(n,K_0,k_*,\nu_2)}{H_{\lambda_0}(1-s)^3}\Big\{\Big[\Big(\fint_{B_R(x_0)}|\overline{\sigma}|^r dx\Big)^{1/r}+\frac{1}{R}\Big(\int_{B_R(x_0)}|\overline{u}|^{r^*}dx\Big)^{1/r^*}\Big]^2$$

$$+R^2\|f\|^2_{W^1_\infty(\Omega_0;\mathbf{R}^n)}+R^{(3-\frac{n}{\overline{n}})2}\Big(\int_{\Omega_0}|\operatorname{div}\varepsilon(f)|^{\overline{n}}dx\Big)^{\frac{2}{\overline{n}}}\Big\}$$

$$\leq\frac{c_{16}(n,K_0,k_*,\nu_2,\Omega_0,\|f\|_{W^2_{\overline{n}}(\Omega_0;\mathbf{R}^n)})}{H_{\lambda_0}(1-s)^3}$$

$$\times\Big\{R^2+\Big[\Big(\fint_{B_R(x_0)}|\overline{\sigma}|^r dx\Big)^{1/r}+\frac{1}{R}\Big(\fint_{B_R(x_0)}|\overline{u}|^{r^*}dx\Big)^{1/r^*}\Big]^2\Big\}.$$

It remains to put $c_9=c_{16}$, and we are done. Lemma 2.3.1 is proved.

$\square$

2.4 Estimates for solutions of certain systems of PDE's with constant coefficients

Suppose that we are given functions

$$\tau \in L^r(B; \mathbb{M}^{n \times n}_{\text{sym}}) \text{ and } v \in L^{r^*}(B; \mathbb{R}^n),$$

where $r > 1$ and $r^* = \frac{r}{r-1}$, satisfying the following system of partial differential equations

$$(4.1) \quad \begin{cases} \varepsilon(v) = \frac{\partial^2 g^*}{\partial \tau^2}(\sigma^0)\tau, \\ \\ \text{div } \tau = 0 \end{cases}$$

on the unit ball $B \subset \mathbb{R}^n$. Here the tensor $\sigma^0 \in \mathbb{M}^{n \times n}_{\text{sym}}$ is chosen according to

$$(4.2) \quad |\sigma^{0D}| \leq \lambda < \sqrt{2}k_*.$$

LEMMA 2.4.1 *For any* $\lambda \in [0, \sqrt{2}k_*[$ *and any* $t_* \in]0, 1[$ *a constant* $c_{17} = c_{17}(t_*, \lambda)$ *exists such that*

$$(4.3) \quad W(t) \leq c_{17}tW(1)$$

for all $0 < t \leq t_*$.

In Lemma 2.4.1 we used the notation

$$W(t) = \left(\fint_{B_t} |\tau - (\tau)_t|^r dy \right)^{1/r}$$

$$+ \frac{1}{t} \inf_{v_\times \in V_\times} \left(\fint_{B_t} |v - \frac{\partial^2 g^*}{\partial \tau^2}(\sigma^0)(\tau)_t y - v_\times|^{r^*} dy \right)^{1/r^*},$$

B_t denoting the ball of radius t concentric to B and $(\tau)_t = \fint_{B_t} \tau \, dy$ is the mean value of τ w.r.t. B_t.

Proof. We put $\varkappa^0 = \frac{\partial g^*}{\partial \tau}(\sigma^0)$. By condition (4.2) we have

$$(4.4) \quad \frac{\partial^2 g}{\partial \varkappa^2}(\varkappa^0) = \left(\frac{\partial^2 g^*}{\partial \tau^2}(\sigma^0) \right)^{-1} : \mathbb{M}^{n \times n}_{\text{sym}} \to \mathbb{M}^{n \times n}_{\text{sym}},$$

and therefore system (4.1) is equivalent to the system of linear elasticity

$$
(4.5) \qquad
\begin{cases}
\operatorname{div} \tau = 0, \\[2mm]
\tau = \frac{\partial^2 g}{\partial \varkappa^2}(\varkappa^0)\varepsilon(v)
\end{cases}
\quad \text{in } B.
$$

Let

$$
E_0(\varkappa; \varepsilon, \varepsilon') = \left(\frac{\partial^2 g}{\partial \varkappa^2}(\varkappa)\varepsilon \right) : \varepsilon' = K_0 tr\,\varepsilon\, tr\,\varepsilon' + E(\varkappa^D; \varepsilon^D, \varepsilon'^D)
$$

$$
\forall \varkappa, \varepsilon, \varepsilon' \in \mathbb{M}^{n \times n}_{\text{sym}}.
$$

Here the bilinear form E is defined by formula (3.9), i.e.

$$
E(\tau; \varepsilon, \varkappa) = \left(\frac{\partial^2 g_0}{\partial \tau^2}(|\tau|)\varepsilon \right) : \varkappa
$$

$$
= \frac{g_0'(|\tau|)}{|\tau|} \varepsilon : \varkappa + \left(g_0''(|\tau|) - \frac{g_0'(|\tau|)}{|\tau|} \right) \frac{\tau : \varepsilon\; \tau : \varkappa}{|\tau|^2}.
$$

Clearly the form E satisfies the estimates (3.5), i.e.

$$
h_-(|\varkappa^D|)|\varepsilon^D|^2 \leq E(\varkappa^D; \varepsilon^D, \varepsilon^D) \leq h_+(|\varkappa^D|)|\varepsilon^D|^2.
$$

From condition (4.2) we deduce that $h_-(|\varkappa^{0D}|)$, $h_+(|\varkappa^{0D}|)$ are positive and finite. Thus, by letting

$$
\bar{\nu}_- = \min\{K_0 n, h_-(|\varkappa^{0D}|)\}, \quad \bar{\nu}_+ = \max\{K_0 n, h_+(|\varkappa^{0D}|)\},
$$

we find that

$$
(4.6) \qquad \bar{\nu}_-|\varepsilon|^2 \leq E_0(\varkappa^0; \varepsilon, \varepsilon) \leq \bar{\nu}_+|\varepsilon|^2 \quad \forall \varepsilon \in \mathbb{M}^{n \times n}_{\text{sym}}.
$$

The ellipticity estimate (4.6) guarantees that the functions v and τ are infinitely differentiable inside of B. Let us rewrite (4.5) as

$$
(4.7) \qquad \int_B E(\varkappa^0; \varepsilon(v), \varepsilon(w))dx = \int_B \bar{\tau} : \varepsilon(w)dx = 0 \quad \forall w \in C_0^\infty(B; \mathbb{R}^n),
$$

where $\bar{\tau} = \tau - (\tau)_1$. We also put $\bar{v} = v - (\varepsilon(v))_1 y - v_\times$ for a rigid displacement $v_\times$ which is chosen according to (see Lemma 3.0.3)

$$
\int_B |\bar{v}|^{r^*} dy = \inf_{\tilde{v}_\times \in V_\times} \int_B |v - \varepsilon(v)_1 y - \tilde{v}_\times|^{r^*} dy.
$$

Remarking that

$$(4.8) \qquad \big(\varepsilon(v)\big)_t = \frac{\partial^2 g^*}{\partial x^2}(\sigma^0)(\tau)_t, \quad 0 < t < 1,$$

we get

$$(4.9) \qquad \overline{\tau} = \frac{\partial^2 g}{\partial x^2}(x^0)\varepsilon(\overline{v})$$

and

$$(4.10) \qquad \overline{v} = v - \frac{\partial^2 g^*}{\partial \tau^2}(\sigma^0)(\tau)_1(y) - v_\times.$$

Let us take a cut-off function $\varphi \in C_0^1(B)$ having the properties

$$(4.11) \qquad \begin{cases} 0 \le \varphi \le 1 \text{ in } B, \; \varphi \equiv 1 \text{ in } B_{t_1}, \; 0 < t_* < t_1 < 1, \\ |\nabla \varphi| \le \frac{2}{1-t_1}. \end{cases}$$

If we insert $w = \varphi \overline{v}$ into (4.7), we arrive at the identity

$$(4.12) \qquad \int_\Omega \varphi \overline{\tau} : \varepsilon(\overline{v})dy = - \int_\Omega \overline{\tau} : (\nabla \varphi \odot \overline{v})dy.$$

From (4.9), (4.12) and (4.6) it follows that

$$(4.13) \qquad \int_{B_{t_1}} |\varepsilon(\overline{v})|^2 dy \le \frac{2}{\overline{\nu}_-(1-t_1)} \int_B |\overline{v}||\overline{\tau}|dy.$$

Further we have

$$(4.14) \qquad \int_{B_{t_1}} \tau_{,k} : \varepsilon(w)dy = 0 \quad \forall w \in C_0^\infty(B_{t_1}; \mathbb{R}^n).$$

Now we take $\varphi \in C_0^1(B_{t_1})$ so that $0 \le \varphi \le 1$ in B_{t_1}, $|\nabla \varphi| \le \frac{2}{t_1-t_2}$ in B_{t_1}, $\varphi \equiv 1$ in B_{t_2}, where $t_2 \in]t_*, t_1[$. Inserting $w = \varphi^2 \tilde{v}_{,k}$ into (4.14) with $\tilde{v} = \overline{v} + w_\times$,

$w_\times \in V_\times$, we get

$$\int_{B_{t_1}} \varphi^2 \tau_{,k} : \varepsilon(v_{,k}) dy = \int_{B_{t_1}} \varphi^2 E_0(\varkappa^0; \varepsilon(v_{,k}), \varepsilon(v_{,k})) dy$$

$$= -2 \int_{B_{t_1}} \varphi \tau_{,k} : (\tilde{v}_{,k} \odot \nabla \varphi) dy = -2 \int_{B_{t_1}} \varphi E_0(\varkappa^0; \varepsilon(v_{,k}), \tilde{v}_{,k} \odot \nabla \varphi) dy$$

$$\leq 2 \left(\int_{B_{t_1}} \varphi^2 E_0(\varkappa^0; \varepsilon(v_{,k}), \varepsilon(v_{,k})) dy \right)^{1/2}$$

$$\times \left(\int_{B_{t_1}} E_0(\varkappa^0; \tilde{v}_{,k} \odot \nabla \varphi, \tilde{v}_{,k} \odot \nabla \varphi) dy \right)^{1/2}$$

and by (4.6)

$$\bar{\nu}_- \int_{B_{t_2}} |\nabla \varepsilon(v)|^2 dy \leq \int_{B_{t_2}} \varphi^2 E_0(\varkappa^0; \varepsilon(v_{,k}), \varepsilon(v_{,k})) dy$$

$$\leq 4 \int_{B_{t_1}} E_0(\varkappa^0; \tilde{v}_{,k} \odot \nabla \varphi, \tilde{v}_{,k} \odot \nabla \varphi) dy \leq \frac{16 \bar{\nu}_+}{(t_1 - t_2)^2} \int_{B_{t_1}} |\nabla \tilde{v}|^2 dy.$$

Choosing the rigid displacement $w_\times$ in a suitable way, we find

$$\int_{B_{t_1}} |\nabla \tilde{v}|^2 dy \leq c_{18}(n) \int_{B_{t_1}} |\varepsilon(\bar{v})|^2 dy,$$

and therefore

$$\int_{B_{t_2}} |\nabla \varepsilon(v)|^2 dy \leq \frac{16 \bar{\nu}_+ c_{18}}{\bar{\nu}_-(t_1 - t_2)^2} \frac{2}{\bar{\nu}_-} \frac{1}{1 - t_1} \int_B |\bar{v}||\bar{\tau}| dy.$$

After a finite number of such steps we can apply Sobolev's inequality and arrive at

$$\sup_{B_{t_*}} |\nabla \varepsilon(v)|^2 \leq c_{19}(t_*, \lambda) \int_B |\bar{\tau}||\bar{v}| dy.$$

Now, for $t \leq t_*$, we can write

$$\left(\fint_{B_t} |\tau - (\tau)_t|^r dy\right)^{1/r} = \left(\fint_{B_t} \left|\frac{\partial^2 g}{\partial \varkappa^2}(\varkappa^0)(\varepsilon(v) - (\varepsilon(v))_t)\right|^r dy\right)^{1/r}$$

$$\leq c_{20}(\lambda)\left(\fint_{B_t} |\varepsilon(v) - (\varepsilon(v))_t|^r dy\right)^{1/r}$$

$$\leq c_{21}(n, \lambda, r)t\left(\fint_{B_t} |\nabla\varepsilon(v)|^r dy\right)^{1/r} \leq c_{21}t\sqrt{c_{19}}\left(\fint_{B} |\bar{\tau}||\bar{v}| dy\right)^{1/2}.$$

Further, by Korn's inequality, we have

$$\frac{1}{t} \inf_{v_\varkappa \in V_\varkappa} \left(\fint_{B_t} |v - (\varepsilon(v))_t y - v_\varkappa|^{r^*} dy\right)^{1/r^*}$$

$$\leq c_{22}(n, r)\left(\fint_{B_t} |\nabla\varepsilon(v)|^{r^*} dy\right)^{1/r^*} \leq c_{22}t\sqrt{c_{19}}\left(\fint_{B} |\bar{\tau}||\bar{v}| dy\right)^{1/2}.$$

Now, from (4.8) and the last two estimates it follows that

$$W(t) \leq c_{23}(t_*, \lambda)t\left(\fint_{B} |\bar{\tau}||\bar{v}| dy\right)^{1/2} \leq c_{23}t\left(\fint_{B} |\bar{\tau}|^r dy\right)^{1/2r}$$

$$\times \left(\fint_{B} |\bar{v}|^{r^*} dy\right)^{1/2r^*} \leq c_{23}\frac{t}{2}\left[\left(\fint_{B} |\bar{\tau}|^r dy\right)^{1/r} + \left(\fint_{B} |\bar{v}|^{r^*} dy\right)^{1/r^*}\right]$$

$$= \tfrac{1}{2}c_{23}tW(1),$$

and Lemma 2.4.1 is proved.

$\square$

2.5   The main lemma and its iteration

Let us introduce another bilinear form

$$E^*(\tau; \sigma, \varkappa) = \left(\frac{\partial^2 g_0^*}{\partial \tau^2}(|\tau|)\sigma \right) : \varkappa$$

$$= \frac{(g_0^*)'(|\tau|)}{|\tau|}\sigma : \varkappa + \left((g_0^*)''(|\tau|) - \frac{(g_0^*)'(|\tau|)}{|\tau|} \right) \frac{\tau : \varkappa \, \tau : \sigma}{|\tau|^2},$$

$\tau, \varkappa, \sigma \in \mathbb{M}_{\mathrm{sym}}^{n \times n}$, $|\tau| < \sqrt{2}k_*$.

For the corresponding quadratic form we have the bounds

(5.1) $h_-^*(|\tau|)|\sigma|^2 \le E^*(\tau; \sigma, \sigma) \le h_+^*(|\tau|)|\sigma|^2$

where

$$h_-^*(s) = \min \left\{ \frac{(g_0^*)'(s)}{s}, (g_0^*)''(s) \right\}, \; h_+^*(s) = \max \left\{ \frac{(g_0^*)'(s)}{s}, (g_0^*)''(s) \right\}$$

and $|s| < \sqrt{2}k_*$.

Since the function $g_0^* :] - \sqrt{2}k_*, \sqrt{2}k_*[\to \mathbb{R}$ is the conjugate function of $g_0 :$ $] - t_0, t_0[\to \mathbb{R}$ (see formulas (1.32)) we have

$$(g_0^*)''(s) = \frac{1}{g_0''(t)} \text{ for } s = g_0'(t) \; (\Leftrightarrow t = (g_0^*)'(s)),$$

and therefore

$$h_+^*(g_0'(t)) = \max \left\{ \frac{t}{g_0'(t)}, \frac{1}{g_0''(t)} \right\}, \; |t| < t_0.$$

If condition (1.21) holds, then we have the estimate

$$h_+^*(s) = h_+^*(g_0'(t)) \le \max\{1, c_*\} \frac{1}{g_0''(t)} = \frac{c_*}{g_0''(t)}, \; |t| < t_0,$$

and thus

(5.2) $h_+^*(s) \le c_*(g_0^*)''(s), \quad |s| < \sqrt{2}k_*.$

MAIN LEMMA 2.5.1 *Suppose that all the conditions of Theorem 2.1.2 hold. Let the number r satisfy*

$$(5.3) \qquad r \in]n, \frac{2n}{n-2}[, \ r \leq \bar{n},$$

and let $r^* = \frac{r}{r-1}$. Moreover, fix $\Omega_0 \subset\subset \Omega$ and some numbers $\varepsilon_3 \in]0,1[, \ t_* \in]0,1[$.

Let $(u, \sigma) \in V_+ \times (Q \cap K)$ denote the weak solution of the minimax–problem (1.1) which is produced via the approximation procedure from section 2, Lemma 2.2.1. Then, for any $t \in]0, t_*[$ and $\lambda \in [0, \sqrt{2}k_*[$, there are positive numbers

$$\varepsilon_0 = \varepsilon_0(t, t_*, \lambda, \varepsilon_3), \ R_0 = R_0(t, t_*, \lambda, \varepsilon_3)$$

such that the conditions

$$(5.4) \qquad B_R(x_0) \Subset \Omega_0,$$

$$(5.5) \qquad |(\sigma^D)_{x_0,R}| \leq \lambda,$$

$$(5.6) \qquad U(x_0, R) + R^{\varepsilon_3} < \varepsilon_0,$$

for some ball $B_R(x_0)$, $0 < R < R_0$, imply the decay estimate

$$(5.7) \qquad U(x_0, tR) \leq 2c_{17}(t_*, \lambda)t\big(U(x_0, R) + R^{\varepsilon_3}\big).$$

Here c_{17} is the constant of Lemma 2.4.1 and the excess $U(x_0, R)$ is defined by the formulas

$$U(x_0, R) = \overline{U}(x_0, R) + \tilde{U}(x_0, R),$$

$$\overline{U}(x_0, R) = \left(\fint_{B_R(x_0)} |\sigma - (\sigma)_{x_0,R}|^r dx \right)^{1/r}, \quad (\sigma)_{x_0,R} = \fint_{B_R(x_0)} \sigma \, dx,$$

$$\tilde{U}(x_0, R) = \frac{1}{R} \inf_{v_\times \in V_\times} \left(\fint_{B_R(x_0)} |u - (\sigma)^{x_0,R}(x - x_0) - v_\times|^{r^*} dx \right)^{1/r^*},$$

$$(\sigma)^{x_0,R} = \frac{\partial g^*}{\partial \tau}((\sigma)_{x_0,R}).$$

Proof.   We note first that $(u, \sigma) \in V_+ \times (Q \cap K)$ satisfies the variational inequality

$$(5.8) \quad \begin{cases} -\displaystyle\int_\Omega \Big\{ u \cdot \mathrm{div}\big(\psi(\tau - \sigma)\big) + \psi\big(g^*(\tau) - g^*(\sigma)\big) \Big\} dy \\[2mm] = -\displaystyle\int_\Omega \Big\{ \psi u \cdot \mathrm{div}(\tau - \sigma) + (\tau - \sigma) : (u \odot \nabla\psi) \\[2mm] \quad + \psi\big(g^*(\tau) - g^*(\sigma)\big) \Big\} dy \leq 0 \end{cases}$$

for all $\tau \in \Sigma_n \cap K$ and all $\psi \in C_0^1(\Omega)$ such that $0 \leq \psi \leq 1$.

Suppose that the claim of the lemma is false. Then there exist numbers t, λ and sequences x^h, R_h, ε_h such that:

$$(5.9) \quad B_{R_h}(x^h) \Subset \Omega_0,$$

$$(5.10) \quad \varepsilon_h \equiv U(x^h, R_h) + R_h^{\varepsilon_3} \to 0 \ \text{ as } h \to +\infty,$$

$$(5.11) \quad |(\sigma^D)_{x^h, R_h}| \leq \lambda,$$

$$(5.12) \quad U(x^h, tR_h) > 2c_{17}t\big(U(x^h, R_h) + R_h^{\varepsilon_3}\big) = 2c_{17}t\varepsilon_h.$$

From (5.8) we get

$$(5.13) \quad \begin{cases} \displaystyle\int_{B_{R_h}(x^h)} \Big\{ -\bar{u}^h \cdot \mathrm{div}\big(\psi(\tau - \sigma)\big) + \psi\big((\sigma)^{x^h, R_h} : (\tau - \sigma) \\[2mm] \quad -g^*(\tau) + g^*(\sigma)\big) \Big\} dx \leq 0 \end{cases}$$

being valid for all $\tau \in C^1(\overline{B}_{R_h}(x^h); \mathbb{M}_{\mathrm{sym}}^{n \times n})$ and $\psi \in C_0^1(B_{R_h}(x^h))$ such that $|\tau^D| \leq \sqrt{2}k_*$ and $0 \leq \psi \leq 1$ where we have abbreviated

$$(5.14) \quad \begin{cases} \bar{u}^h = u - (\sigma)^{x^h, R_h}(x - x^h) - u_\times^h, \quad u_\times^h \in V_\times, \\[2mm] (\sigma)^{x^h, R_h} = \dfrac{\partial g^*}{\partial \tau}\big((\sigma)_{x^h, R_h}\big), \\[2mm] \bar{\sigma}^h = \sigma - (\sigma)_{x^h, R_h}. \end{cases}$$

Here the rigid displacement $u_\times^h$ is chosen according to

$$(5.15) \qquad \int_{B_{R_h}(x^h)} |\bar{u}^h|^{r^*} dx = \inf_{v_\times \in V_\times} \int_{B_{R_h}(x^h)} |u - (\sigma)^{x^h,R_h}(x - x^h) - v_\times|^{r^*} dx.$$

Now we use Lemma 2.3.1 with $x_0 = x^h$, $R = R_h$, $\varkappa^0 = (\sigma)^{x^h,R_h}$. As a result we have the estimate

$$(5.16) \qquad \frac{1}{R_h^{n-2}} \int_{B_{sR_h}(x^h)} |\nabla\bar{\sigma}^h|^2 dx \le \frac{c_{24}(\Omega_0, f, \lambda)}{(1-s)^3} \left\{ U^2(x^h, R_h) + R_h^2 \right\}.$$

Let us introduce the coordinate transformation $x = x^h + R_h y$, $y \in B$, and define

$$(5.17) \qquad v^h(y) = \frac{1}{\varepsilon_h R_h} \bar{u}^h(x), \quad \sigma^h(y) = \bar{\sigma}^h(x)/\varepsilon_h.$$

It is easy to check that

$$(5.18) \qquad \bar{U}(x^h, \rho) = \varepsilon_h \overline{W}^h(\rho/R_h),$$

where

$$\overline{W}^h(t) = \left(\fint_{B_t} |\sigma^h - (\sigma^h)_t|^r dy \right)^{1/r}.$$

We have

$$\tilde{U}(x^h, \rho) = \frac{1}{\rho} \inf_{v_\times \in V_\times} \left(\fint_{B_\rho(x^h)} |\bar{u}^h + (\sigma)^{x^h,R_h}(x - x^h) + u_\times^h \right.$$
$$\left. -(\sigma)^{x^h,\rho}(x - x^h) - v_\times|^{r^*} dx \right)^{1/r^*}$$

and

$$(\sigma)^{x^h,\rho} - (\sigma)^{x^h,R_h} = \int_0^1 \frac{\partial^2 g^*}{\partial \tau^2} \left((\sigma)_{x^h,R_h} + \theta \left((\sigma)_{x^h,\rho} \right. \right.$$
$$\left. \left. - (\sigma)_{x^h,R_h} \right) \right) d\theta \left((\sigma)_{x^h,\rho} - (\sigma)_{x^h,R_h} \right).$$

As a result we get

$$(5.19) \qquad (\sigma)^{x^h,\rho} - (\sigma)^{x^h,R_h} = \varepsilon_h \int_0^1 \frac{\partial^2 g^*}{\partial \tau^2} \left((\sigma)_{x^h,R_h} + \theta(\sigma^h)_{\rho/R_h} \right) (\sigma^h)_{\rho/R_h} d\theta,$$

$$(5.20) \quad \begin{cases} \tilde{U}(x^h, \rho) = \varepsilon_h \frac{1}{\rho/R_h} \inf_{v_x \in V_x} \left(\fint_{B_{\rho/R_h}} \left| v^h(y) \right. \right. \\[2ex] \left. -\left(\int_0^1 \frac{\partial^2 g^*}{\partial \tau^2}((\sigma)_{x^h, R_h} + \theta(\sigma^h)_{\rho/R_h}) d\theta(\sigma^h)_{\rho/R_h} \right) y - v_x \left.\Big|^{r^*} dy \right)^{1/r^*} \\[2ex] =: \varepsilon_h \tilde{W}^h(\rho/R_h). \end{cases}$$

Now, if we set

$$W^h(t) = \overline{W}^h(t) + \tilde{W}^h(t),$$

we arrive at the formula

$$(5.21) \quad U(x^h, \rho) = \varepsilon_h W^h(\rho/R_h).$$

We remark that

$$(5.22) \quad (\sigma^h)_1 = 0,$$

$$(5.23) \quad W^h(1) = \left(\fint_B |\sigma^h|^r dy \right)^{1/r} + \left(\fint_B |v^h|^{r^*} dy \right)^{1/r^*}.$$

From the above relations it follows that

$$(5.24) \quad W^h(1) + R_h^{\varepsilon_3}/\varepsilon_h = 1,$$

$$(5.25) \quad W^h(t) > 2c_{17}t,$$

and (5.16) reads after scaling

$$(5.26) \quad \int_{B_s} |\nabla\sigma^h|^2 dy \le \frac{c_{24}}{(1-s)^3} \{W^h(1) + R_h/\varepsilon_h\}^2.$$

Consider $\varphi \in C_0^1(B)$ such that $0 \le \varphi \le 1$ in B and let $\psi(x) = \varphi(\frac{x-x^h}{R_h})$.

Using (5.13) with ψ from above we find that

$$(5.27) \quad \begin{cases} \int_B \Big\{ -v^h \cdot \mathrm{div}\left(\varphi(\tilde{\tau} - (\sigma)_{x^h, R_h} - \varepsilon_h \sigma^h) \right) + \frac{\varphi}{\varepsilon_h}((\sigma)^{x^h, R_h} : \\[2ex] (\tilde{\tau} - (\sigma)_{x^h, R_h} - \varepsilon_h \sigma^h) - g^*(\tilde{\tau}) + g^*((\sigma)_{x^h, R_h} + \varepsilon_h \sigma^h)) \Big\} dy \le 0 \end{cases}$$

for all $\tilde{\tau} \in C^1(\overline{B}; \mathbb{M}^{n\times n}_{\text{sym}})$ and $\varphi \in C^1_0(B)$ such that $|\tilde{\tau}^D| \le \sqrt{2}k_*$ and $0 \le \varphi \le 1$.

Let us fix an arbitrary number $m > 0$. Then, for all h large enough, the inequality

$$\lambda + \varepsilon_h m \le \frac{\sqrt{2}k_* + \lambda}{2} = \lambda_0 < \sqrt{2}k_*$$

is valid.

Let

$$K_m = \left\{ \tau \in C^1(\overline{B}; \mathbb{M}^{n\times n}_{\text{sym}}) : |\tau^D| \le m \right\},$$

and observe that the tensor $\tilde{\tau} = (\sigma)_{x^h, R_h} + \varepsilon_h \tau$ satisfies $|\tilde{\tau}^D| \le \lambda + \varepsilon_h m \le \lambda_0 < \sqrt{2}k_*$ for any τ in K_m. Hence we can use inequality (5.27). Taylor's formula then implies

(5.28)
$$\begin{cases} \displaystyle \int_B \left\{ -v^h \cdot \text{div}(\varphi(\tau - \sigma^h)) - \frac{\varphi}{2}\left(\int_0^1 \left(\frac{\partial^2 g^*}{\partial \tau^2}((\sigma)_{x^h, R_h} + \theta_1 \varepsilon_h \tau) \tau \right) : \right. \\ \\ \displaystyle \left. \tau \, d\theta_1 - \int_0^1 \left(\frac{\partial^2 g^*}{\partial \tau^2}((\sigma)_{x^h, R_h} + \theta_2 \varepsilon_h \sigma^h) \sigma^h \right) : \sigma^h d\theta_2 \right) \right\} dy \le 0 \end{cases}$$

for all $\tau \in K_m$ and $\varphi \in C^1_0(B)$ such that $0 \le \varphi \le 1$.

By (5.23) and (5.24) we have (at least for a subsequence)

(5.29) $\sigma^h \rightharpoonup \hat{\sigma}$ in $L^r(B; \mathbb{M}^{n\times n}_{\text{sym}})$,

(5.30) $v^h \rightharpoonup \hat{v}$ in $L^{r^*}(B; \mathbb{R}^n)$.

Now let us take arbitrary functions $\tau \in K_m$ and $\varphi \in C^1_0(B)$ with the restriction $0 \le \varphi \le 1$. Next we choose $s \in [t, 1[$ so that $\text{spt}\,\varphi \subset B_s$.

By (5.26) and (5.29) we can arrange (at least for another subsequence)

(5.31) $\sigma^h \to \hat{\sigma}$ in $L^r(B_s; \mathbb{M}^{n\times n}_{\text{sym}})$,

(5.32) $\sigma^h \to \hat{\sigma}$ a.e. in B_s.

Further we have

$$\left(\int_B |\mathrm{div}\sigma^h|^r\,dy\right)^{1/r} = \left(\int_{B_{R_h}(x^h)} |\mathrm{div}\sigma|^r\,dx\right)^{1/r} \frac{R_h}{\varepsilon_h R_h^{n/r}}$$

$$= \frac{R_h^{1-n/r}}{\varepsilon_h}\left(\int_{B_{R_h}(x^h)} |f|^r\,dx\right)^{1/r} \le \frac{R_h}{\varepsilon_h}\sup_{x\in\Omega_0}|f|$$

$$\le \frac{R_h}{\varepsilon_h} c_{25}\left(\|f\|_{W_{\tilde{n}}^2(\Omega_0;\mathbf{R}^n)}\right) \longrightarrow 0 \text{ as } h \to +\infty$$

(see (5.24)). So we have shown

(5.33) $\mathrm{div}\sigma^h \to \mathrm{div}\hat{\sigma} = 0$ in $L^r(B;\mathbf{R}^n)$.

The relations (5.29) – (5.32) allow us to take the limit in inequality (5.28) for given φ and τ. In particular, from (5.30), (5.31), (5.33) we derive

(5.34) $\displaystyle\int_{B_s} v^h \cdot \mathrm{div}\left(\varphi(\tau - \sigma^h)\right)dy \to \int_{B_s} \hat{v} \cdot \mathrm{div}\left(\varphi(\tau - \hat{\sigma})\right)dy.$

For the last relation we took into account that

(5.35) $\mathrm{div}\hat{\sigma} = 0$ in B.

By (5.11) we find a tensor σ^0 whose trace might be taken to be zero such that

$$(\sigma^D)_{x^h,R_h} \to \sigma^{0D} \quad \text{and} \quad |\sigma^{0D}| \le \lambda.$$

We have

(5.36)
$$\begin{cases} \left(\frac{\partial^2 g^*}{\partial\tau^2}\left((\sigma)_{x^h,R_h} + \theta_1\varepsilon_h\tau\right)\tau\right) : \tau \\[2mm] = \left(\frac{\partial^2 g^*}{\partial\tau^2}\left(\sigma^{0D} + \theta_1\varepsilon_h\tau^D\right)\tau\right) : \tau \longrightarrow \left(\frac{\partial^2 g^*}{\partial\tau^2}(\sigma^0)\hat{\sigma}\right) : \hat{\sigma} \end{cases}$$

and using the same arguments

(5.37) $\left(\dfrac{\partial^2 g^*}{\partial\tau^2}\left((\sigma)_{x^h,R_h} + \theta_2\varepsilon_h\sigma^h\right)\sigma^h\right) : \sigma^h \to \left(\dfrac{\partial^2 g^*}{\partial\tau^2}(\sigma^0)\hat{\sigma}\right) : \hat{\sigma}$

a.e. in $B_s \times [0,1]$.

Since the quadratic form $\varkappa \mapsto \left(\frac{\partial^2 g^*}{\partial \tau^2}(\sigma^0)\varkappa\right) : \varkappa$ is non–negative, we conclude with the help of Fatou's lemma that

(5.38)
$$
\begin{cases}
\liminf\limits_{h \to +\infty} \int\limits_B \varphi \int\limits_0^1 \left(\frac{\partial^2 g^*}{\partial \tau^2}\big((\sigma)_{x^h,R_h} + \theta_2 \varepsilon_h \sigma^h\big)\sigma^h\right) : \sigma^h \, d\theta_2 dy \\[4mm]
\geq \int\limits_B \varphi \left(\frac{\partial^2 g^*}{\partial \tau^2}(\sigma^0)\hat{\sigma}\right) : \hat{\sigma} dy.
\end{cases}
$$

From (5.2) it follows also that

$$
\left(\frac{\partial^2 g^*}{\partial \tau^2}\big((\sigma)_{x^h,R_h} + \theta_1 \varepsilon_h \tau\big)\tau\right) : \tau
$$

$$
= \frac{1}{n^2 K_0} tr^2 \tau + E^*\big((\sigma^D)_{x^h,R_h} + \theta_1 \varepsilon_h \tau^D; \tau, \tau\big)
$$

$$
\leq \frac{1}{n^2 K_0} tr^2 \tau + c_*(g_0^*)''\big(|(\sigma^D)_{x^h,R_h} + \theta \varepsilon_h \tau^D|\big)|\tau^D|^2
$$

$$
\leq \frac{1}{n^2 K_0} tr^2 \tau + c_* \sup_{-\lambda_0 \leq s \leq \lambda_0} (g_0^*)''(s)|\tau^D|^2
$$

$$
\leq c_{26}(\lambda_0, n, K_0, c_*)|\tau|^2.
$$

We recall to the reader that, as it was shown in the proof of Lemma 2.3.1, $c_* = \frac{1}{H_{k_*}}$.

But then, by Lebesgue's theorem, (5.36) and the above estimates

(5.39)
$$
\begin{cases}
\lim\limits_{h \to +\infty} \int\limits_B \varphi \int\limits_0^1 \left(\frac{\partial^2 g^*}{\partial \tau^2}\big((\sigma)_{x^h,R_h} + \theta_1 \varepsilon_h \tau\big)\tau\right) : \tau d\theta_1 dy \\[4mm]
= \int\limits_B \varphi \left(\frac{\partial^2 g^*}{\partial \tau^2}(\sigma^0)\tau\right) : \tau dy.
\end{cases}
$$

Now from (5.34), (5.38), (5.39) and (5.28) we derive

(5.40)
$$
\begin{cases}
\int\limits_{B_s} \int \Big\{ -\hat{v} \cdot \operatorname{div}\big(\varphi(\tau - \hat{\sigma})\big) \\[4mm]
-\frac{\varphi}{2}\left(\big(\frac{\partial^2 g^*}{\partial \tau^2}(\sigma^0)\tau\big) : \tau - \big(\frac{\partial^2 g^*}{\partial \tau^2}(\sigma^0)\hat{\sigma}\big) : \hat{\sigma}\right) \Big\} dy \leq 0
\end{cases}
$$

for all $\tau \in K_m$.

By arbitrariness of m it follows that inequality (5.40) holds for all $\tau \in C^1(\overline{B}; M_{\text{sym}}^{n\times n})$. We know that $\hat{\sigma} \in L^r(B; M_{\text{sym}}^{n\times n})$ and $\operatorname{div}\hat{\sigma} \in L^r(B; \mathbb{R}^n)$. It is clear that, for any $\tau \in L^r(B; M_{\text{sym}}^{n\times n})$ with the property $\operatorname{div}\tau \in L^r(B; \mathbb{R}^n)$, there exists a sequence $\tau^m \in C^1(\overline{B}; M_{\text{sym}}^{n\times n})$ such that

$$\tau^m \to \tau \text{ in } L^r(B; M_{\text{sym}}^{n\times n}) \text{ and } \operatorname{div}\tau^m \to \operatorname{div}\tau \text{ in } L^r(B; \mathbb{R}^n).$$

But then it follows from (5.40) that

$$\int_{B_s} \left[-\hat{v} \cdot \operatorname{div}(\varphi\tau) - \varphi\left(\frac{\partial^2 g^*}{\partial \tau^2}(\sigma^0)\hat{\sigma}\right) : \tau \right] dy \le 0 \quad \forall \tau \in C^1(\overline{B}; M_{\text{sym}}^{n\times n}).$$

Since the function $\varphi \in C_0^1(B)$ is arbitrary (satisfying the condition $0 \le \varphi \le 1$), we finally get

$$\varepsilon(\hat{v}) = \frac{\partial^2 g^*}{\partial \tau^2}(\sigma^0)\hat{\sigma} \quad \text{in } B.$$

Taking into account identity (5.35), we conclude that the pair $\hat{v}$, $\hat{\sigma}$ satisfies all conditions of Lemma 2.4.1. According to this lemma we have estimate (4.3) in which the functions v and τ have to be replaced by the functions $\hat{v}$ and $\hat{\tau}$, respectively. Now, if we can show that

$$(5.41) \quad v^h \to \hat{v} \text{ in } L^{r^*}(B_t; \mathbb{R}^n),$$

$$(5.42) \quad \sigma^h \to \hat{\sigma} \text{ in } L^r(B_t; M_{\text{sym}}^{n\times n}),$$

then the proof of the lemma is complete. Indeed, in this case, by (5.25), we have

$$W(t) \ge 2c_{17}t,$$

and the lower semicontinuity of the norms with respect to weak convergence implies

$$W(1) \le 1.$$

But then

$$W(t) \le c_{17}t\, W(1) \le c_{17}t \le W(t)/2,$$

which is a contradiction.

We are now going to prove (5.41) and (5.42). (5.42) follows from (5.31) since $s \geq t$.

Let $\tau \in C_0^1(B_t; \mathbb{M}_{\text{sym}}^{n \times n})$, $|\tau| \leq 1$, and consider a cut–off function φ such that $\varphi \equiv 1$ in B_t, $\text{spt}\varphi \subset B_s$, $t < s < 1$. We insert these functions into (5.28) and arrive at the relation

$$- \int_{B_t} v^h \cdot \text{div}\tau \, dy \leq - \int_{B_s} \sigma^h : (v^h \odot \nabla\varphi) dy$$

$$+ \int_{B_s} \frac{\varphi}{2} \int_0^1 \left(\frac{\partial^2 g^*}{\partial \tau^2} \left((\sigma)_{x^h, R_h} + \theta_1 \varepsilon_h \tau \right) \tau \right) :$$

$$\tau d\theta_1 dy \leq \int_{B_s} |\sigma^h| |\nabla\varphi| |v^h| dy + c_{26} \int_{B_s} |\tau|^2 dy.$$

Using $|\tau| \leq 1$ we get for h sufficiently large

$$|(\sigma^D)_{x^h, R_h} + \varepsilon_h \theta_1 \tau^D| \leq \lambda + \varepsilon_h \leq \frac{\sqrt{2}k_* + \lambda}{2} = \lambda_0 < \sqrt{2}k_*.$$

From this we obtain

$$\int_{B_t} |\varepsilon(v^h)| dy = \sup_{\substack{\tau \in C_0^\infty(B_t; \mathbb{M}_{\text{sym}}^{n \times n}), \\ |\tau| \leq 1}} \left(- \int_{B_t} v^h \cdot \text{div}\tau dy \right)$$

$$\leq c_{26}' \left\{ \left(\int_B |\sigma^h|^r dy \right)^{1/r} \left(\int_B |v^h|^{r^*} dy \right)^{1/r^*} + 1 \right\}.$$

By (5.23), (5.24) the sequence v^h is bounded in $BD(B_t; \mathbb{R}^n)$ and hence precompact in $L^{r^*}(B_t; \mathbb{R}^n)$ (see Appendix A3). This implies (5.41) and Lemma 2.5.1 is established.

□

We are now going to iterate Lemma 2.5.1. First observe the inequalities

$$(5.43) \quad \begin{cases} |(\sigma^D)_{x_0, t^k R} - (\sigma^D)_{x_0, R}| \leq \sum_{i=0}^{k-1} |(\sigma^D)_{x_0, t^{i+1}R} - (\sigma^D)_{x_0, t^i R}|, \\ \\ |(\sigma^D)_{x_0, t^{i+1}R} - (\sigma^D)_{x_0, t^i R}| \leq t^{-\frac{n}{r}} \overline{U}(x_0, t^i R) \leq t^{-\frac{n}{r}} U(x_0, t^i R), \end{cases}$$

from which it is easy to get the estimates

(5.44)

$$
\begin{cases}
|(\sigma^D)_{x_0,R}| - t^{-\frac{n}{r}} \sum_{i=0}^{k-1} U(x_0, t^i R) \le |(\sigma^D)_{x_0, t^k R}| \\[2ex]
\le |(\sigma^D)_{x_0, R}| + t^{-\frac{n}{r}} \sum_{i=0}^{k-1} U(x_0, t^i R).
\end{cases}
$$

LEMMA 2.5.2 *Let the numbers* $\nu \in]0,1[,\ t_* \in]0,1[,\ t \in]0, t_*[,\ \lambda \in]0, \sqrt{2}k_*[$ *be chosen according to*

(5.45) $2c_{17}(t_*, \lambda) t^{1-\nu} = 1$

and define ε_0 *and* R_0 *as in Lemma 2.5.1. Suppose further that the following conditions hold*

(5.46) $0 < R < R_0, \quad B_R(x_0) \subset\subset \Omega_0,$

(5.47) $|(\sigma^D)_{x_0, R}| < \lambda_1 < \lambda < \sqrt{2}k_*,$

(5.48) $U(x_0, R) + R^{\varepsilon_3} < \bar{\varepsilon}_0 = (1 - t^{\varepsilon_3 - \nu}) \min\{\varepsilon_0, (1 - t^\nu) t^{\frac{n}{r}} (\lambda - \lambda_1)\}$

where $\varepsilon_3 \in]\nu, 1[$. *Then, for any* $k \in \mathbb{N}$, *we have the estimates*

(5.49) $|(\sigma^D)_{x_0, t^k R}| < \lambda,$

(5.50) $U(x_0, t^k R) \le t^{\nu k} \left(U(x_0, R) + R^{\varepsilon_3} \dfrac{1 - t^{(\varepsilon_3 - \nu)k}}{1 - t^{\varepsilon_3 - \nu}} \right).$

Remark: Here we use the same notation as in Lemma 2.5.1.

Proof.   We will prove the lemma by induction. Let $k = 1$. Then inequality (5.50) follows from Lemma 2.5.1. In view of (5.44), (5.47), (5.48) we have

$$
|(\sigma^D)_{x_0, tR}| \le |(\sigma^D)_{x_0, R}| + t^{-\frac{n}{r}} U(x_0, R) < \lambda_1
$$

$$
+ \frac{t^{-\frac{n}{r}} (U(x_0, R) + R^{\varepsilon_3})}{1 - t^{\varepsilon_3 - \nu}} < \lambda_1 + \frac{t^{-\frac{n}{r}} \bar{\varepsilon}_0}{1 - t^{\varepsilon_3 - \nu}}
$$

$$
< \lambda_1 + \frac{t^{-\frac{n}{r}}}{1 - t^{\varepsilon_3 - \nu}} (1 - t^\nu) t^{\frac{n}{r}} (\lambda - \lambda_1)(1 - t^{\varepsilon_3 - \nu}).
$$

Therefore estimate (5.49) for $k = 1$ is valid.

Suppose that all claims of the lemma are valid for $k = 1, 2, \ldots, s$. We wish to prove them for $k = s + 1$. Since

(5.51) $\quad |(\sigma^D)_{x_0, t^k R}| < \lambda,$

(5.52) $\quad U(x_0, t^k R) \leq t^{\nu k} \left(U(x_0, R) + R^{\varepsilon_3} \dfrac{1 - t^{(\varepsilon_3 - \nu)k}}{1 - t^{\varepsilon_3 - \nu}} \right) < \varepsilon_0, \quad k = 1, \ldots, s,$

we may use Lemma 2.5.1 replacing R there by $t^s R$ which gives

$$U(x_0, t^{s+1} R) \leq t^{\nu} \left(U(x_0, t^s R) + (t^s R)^{\varepsilon_3} \right)$$

$$\leq t^{\nu(s+1)} \left(U(x_0, R) + R^{\varepsilon_3} \dfrac{1 - t^{(\varepsilon_3 - \nu)k}}{1 - t^{\varepsilon_3 - \nu}} + R^{\varepsilon_3} t^{(\varepsilon_3 - \nu)k} \right)$$

$$= t^{\nu(s+1)} \left(U(x_0, R) + R^{\varepsilon_3} \dfrac{1 - t^{(\varepsilon_3 - \nu)(s+1)}}{1 - t^{\varepsilon_3 - \nu}} \right).$$

So (5.52) for $k = s + 1$ is proved. Moreover

$$|(\sigma^D)_{x_0, t^{s+1} R}| \leq |(\sigma^D)_{x_0, R}| + t^{-\frac{n}{r}} \sum_{k=0}^{s} U(x_0, t^k R)$$

$$< \lambda_1 + \dfrac{t^{-\frac{n}{r}}}{1 - t^{\varepsilon_3 - \nu}} \left(U(x_0, R) + R^{\varepsilon_3} \right) \sum_{k=0}^{s} t^{\nu k}$$

$$< \lambda_1 + \dfrac{t^{-\frac{n}{r}} \overline{\varepsilon}_0}{(1 - t^{\varepsilon_3 - \nu})(1 - t^{\nu})} < \lambda$$

which completes the proof.

$\square$

LEMMA 2.5.3 *Suppose that all the conditions of Lemma 2.5.2 are satisfied. Then we have the following estimates:*

(5.53) $\quad \overline{U}(x_0, \rho) \leq 2(\dfrac{1}{t})^{\frac{n}{r} + \nu} (\dfrac{\rho}{R})^{\nu} (U(x_0, R) + R^{\varepsilon_3})(1 - t^{\varepsilon_3 - \nu})^{-1},$

(5.54) $\quad |(\sigma^D)_{x_0, \rho}| < \lambda, \quad 0 < \rho \leq R, \quad B_R(x_0) \subset\subset \Omega_0.$

Proof.   Let $\rho \in]0, R]$. We take k so that

$$t^{k+1} < \rho/R \leq t^k \Rightarrow \frac{R}{\rho}t^k < \frac{1}{t}.$$

Then we have

$$\overline{U}(x_0, \rho) = \left(\fint_{B_\rho(x_0)} |\sigma - (\sigma)_{x_0,\rho}|^r dx \right)^{1/r} \leq \left(\fint_{B_\rho(x_0)} |\sigma - (\sigma)_{x_0,t^k R}|^r dx \right)^{1/r}$$

$$+|(\sigma)_{x_0,\rho} - (\sigma)_{x_0,t^k R}| \leq 2\left(\fint_{B_\rho(x_0)} |\sigma - (\sigma)_{x_0,t^k R}|^r dx \right)^{1/r}$$

$$\leq 2\left(\frac{t^k R}{\rho} \right)^n \left(\fint_{B_{t^k R}(x_0)} |\sigma - (\sigma)_{x_0,t^k R}|^r dx \right)^{1/r}$$

$$\leq 2(\tfrac{t^k R}{\rho})^{\frac{n}{r}} \overline{U}(x_0, t^k R),$$

hence

$$\overline{U}(x_0, \rho) \leq 2(\tfrac{1}{t})^{\frac{n}{r}} \overline{U}(x_0, t^k R) \leq 2(\tfrac{1}{t})^{\frac{n}{r}} t^{\nu k}$$

$$\times \frac{U(x_o, R) + R^{\varepsilon_3}}{1 - t^{\varepsilon_3 - \nu}} \leq 2(\tfrac{1}{t})^{\frac{n}{r}} (\tfrac{1}{t}\tfrac{\rho}{R})^\nu \frac{U(x_o, R) + R^{\varepsilon_3}}{1 - t^{\varepsilon_3 - \nu}},$$

and (5.53) follows. Next we observe

$$|(\sigma^D)_{x_0,\rho} - (\sigma^D)_{x_0,R}| \leq |(\sigma^D)_{x_0,\rho} - (\sigma^D)_{x_0,t^k R}|$$

$$+ \sum_{i=0}^{k-1} |(\sigma^D)_{x_0,t^{i+1}R} - (\sigma^D)_{x_0,t^i R}|$$

$$\leq (\tfrac{1}{t})^{\frac{n}{r}} U(x_0, t^k R) + (\tfrac{1}{t})^{\frac{n}{r}} \sum_{i=0}^{k-1} U(x_0, t^i R)$$

$$= (\tfrac{1}{t})^{\frac{n}{r}} \sum_{i=0}^{k} U(x_0, t^i R) < (\tfrac{1}{t})^{\frac{n}{r}} \frac{1}{1 - t^{\varepsilon_3 - \nu}} \frac{U(x_0, R) + R^{\varepsilon_3}}{1 - t^\nu}$$

$$< \frac{\overline{\varepsilon}_0 t^{-\frac{n}{r}}}{(1 - t^\nu)(1 - t^{\varepsilon_3 - \nu})} \leq \frac{t^{-\frac{n}{r}}(1 - t^{\varepsilon_3 - \nu})(1 - t^\nu)t^{\frac{n}{r}}(\lambda - \lambda_1)}{(1 - t^\nu)(1 - t^{\varepsilon_3 - \nu})}$$

$$= \lambda - \lambda_1.$$

This inequality implies (5.54).

$\square$

2.6   Proof of Theorem 2.1.2

As in the proof of the Main Lemma 2.5.1 we are going to consider the saddle point (u, σ) obtained in Lemma 2.2.1.

We first deduce from inequality (5.8) the relation

$$(6.1) \qquad \int_{B_R(x_0)} \left\{ -u_R \cdot \operatorname{div}\big(\psi(\tau - \sigma)\big) + \psi\big(\sigma^R : (\tau - \sigma) - g^*(\tau) + g^*(\sigma)\big) \right\} dx \le 0$$

being valid for all $\tau \in C^1(\overline{B}_R(x_0); M^{n\times n}_{\text{sym}})$ and $\psi \in C^1_0(B_R(x_0))$ such that $|\tau^D| \le \sqrt{2}k_*$ and $0 \le \psi \le 1$. Here

$$u_R = u - \sigma^R(x - x_0) - v_{\times R}, \ v_{\times R} \in V_\times, \qquad \sigma^R = \frac{\partial g^*}{\partial \tau}\big((\sigma)_{x_0,R}\big)$$

and it is assumed that

$$(6.2) \qquad |(\sigma^D)_{x_0,R}| \le \lambda < \sqrt{2}k_*.$$

We now use the same arguments which led to (5.27) and (5.28). We write (6.1) in the form

$$\int_{B_R(x_0)} \left\{ -u_R \cdot \operatorname{div}\big(\psi(\tau - (\sigma)_{x_0,R} + (\sigma)_{x_0,R} - \sigma)\big) + \psi\Big(\sigma^R : \big(\tau - (\sigma)_{x_0,R}\big) \right.$$
$$-g^*\big(\tau - (\sigma)_{x_0,R} + (\sigma)_{x_0,R}\big) - \sigma^R : \big(\sigma - (\sigma)_{x_0,R}\big)$$
$$\left. +g^*\big(\sigma - (\sigma)_{x_0,R} + (\sigma)_{x_0,R}\big)\Big) \right\} dx \le 0,$$

and then, by applying Lagrange's formula, estimates (5.1), (5.2) and also imposing the additional restriction

$$(6.3) \qquad |\tau^D - (\sigma^D)_{x_0,R}| \le \frac{\sqrt{2}k_* - \lambda}{2},$$

we have

$$(6.4) \quad \begin{cases} 0 \geq \displaystyle\int_{B_R(x_0)} \Big\{ -\psi u_R \cdot \operatorname{div}\big(\tau - (\sigma)_{x_0,R}\big) - (u_R \odot \nabla\psi) : \big(\tau - (\sigma)_{x_0,R}\big) \\ \quad + \psi u_R \cdot \operatorname{div}\big(\sigma - (\sigma)_{x_0,R}\big) + (u_R \odot \nabla\psi) : \big(\sigma - (\sigma)_{x_0,R}\big) \\[2mm] \quad -\tfrac{\psi}{2}\Big\langle \displaystyle\int_0^1 \Big(\frac{\partial^2 g^*}{\partial\tau^2}\big((\sigma)_{x_0,R} + \theta_1\big(\tau - (\sigma)_{x_0,R}\big)\big)\big(\tau - (\sigma)_{x_0,R}\big)\Big) : \\[2mm] \quad \big(\tau - (\sigma)_{x_0,R}\big)d\theta_1 - \displaystyle\int_0^1 \Big(\frac{\partial^2 g^*}{\partial\tau^2}\big((\sigma)_{x_0,R} \\[2mm] \quad + \theta_2\big(\sigma - (\sigma)_{x_0,R}\big)\big)\big(\sigma - (\sigma)_{x_0,R}\big)\Big) : \big(\sigma - (\sigma)_{x_0,R}\big)d\theta_2 \Big\rangle \Big\}\, dx \\[3mm] \quad \geq \displaystyle\int_{B_R(x_0)} \Big\{ -\psi u_R \cdot \operatorname{div}\big(\tau - (\sigma)_{x_0,R}\big) - (u_R \odot \nabla\psi) : \big(\tau - (\sigma)_{x_0,R}\big) \\ \quad -\psi u_R \cdot f - |u_R| \|\nabla\psi\| |\sigma - (\sigma)_{x_0,R}| \\[2mm] \quad -\tfrac{\psi}{2}\Big[\frac{tr^2\big(\tau - (\sigma)_{x_0,R}\big)}{n^2 K_0} + \sup_{|s| \leq \frac{\sqrt{2}k_* + \lambda}{2}} (g_0^*)''(s)|\tau^D - (\sigma^D)_{x_0,R}|^2\Big]\Big\}\, dx. \end{cases}$$

For the second inequality we have used (5.3) and the following elementary estimate

$$\big|(\sigma^D)_{x_0,R} + \theta_1\big(\tau^D - (\sigma^D)_{x_0,R}\big)\big| \leq |(\sigma^D)_{x_0,R}| + |\tau^D - (\sigma^D)_{x_0,R}|$$

$$\leq \frac{\sqrt{2}k_* + \lambda}{2}.$$

Let the cut–off function ψ satisfy $\psi \equiv 1$ on $B_{R/2}(x_0)$, $|\nabla\psi| \leq \frac{3}{R}$ in $B_R(x_0)$. We insert $\tau = (\sigma)_{x_0,R} + \frac{\sqrt{2}k_* - \lambda}{2}\varkappa$ into (6.4), assuming that $\varkappa \in C_0^1\big(B_R(x_0); \mathbb{M}^{n\times n}_{\mathrm{sym}}\big)$, $|\varkappa| \leq 1$ on $B_R(x_0)$ and $\operatorname{spt}\varkappa \subset B_{R/2}(x_0)$. Then the tensor τ satisfies restriction (6.3) and we can derive from (6.4) the following inequality

$$\int_{B_R(x_0)} \Big\{ -u_R \cdot \operatorname{div}\varkappa \frac{\sqrt{2}k_* - \lambda}{2} - \frac{(\sqrt{2}k_* - \lambda)^2}{8} c_{27}(n, K_0, \lambda, k_*)|\varkappa|^2 \Big\}\, dx$$
$$\leq \int_{B_R(x_0)} |u_R| |f| dx + \frac{3}{R}\int_{B_R(x_0)} |u_R| |\sigma - (\sigma)_{x_0,R}| dx$$

being valid for all $\varkappa \in C_0^1\big(B_{R/2}(x_0); \mathbb{M}^{n\times n}_{\mathrm{sym}}\big)$ such that $|\varkappa| \leq 1$ in $B_{R/2}(x_0)$. Here

$$c_{27} = \max\Big\{ \frac{1}{nK_0},\ \max_{|s| \leq \frac{\sqrt{2}k_* + \lambda}{2}} (g_0^*)''(s) \Big\}.$$

Choosing a rigid displacement $v_{\times R} \in V_\times$ so that

$$(6.5) \qquad \tilde{U}(x_0, R) = \frac{1}{R}\left(\fint_{B_R(x_0)} |u_R|^{r^*} dx \right)^{1/r^*}$$

and recalling the definition of how to apply a convex function to a measure, we obtain

$$\int_{B_{R/2}(x_0)} H(|\varepsilon(u_R)|) \equiv$$

$$\sup_{\substack{\varkappa \in C_0^1(B_{R/2}(x_0);!A_{\mathrm{sym}}^{n \times n}), \\ |\varkappa| \le 1 \text{ in } B_{R/2}(x_0)}} \int_{B_{R/2}(x_0)} \left\{ -u_R \cdot \operatorname{div} \varkappa - c_{27} \frac{(\sqrt{2}k_* - \lambda)}{4} |\varkappa|^2 \right\} dx$$

$$\le \frac{2}{\sqrt{2}k_* - \lambda} \int_{B_R(x_0)} |u_R| \left(|f| + \frac{3|\sigma - (\sigma)_{x_0,R}|}{R} \right) dx \le |B_{R/2}| S(x_0, R),$$

where

$$S(x_0, R) \equiv \frac{2^{n+1}}{\sqrt{2}k_* - \lambda} \left[R\left(\fint_{B_R(x_0)} |f|^r dx \right)^{1/r} + 3\overline{U}(x_0, R) \right] \tilde{U}(x_0, R)$$

$$\le S_1(x_0, R) \equiv \frac{2^{n+3}}{\sqrt{2}k_* - \lambda} c_{28}\left(\|f\|_{W_{\overline{n}}^2(\Omega_0; \mathbf{R}^n)} \right) \left[R + \overline{U}(x_0, R) \right] \tilde{U}(x_0, R)$$

and

$$H(t) = \begin{cases} \dfrac{t^2}{c_{27}(\sqrt{2}k_* - \lambda)} & \text{if } |t| \le \dfrac{\sqrt{2}k_* - \lambda}{2} c_{27} \\ |t| - \dfrac{\sqrt{2}k_* - \lambda}{4} c_{27} & \text{if } |t| > \dfrac{\sqrt{2}k_* - \lambda}{2} c_{27} \end{cases}$$

$$= \sup\left\{ st - \frac{\sqrt{2}k_* - \lambda}{4} c_{27} s^2 : |s| \le 1 \right\}.$$

After application of Jensen's inequality (see Appendix A.4) we arrive at the relation

$$(6.6) \qquad H\left(\fint_{B_{R/2}(x_0)} |\varepsilon(u_R)| \right) \le \fint_{B_{R/2}(x_0)} H(|\varepsilon(u_R)|) \le S_1(x_0, R).$$

Using the imbedding theorem (see Appendix A.3) it is easy to establish the estimate

$$(6.7) \qquad \begin{cases} \dfrac{2}{R}\left(\fint_{B_{R/2}(x_0)} |\tilde{u}_R|^{r^*} dx \right)^{1/r^*} \le c_{29}(n, r) \fint_{B_{R/2}(x_0)} |\varepsilon(\tilde{u}_R)| \\ = c_{29} \fint_{B_{R/2}(x_0)} |\varepsilon(u_R)|, \end{cases}$$

where

$$\tilde{u}_R = u - \sigma^R(x - x_0) - \tilde{v}_{\times R}, \quad \tilde{v}_{\times R} \in V_\times,$$

and

(6.8)
$$\fint_{B_{R/2}(x_0)} |\tilde{u}_R|^{r^*} dx = \inf_{v_\times \in V_\times} \fint_{B_{R/2}(x_0)} |u - \sigma^R(x - x_0) - v_\times|^{r^*} dx.$$

So we can replace (6.6) by

(6.9)
$$H\left(\frac{2}{c_{29}R}\left(\fint_{B_{R/2}(x_0)} |\tilde{u}_R|^{r^*} dx\right)^{1/r^*}\right) \le S_1(x_0, R).$$

Let us consider two cases. Suppose first that

(6.10)
$$\frac{2}{R c_{29}}\left(\fint_{B_{R/2}(x_0)} |\tilde{u}_R|^{r^*} dx\right)^{1/r^*} \le \frac{\sqrt{2}k_* - \lambda}{2} c_{27}.$$

Inequality (6.9) in this case takes the form

$$\frac{1}{(\sqrt{2}k_* - \lambda)c_{27}c_{29}^2}\left(\frac{2}{R}\fint_{B_{R/2}(x_0)} |\tilde{u}_R|^{r^*} dx\right)^{2/r^*} \le S_1(x_0, R).$$

From this we get

(6.11)
$$\begin{cases}
\frac{2}{R}\left(\fint_{B_{R/2}(x_0)} |\tilde{u}_r|^{r^*} dx\right)^{1/r^*} \le \sqrt{(\sqrt{2}k_* - \lambda)c_{27}c_{29}^2 S_1(x_0, R)} \\
= \left(\frac{2^{n+3}}{\sqrt{2}k_* - \lambda}c_{28}(R + \overline{U}(x_0, R))\tilde{U}(x_0, R)(\sqrt{2}k_* - \lambda)c_{27}c_{29}^2\right)^{1/2} \\
= \sqrt{c_{27}c_{28}c_{29}^2 2^{n+3}}(R + \overline{U}(x_0, R))^{1/2}\tilde{U}^{1/2}(x_0, R) \\
\le \frac{1}{2}\tilde{U}(x_0, R) + c_{27}c_{28}c_{29}^2 2^{n+2}(R + \overline{U}(x_0, R)).
\end{cases}$$

In the opposite case we first note that

$$\frac{1}{2}\frac{2}{c_{29}R}\left(\fint_{B_{R/2}(x_0)} |\tilde{u}_R|^{r^*} dx\right)^{1/r^*}$$

$$\le H\left(\frac{2}{Rc_{29}}\left(\fint_{B_{R/2}(x_0)} |\tilde{u}_R|^{r^*} dx\right)^{1/r^*}\right) \le S_1(x_0, R),$$

and therefore

$$(6.12) \quad \begin{cases} \dfrac{2}{R}\left(\fint_{B_{R/2}(x_0)} |\tilde{u}_R|^{r^*}\,dx \right)^{1/r^*} \leq 2c_{29}S_1(x_0, R) \\[4mm] \leq \dfrac{2^{n+4}}{\sqrt{2k_* - \lambda}}\,c_{29}c_{28}\big[R + \overline{U}(x_0, R)\big]\tilde{U}(x_0, R). \end{cases}$$

Putting together estimates (6.11) and (6.12) we get the final result

$$(6.13) \quad \begin{cases} \dfrac{2}{R}\left(\fint_{B_{R/2}(x_0)} |\tilde{u}_R|^{r^*}\,dx \right)^{1/r^*} \leq \max\left\{\dfrac{1}{2}, c_{30}(R + \overline{U}(x_0, R))\right\}\tilde{U}(x_0, R) \\[4mm] + c_{31}(R + \overline{U}(x_0, R)), \end{cases}$$

where

$$c_{30} = \frac{2^{n+4}}{\sqrt{2k_* - \lambda}}\,c_{29}c_{28}, \quad c_{31} = c_{27}c_{28}c_{29}^2 2^{n+2}.$$

Now, we can prove the following

LEMMA 2.6.1 *Let $B_R(x_0) \Subset \Omega_0$ and suppose that*

$$(6.14) \quad |(\sigma^D)_{x_0, R}| < \lambda - 2^{\frac{n}{r}}\overline{U}(x_0, R) \quad (0 \leq \lambda < \sqrt{2}k_*).$$

Then we have the estimate

$$(6.15) \quad \tilde{U}(x_0, R/2) \leq \max\left\{\frac{1}{2}, c_{30}(R + \overline{U}(x_0, R))\right\}\tilde{U}(x_0, R) + c_{32}(R + \overline{U}(x_0, R)),$$

where the constant c_{32} depends only on K_0, n, λ, r, k_*, $\|f\|_{W^2_{\frac{n}{n}}(\Omega_0; \mathbf{R}^n)}$.

Proof. We will use estimate (6.13). Taking into account formula (6.5) in which R is replaced by $R/2$ (see also (6.8)), we have

$$\tilde{U}(x_0, R/2) = \frac{2}{R}\left(\fint_{B_{R/2}(x_0)} |u - \sigma^{R/2}(x - x_0) - v_{\times R/2}|^{r^*}\,dx \right)^{1/r^*}$$

$$\leq \frac{2}{R}\left(\fint_{B_{R/2}(x_0)} |u - \sigma^{R/2}(x - x_0) - \tilde{v}_{\times R}|^{r^*}\,dx \right)^{1/r^*}$$

$$\leq \frac{2}{R}\left(\fint_{B_{R/2}(x_0)} |\tilde{u}_R|^{r^*} \right)^{1/r^*} + |\sigma^{R/2} - \sigma^R|$$

$$\leq \max\left\{\frac{1}{2}, c_{30}(R + \overline{U}(x_0, R))\right\}\tilde{U}(x_0, R) + c_{31}(R + \overline{U}(x_0, R)) + |\sigma^{R/2} - \sigma^R|.$$

On the other hand, we also have

(6.16) $|(\sigma)_{x_0,R/2} - (\sigma)_{x_0,R}| \leq 2^{\frac{n}{r}}\overline{U}(x_0,R).$

Due to (6.14) and (6.16) we get

(6.17) $|(\sigma^D)_{x_0,R} + \theta((\sigma^D)_{x_0,R/2} - (\sigma^D)_{x_0,R})| \leq \lambda$

for any $\theta \in [0,1]$. Lagrange's formula gives

$$|\sigma^{R/2} - \sigma^R| = |\tfrac{\partial g^*}{\partial \tau}((\sigma)_{x_0,R/2}) - \tfrac{\partial g^*}{\partial \tau}((\sigma)_{x_0,R})|$$

$$\leq \int_0^1 |\tfrac{\partial^2 g^*}{\partial \tau^2}((\sigma)_{x_0,R} + \theta((\sigma)_{x_0,R/2} - (\sigma)_{x_0,R}))((\sigma)_{x_0,R/2} - (\sigma)_{x_0,R})|\, d\theta$$

$$\leq \max\left\{\tfrac{1}{nK_0},\ \sup_{|s|\leq\lambda}(g_0^*)''(s)\right\}|(\sigma)_{x_0,R/2} - (\sigma)_{x_0,R}|$$

$$\leq 2^{\frac{n}{r}}\max\left\{\tfrac{1}{nK_0},\ \sup_{|s|\leq\lambda}(g_0^*)''(s)\right\}\overline{U}(x_0,R).$$

If we put

$$c_{32} = c_{31} + 2^{\frac{n}{r}}\max\left\{\frac{1}{nK_0},\ \sup_{|s|\leq\lambda}(g_0^*)''(s)\right\},$$

then we arrive at (6.15). Lemma 2.6.1 is proved.

$\square$

LEMMA 2.6.2 *Let $x_0 \in \Omega_0$ be such that*

(6.18) $\lim_{R\to 0}\overline{U}(x_0,R) = 0,$

(6.19) $\lim_{R\to 0}|(\sigma^D)_{x_0,R}| = \lambda_0 < \sqrt{2}k_*.$

Then

(6.20) $\liminf_{R\to 0}\tilde{U}(x_0,R) = 0.$

Proof.   By conditions (6.18) and (6.19) there is a number $R_1 > 0$ such that $|(\sigma^D)_{x_0,R}| < \frac{\sqrt{2}k_* + \lambda_0}{2}$ and $2^{\frac{n}{r}}\overline{U}(x_0,R) < \frac{\sqrt{2}k_* - \lambda_0}{4}$ for all $R \in]0, R_1[$. Therefore

we have

(6.21) $\quad |(\sigma^D)_{x_0,R}| < \dfrac{3\sqrt{2}k_* + \lambda_0}{4} - 2^{\frac{n}{r}}\overline{U}(x_0, R)$

for all $R \in]0, R_1[$.

According to (6.18) there exists a number $R_2 > 0$ having the property

(6.22) $\quad c_{30}\left(n, K_0, r, \|f\|_{W^2_{\frac{n}{r}}(\Omega_0;\mathbf{R}^n)}, \dfrac{3\sqrt{2}k_* + \lambda_0}{4}\right)\overline{U}(x_0, R) < \dfrac{1}{2}$

for any $R \in]0, R_2[$. But then, by (6.21), (6.22) and Lemma 2.6.1, we have

(6.23) $\quad \begin{cases} \tilde{U}(x_0, R/2) \leq \frac{1}{2}\tilde{U}(x_0, R) + c_{32}\left(n, K_0, r, \|f\|_{W^2_{\frac{n}{r}}(\Omega_0;\mathbf{R}^n)}, \frac{3\sqrt{2}k_*+\lambda_0}{4}\right) \\ (R + \overline{U}(x_0, R)) \text{ for any } R \in]0, \min\{R_1, R_2\}[. \end{cases}$

In (6.22) and (6.23) we used the following notation: the constants c_{30} and c_{32} defined before depend on a parameter λ and we now replace λ by the quantity $\frac{3\sqrt{2}k_*+\lambda_0}{4}$.

Iteration of (6.23) gives

(6.24) $\quad \tilde{U}(x_0, R/2^k) \leq \dfrac{1}{2^k}\tilde{U}(x_0, R) + c_{32}\displaystyle\sum_{i=0}^{k-1}\dfrac{1}{2^{k-1-i}}\left(\overline{U}\left(x_0, \dfrac{R}{2^i}\right) + \dfrac{R}{2^i}\right)$

for any $R \in]0, \min\{R_1, R_2\}[$.

It is known that

$$\lim_{k\to+\infty}\dfrac{\displaystyle\sum_{i=0}^{k-1} 2^i\left(\overline{U}(x_0, \frac{R}{2^i}) + \frac{R}{2^i}\right)}{2^{k-1}} = \lim_{k\to+\infty}\dfrac{2^k\left(\overline{U}(x_0, \frac{R}{2^k}) + \frac{R}{2^k}\right)}{2^k - 2^{k-1}}$$

provided the limit of the right hand side exists. But by (6.18) this limit is equal to zero. Now, it follows from (6.24) that

$$\tilde{U}(x_0, \dfrac{R}{2^k}) \to 0 \text{ as } k \to +\infty.$$

Lemma 2.6.2 is proved.

$\qquad\qquad\qquad\qquad\qquad\qquad\qquad\qquad\qquad\qquad\qquad\qquad\qquad\qquad\qquad$ □

Proof of Theorem 2.1.2.   Let us choose a number $r \in]n, \frac{2n}{n-2}[$ such that $r \leq \bar{n}$ and introduce the sets

$$\Sigma_1 = \left\{ x_0 \in \Omega : \limsup_{R \to 0} \overline{U}(x_0, R) > 0 \right\},$$

$$\Sigma_2 = \left\{ x_0 \in \Omega : \not\exists \lim_{R \to 0} |(\sigma^D)_{x_0, R}| \right\}.$$

It is clear that $|\Sigma| = 0$ where $\Sigma = \Sigma_1 \cup \Sigma_2$. Let us represent the set $\Omega \setminus \Sigma$ as the union of the two disjoint subsets

$$\Omega_1 = \{ x_0 \in \Omega \setminus \Sigma : |\sigma^D(x_0)| < \sqrt{2}k_* \}$$

$$\Omega_2 = \{ x_0 \in \Omega \setminus \Sigma : |\sigma^D(x_0)| = \sqrt{2}k_* \}.$$

We are going to show that the set Ω_1 is open and $\sigma \in C^\nu(\Omega_1; \mathbb{M}^{n \times n}_{\text{sym}})$ for any $\nu \in]0, 1[$.

To begin with, we take an arbitrary number $\nu \in]0, 1[$ and define

$$t_* = (1/2)^{\frac{1}{1-\nu}} < 1.$$

Let $x_0 \in \Omega_0$. We choose some domain Ω_0 such that $x_0 \in \Omega_0 \subset\subset \Omega$ and let

$$\lambda = \frac{|\sigma^D(x_0)| + \sqrt{2}k_*}{2}.$$

Now we can determine the constant $c_{17}(t_*, \lambda)$ from Lemma 2.4.1. Without loss of generality let c_{17} be greater then $3/2$.

If we set

$$t = (1/2c_{17})^{\frac{1}{1-\nu}},$$

then $t \in]0, t_*[$ and identity (5.45) holds. So we can determine the constants ε_0 and R_0 of Lemma 2.5.1 by choosing $\varepsilon_3 = \frac{1+\nu}{2}$.

By Lemma 2.6.2,

$$\liminf_{R \to 0} \tilde{U}(x_0, R) = 0$$

and since

$$\lim_{R \to 0} |(\sigma^D)_{x_0, R}| < \lambda_1 = \frac{|\sigma^D(x_0)| + \lambda}{2} < \lambda < \sqrt{2}k_*,$$

a number $R > 0$ exists such that conditions (5.46)–(5.48) hold.

The functions $x \mapsto |(\sigma^D)_{x,R}|$ and $x \mapsto U(x, R)$ are upper semicontinuous at the point x_0, and so a nonempty ball $B_{r_1}(x_0)$ exists such that at each point of $B_{r_1}(x_0)$ conditions (5.46)–(5.48) are satisfied, i.e.

$$B_R(x) \subset\subset \Omega_0 \ (0 < R < R_0), \quad |(\sigma^D)_{x,R}| < \lambda_1,$$
$$U(x, R) + R^{\varepsilon_3} < \bar{\varepsilon}_0 \quad \text{for any } x \in B_{r_1}(x_0).$$

But then, by Lemma 2.5.3, we have the estimates

$$\overline{U}(x, \rho) \leq 2(\tfrac{1}{t})^{\frac{n}{r}+\nu}(\tfrac{\rho}{R})^\nu \frac{U(x, R) + R^{\varepsilon_3}}{1 - t^{\varepsilon_3 - \nu}}$$

$$\leq 2(\tfrac{1}{t})^{\frac{n}{r}+\nu} \frac{\bar{\varepsilon}_0}{1 - t^{\varepsilon_3 - \nu}}(\tfrac{\rho}{R})^\nu, \quad 0 < \rho \leq R, \quad x \in B_{r_1}(x_0).$$

It follows that the function $x \mapsto \sigma(x)$ is Hölder continuous with exponent ν in some neighborhood of the point x_0, and moreover, all points of this neighborhood belong to Ω_1. This means that Ω_1 is an open set and $\sigma \in C^\nu(\Omega_1; M_{\text{sym}}^{n \times n})$. Theorem 2.1.2 is proved.

$\square$

2.7   Open Problems

The following problems should be investigated:

a) To prove the statement of Theorem 2.1.2 by direct methods, and so to get explicit integral conditions under which a given point of an elastoplastic body belongs to the elastic or plastic zone. Partially this problem has been solved in [Se10]. In the twodimensional case it turned out to be possible to apply the so–called "hole–filling trick" and to get the following results: we introduce the function

$$\Phi(x_0, R) = \int_{B_R(x_0)} |\nabla \sigma|^2 dx$$

and the sets

$$\Sigma_1 = \{x_0 \in \Omega : \lim_{R \to 0} |(\sigma^D)_{x_0,R}| = \sqrt{2}k_* \}$$

$$\Sigma_2 = \left\{x_0 \in \Omega : \limsup_{R \to 0} \frac{\Phi^{1/2}(x_0, R) \ln \frac{1}{R}}{\ln(\frac{1}{\Phi(x_0,R)+R^2})} > 0 \right\}.$$

THEOREM 2.7.1 *Suppose that* $n = 2$ *and let conditions*(1.20), (1.21) *of Theorem 2.1.1 and condition* (1.33) *hold. Assume further that the function* $x \mapsto \varepsilon^D(f(x))$ *is continuous in* Ω. *Then the set* $\Omega_0 = \Omega \setminus \Sigma$, *where* $\Sigma = \Sigma_1 \cup \Sigma_2$, *is open. Moreover, the tensor* σ *is continuous in* Ω_0 *and*

$$\mathcal{F}(\sigma(x)) < 0 \text{ for all } x \in \Omega_0.$$

It is easy to see that the twodimensional Lebesgue measure of the set Σ_2 is equal to zero, i.e. $|\Sigma_2| = 0$. However, using a φ–measure of Hausdorff, one can establish more precise estimates for the set Σ_2. Suppose that φ is some nondecreasing positive function on the segment $[0, 1]$, E is an arbitrary set in $\mathbb{R}^n$. We let

$$\mathcal{H}(E, \varphi) = \liminf_{\delta \to 0} \{ \sum_i \varphi(r_i) : E \subset \bigcup_i B_i, \ r_i < \delta \}.$$

Here $\{B_i\}$ is an arbitrary countable cover of the set E by open balls B_i of radius $r_i < \delta$. Applying known arguments (see, for example, [Gi]), one can show that $\mathcal{H}(\Sigma_2, \varphi) = 0$ for the choice $\varphi(t) = (\ln \frac{1}{t})^{-2}$.

If we write $\mathcal{H}^m(E)$ in place of $\mathcal{H}(E, \varphi)$ for the choice $\varphi(t) = t^m$, then $\mathcal{H}^m(\Sigma_2) = 0$ for any $0 < m \le 2$, and therefore the Hausdorff dimension of Σ_2 is equal to zero.

b) To prove a global variant of Theorem 2.1.1, i.e. to show that
$\sigma \in W_2^1(\Omega; M_{sym}^{n \times n})$.

c) To prove that

$$\sigma \in W_{\infty, loc}^1(\Omega; M_{sym}^{n \times n}).$$

We note that in the case of elastoplastic torsion this result is correct (see, for example, [BS], [CR], [F]).

d) To study the properties of the internal free boundary dividing an elastoplastic body into elastic and plastic zones.

2.8 Remarks on the regularity of minimizers of variational functionals from the deformation theory of plasticity with power hardening

For plasticity with hardening (formulated within the framework of the deformation theory), the classical boundary value problem includes the equilibrium equations for the stresses and the deformation relations stated as equations (1.1) and (1.2) of Chapter 1; the constitutive equations (or stress–strain relations) should now be replaced by the more simple relation

$$\sigma = K_0 1 \text{div } u + \gamma(|\varepsilon^D(u)|)\varepsilon^D(u),$$

where γ is a given positive function. The latter equation has the principal form of the Hookean law for an isotropic elastic body, the only difference is that γ is not a constant but a function depending on $|\varepsilon^D(u)|$. Such constitutive equations, in fact, describe rather physically nonlinear elasticity (still in a geometrically linear way) than plasticity. But as it was discussed in the introduction, for so–called simple loadings, one may use these kind of constitutive equations in plasticity with hardening, the reader is referred to, for example, [K], [Kl], [NH]. It is now convenient to introduce the function

$$g_0(t) = \int\limits_0^t \theta\gamma(\theta)\, d\theta.$$

Then the stress–strain relation takes the form

$$\sigma = K_0 1 \text{div } u + \frac{g_0'(|\varepsilon^D(u)|)}{|\varepsilon^D(u)|}\varepsilon^D(u)$$

and, therefore,

$$|\sigma^D| = g_0'(|\varepsilon^D(u)|).$$

A typical example of such a function g_0 is the following one:

$$g_0'(t) = 2\mu \begin{cases} \dfrac{t}{t_0} & , \text{ if } 0 \le t \le t_0 \\ \left(\dfrac{t}{t_0}\right)^\alpha & , \text{ if } t \ge t_0 \end{cases}$$

for some $0 < \alpha \le 1$. So, if $|\sigma^D(x_0)| < 2\mu$, then the point x_0 belongs to this zone of the elastoplastic body which undergoes only elastic deformations. If $|\sigma^D(x_0)| > 2\mu$, then x_0 belongs to the plastic zone where the body undergoes both elastic and plastic deformations. Roughly speaking, in contrast to perfect elastoplasticity, for "plasticity with hardening", one has to increase stresses in order to increase the plastic deformations.

As we shall see below, the solution of the classical boundary value problem describing the equilibrium state of an elastoplastic body with hardening can be interpreted as the minimizer of the variational functional (1.4).

So, in this section we consider the variational functional from (1.4), i.e. we let

$$I(v) = \int_\Omega g(\varepsilon(v))\, dx - \overline{M}(v),$$

$$g(\varkappa) = \tfrac{1}{2}K_0 tr^2 \varkappa + g_0(|\varkappa^D|), \quad \varkappa \in \mathbb{M}^{n\times n}_{sym},$$

with a function $g_0 : \mathbb{R} \to \mathbb{R}$ having superlinear growth. More precisely, we assume that the following conditions hold:

(8.1)
$$\begin{cases}
(i) & g_0 \text{ is of class } C^1,\ g_0(0) = 0,\ g_0'(0) = 0; \\[2mm]
(ii) & \text{the function } t \mapsto g_0'(t) \text{ is continuously differentiable} \\
& \text{for all } t \text{ except perhaps at } \pm t_0. \text{ The left and right} \\
& \text{derivatives } (g_0'')_-(t_0) \text{ and } (g_0'')_+(t_0) \text{ exist. The function} \\
& t \mapsto g_0''(t) \text{ is extended to } t_0 \text{ by means of the value} \\
& (g_0'')_+(t_0); \\[2mm]
(iii) & \text{There are positive numbers } c_1 \text{ and } c_2 \text{ such that} \\
& c_1(1+t^2)^{\frac{\alpha-1}{2}} \le h_-(t) := \min\{g_0''(t),\, g_0'(t)/t\} \le \\
& \le h_+(t) := \max\{g_0''(t),\, g_0'(t)/t\} \le \\
& \le c_2(1+t^2)^{\frac{\alpha-1}{2}}, \quad t \ge 0,
\end{cases}$$

for some

(8.2) $0 < \alpha \le 1.$

We remark that in the deformation theory of plasticity based on the von Mises plasticity test, the range of admissible values for α corresponds to the model of an elasto–plastic body with power hardening. The case $\alpha = 1$ describes linear hardening.

The natural domain of the functional I is the Banach space

$$D^{2,1+\alpha} = \{v : \text{div } v \in L^2(\Omega),\ |v| + |\varepsilon^D(v)| \in L^{1+\alpha}(\Omega)\}.$$

If we assume that we are given functions

(8.3) $f \in L^n(\Omega; \mathbb{R}^n),\ F \in L^\infty(\partial_2\Omega; \mathbb{R}^n),\ u_0 \in W_2^1(\Omega; \mathbb{R}^n),$

then the variational problem

(8.4) $\left\{ \begin{array}{l} \text{to find } u \in V_0 + u_0 = \{v \in D^{2,1+\alpha}(\Omega) : v = 0 \text{ on } \partial_1\Omega\} + u_0 \\ \text{such that } I(u) = \inf\{I(v) : v \in V_0 + u_0\} \end{array} \right.$

has a unique solution.

This follows from the fact the the strictly convex functional I is continuous on $D^{2,1+\alpha}(\Omega)$ and coercive on the affine submanifold $V_0 + u_0$ of the reflexive space $D^{2,1+\alpha}(\Omega)$.

We wish to describe briefly what regularity results for minimizers of (8.4) can be expected. Proofs of the corresponding statements can be given via duality approach and may be obtained by nonessential modifications of the proofs of Theorem 2.1.1 and 2.1.2. For details we refer the reader to the paper [Se7].

Now let us formulate the dual problem of (8.4)

(8.5) $\left\{ \begin{array}{l} \text{to find } \sigma \in Q_f \text{ such that} \\ R(\sigma) = \sup\{R(\tau) : \tau \in Q_f\}, \end{array} \right.$

where

$$Q_f := \{\tau \in \Sigma : \int_\Omega \tau : \varepsilon(v)dx = \overline{M}(v) \quad \forall v \in V_0\}$$

is the set of all tensor fields satisfying the equilibrium equations in terms of stresses,

$$\Sigma := \{\tau \in L^2(\Omega; M^{n\times n}_{sym}) : |\tau^D| \in L^{\frac{1+\alpha}{\alpha}}(\Omega)\}$$

is the space of admissible tensors,

$$R(\tau) := \int_{\partial_1\Omega} (\tau\nu) \cdot u_0 d\ell - \int_\Omega g^*(\tau)dx$$

is the functional of the problem,

$$\int_{\partial_1\Omega} (\tau\nu) \cdot u_0 d\ell := \int_\Omega \tau : \varepsilon(u_0)dx - \overline{M}(u_0),$$

$$g^*(\tau) = \frac{1}{2n^2K_0}tr^2\tau + g_0^*(|\tau^D|),$$

and $g_0^*(s) = \sup\{st - g_0(t) : t \in \mathbb{R}\}$ is the conjugate function of g_0.

It is well known that problem (8.5) also has a unique solution which is connected to that of (8.4) by the relations

$$
(8.6) \quad
\begin{cases}
I(u) = R(\sigma), \\
\sigma = K_0 \operatorname{div} u \, \mathbf{1} + \dfrac{g_0'(|\varepsilon^D(u)|)}{|\varepsilon^D(u)|} \varepsilon^D(u).
\end{cases}
$$

We assume that the following density condition holds:

(8.7) $V_* = V_0 \cap W_2^1(\Omega, \mathbb{R}^n)$ is dense in V_0 w.r.t. the norm of $D^{2,1+\alpha}(\Omega)$,

and refer the reader to Appendix A.2 concerning tests for (8.7).

The main results of this section, announced in [Se6] and proved in [Se7], are gathered in the following assertions.

THEOREM 2.8.1 *If conditions (8.1)–(8.3) and (8.7) hold and if further*

(8.8) $f \in W_{\frac{1+\alpha}{\alpha}}^1(\Omega; \mathbb{R}^n),$

then

$$\sigma \in W_{2,\mathrm{loc}}^1(\Omega; \mathbb{M}_{\mathrm{sym}}^{n \times n}).$$

THEOREM 2.8.2 *If conditions (8.1)–(8.3) and (8.7) hold and if*

(8.9) $f \in W_{\bar{n}}^2(\Omega; \mathbb{R}^n),$

for some

$$(8.10) \quad \bar{n} > \frac{n(1+\alpha)}{n\alpha + 1 + \alpha},$$

then there exist open sets Ω_1 and Ω_2 in Ω and a number $\varepsilon_2 = \varepsilon_2(n, \bar{n}) \in\,]0, 1]$ such that

$$\sigma \in C^\nu(\Omega_1 \cup \Omega_2; \mathbb{M}_{\mathrm{sym}}^{n \times n})$$

for all $\nu \in]0, \varepsilon_2[$ and

$$\begin{cases} \mathcal{F}(\sigma(x)) := |\sigma^D(x)| - \sqrt{2}k_* < 0 \text{ for all } x \in \Omega_1 \\[2mm] \mathcal{F}(\sigma(x)) > 0 \text{ for all } x \in \Omega_2 \\[2mm] \mathcal{F}(\sigma(x)) = 0 \text{ for almost all } x \in \Omega \setminus (\Omega_1 \cup \Omega_2) \end{cases}$$

where $\sqrt{2}k_* = g_0'(t_0)$.

Under conditions (8.1) and (8.2) the dual function g_0^* coincides with the Legendre transform of g_0, i.e.

$$g_0^*(s) = st - g_0(t), \quad s = g_0'(t).$$

Consequently, g_0^* is a function of class C^1; in addition, its second derivative exists and satisfies

$$(g_0^*)''(s) = \frac{1}{g_0''(t)}, \quad s = g_0'(t) \Leftrightarrow t = (g_0^*)'(s),$$
$$((g_0^*)'')_\pm(\sqrt{2}k_*) = \frac{1}{(g_0'')_\pm(t_0)}.$$

From the second identity in (8.6) we obtain

$$\varepsilon(u) = \frac{1}{n^2 K_0} tr\, \sigma \, 1 + \frac{(g_0^*)'(|\sigma^D|)}{|\sigma^D|} \sigma^D.$$

Hence, we deduce that

$$\varepsilon(u) \in C^\nu(\Omega_1 \cup \Omega_2; \mathbb{M}_{\mathrm{sym}}^{n \times n})$$

for every $\nu \in]0, \varepsilon_2[$. The differentiability properties of the minimizer of problem (8.4) on $\Omega_1 \cup \Omega_2$ are further improved by means of a well–known scheme and are determined only by the smoothness of the data g_0 and f.

REMARK 2.8.1 Conditions (8.8), (8.9) and (8.10) are not optimal. In particular, the claim of Theorem 2.8.1 is true if $f \in L_{\mathrm{loc}}^{\frac{1+\alpha}{\alpha}}(\Omega; \mathbb{R}^n)$.

REMARK 2.8.2 In a number of practically meaningful cases the sets Ω_1 and Ω_2 may be interpreted as the elastic and plastic domains.

We comment briefly on the above remark. Let $g_0(t) = \mu t^2$ for $t \in [0, t_0]$ where μ is a positive constant. Then, from Theorem 2.8.2, it follows that the equilibrium equations of linear elasticity are satisfied in Ω_1, i.e.

$$\int_{\Omega_1} \{K_0 \operatorname{div} u \operatorname{div} v + 2\mu \varepsilon^D(u) : \varepsilon^D(v) - f \cdot v\} dx = 0 \quad \forall v \in C_0^\infty(\Omega_1; \mathbb{R}^n),$$

and that the deformation tensor is computed according to the formula for an elastic material

$$\varepsilon(u) = \frac{1}{n^2 K_0} tr\,\sigma\,\mathbf{1} + \frac{1}{2\mu}\sigma^D.$$

Hence it is natural to call Ω_1 the elastic domain.

On the open set Ω_2 we have the quasilinear system of differential equations describing the equilibrium of an elasto–plastic body

$$\int_{\Omega_2} \{K_0\,\mathrm{div}\,u\,\mathrm{div}\,v + \frac{g_0'(|\varepsilon^D(u)|)}{|\varepsilon^D(u)|}\varepsilon^D(u) : \varepsilon^D(v) - f\cdot v\}dx = 0$$

$$\forall v \in C_0^\infty(\Omega_2; \mathbb{R}^n),$$

and the plastic deformation

$$\varepsilon_p = \varepsilon(u) - \varepsilon_e = \varepsilon(u) - \frac{1}{n^2 K_0}tr\,\sigma\,\mathbf{1} - \frac{1}{2\mu}\sigma^D = \left(\frac{(g_0^*)'(|\sigma^D|)}{|\sigma^D|} - \frac{1}{2\mu}\right)\sigma^D$$

is nonzero at every point of this set. Due to this facts Ω_2 may be called the plastic domain.

Unfortunately we have considerably less information on $\Omega_* = \Omega \setminus (\Omega_1 \cup \Omega_2)$. We mention only that at almost all points the plastic deformation is zero and $|\varepsilon^D(u)| = t_0$. Formally the set Ω_* could be called the elasto–plastic boundary, since it separates the elastic and plastic domains. This set could again be partitioned into subsets, each having its own nature. The elements of such a partition could be, for example, $\partial\Omega_1$, $\partial\Omega_2$, and int Ω_*. The last set, generally speaking, is nonempty; even more, it is easy to give an example where this set coincides with the entire domain Ω. Moreover, on the set int Ω_* the equilibrium equations for a linear elastic medium and those for an elasto–plastic medium are satisfied simultaneously, since they coincide there. Finally, Ω_* may contain singular points, where a loss of smoothness may occur since the variational problem (8.4) is considered in a class of vector–valued functions.

THEOREM 2.8.3 *If the conditions of Theorem 2.8.2 hold and, in addition,*

$$g_0''(t) \text{ is continuous at } t_0,$$

then there exist an open set $\tilde{\Omega} \subset \Omega$ and a number $\varepsilon_2 = \varepsilon_2(n,\bar{n}) \in]0,1]$ such that

$$\sigma \in C^\nu(\tilde{\Omega}; \mathbb{M}_{\mathrm{sym}}^{n\times n})$$

for all $\nu \in]0,\varepsilon_2[$ and $|\Omega \setminus \tilde{\Omega}| = 0$.

REMARK 2.8.3 From Theorem 2.8.3 and the second relation in (8.4) it follows that the solution of problem (8.4) is regular on an open set of full measure if the data g_0 and f are sufficiently regular.

At the end of this section we remark that if $n = 2$, then $\varepsilon(u)$ and σ are continuous functions in Ω. The proof of this fact for a more difficult case will be given in Chapter 4.

Appendix A

A.1 Density of smooth functions in spaces of tensor–valued functions

In what follows we assume that Ω is a bounded domain in $\mathbb{R}^n$ whose boundary is Lipschitz continuous.

We denote by $\mathbb{M}^{N \times n}$ the space of all real $N \times n$ matrices. Adopting the convention of summation over repeated Latin and Greek indices running from 1 to N and from 1 to n, respectively, we will use the notation

$$\sigma : \varkappa = \sigma_{i\alpha} \varkappa_{i\alpha}, \ \sigma = (\sigma_{i\alpha}), \ \varkappa = (\varkappa_{i\alpha}) \in \mathbb{M}^{N \times n}, \ |\sigma| = \sqrt{\sigma : \sigma}.$$

LEMMA A.1.1 *Let* $\sigma \in L^\infty(\Omega; \mathbb{M}^{N \times n})$, *div* $\sigma = (\sigma_{i\alpha,\alpha}) \in L^s(\Omega; \mathbb{R}^N)$ *and let* $1 \le s \le t < +\infty$. *Then a sequence* $\sigma^m \in C^\infty(\overline{\Omega}; \mathbb{M}^{N \times n})$ *exists such that*

(1.1) $\sigma^m \to \sigma$ *in* $L^t(\Omega; \mathbb{M}^{N \times n})$;

(1.2)   div $\sigma^m \to$ div σ *in* $L^s(\Omega; \mathbb{R}^N)$;

(1.3) $\sigma^m \overset{*}{\to} \sigma$ *in* $L^\infty(\Omega; \mathbb{M}^{N \times n})$;

(1.4) $\|\sigma^m\|_{L^\infty(\Omega)} = \|\sigma\|_{L^\infty(\Omega)}$.

Proof.   Since the boundary of Ω is Lipschitz continuous, for any point $x \in \partial\Omega$ there is a neighborhood O_x such that the domain $O_x \cap \Omega$ is star–shaped relative to some of its points. All these neighborhoods together with Ω form an open cover of the compact set $\overline{\Omega}$, and so a partition of unity exists such that

$$\begin{cases} \varphi_k \in C_0^\infty(\mathbb{R}^n), \ \varphi_k \ge 0 \text{ in } \mathbb{R}^n, \quad k = 0, 1, 2, \ldots, r; \\[2mm] \text{spt } \varphi_0 \subset \Omega \equiv \Omega^0, \text{ spt } \varphi_k \subset \Omega^k = O_{x_k} \cap \Omega, \quad k = 1, 2, \ldots, r; \\[2mm] \sum_{k=0}^{r} \varphi_k \equiv 1 \text{ in } \overline{\Omega}. \end{cases}$$

We set $\sigma_k = \varphi_k \sigma$, $k = 0, 1, \ldots, r$. Let $x_k^* \in \Omega^k$ be a point with respect to which Ω^k is star–shaped. Now we stretch the domain Ω^k with the help of a similarity transformation relative to the point x_k^*. In the domains

$$\Omega_\lambda^k = \{x \in \mathbb{R}^n : x_k^* + \lambda(x - x_k^*) \in \Omega^k\}, \quad 0 < \lambda < 1,$$

we define the functions

$$\sigma_k^\lambda(x) = \sigma_k(x_k^* + \lambda(x - x_k^*)), \quad x \in \Omega_\lambda^k.$$

Further we let

$$\sigma_k^{\lambda,\rho}(x) = \int_{\Omega_\lambda^k} \omega_\rho(x - y)\sigma_k^\lambda(y)dy, \quad \sigma_0^\rho(x) = \int_\Omega \omega_\rho(x - y)\sigma_0(y)dy,$$

where ω_ρ is the standard smoothing kernel, i.e.

$$\omega(x) = c_n \begin{cases} e^{\frac{1}{|x|^2 - 1}} & \text{if } |x| < 1 \\ 0 & \text{if } |x| \geq 1 \end{cases}, \quad \int_{\mathbb{R}^n} \omega(x)dx = 1,$$

ρ a positive parameter and $\omega_\rho(z) = \rho^{-n}\omega(z/\rho)$.

Under the conditions of the lemma div $\sigma_k \in L^s(\Omega; \mathbb{R}^N)$, and, using known arguments (see, for example, [So1], [So2]), we get that, for any $\varepsilon > 0$, a number $\delta(\varepsilon)$ exists such that, for any $\lambda \in]1 - \delta(\varepsilon), 1[$, there is a number $\mu(\lambda, \varepsilon)$ such that

$$\begin{cases} \|\sigma^{\lambda,\rho} - \sigma\|_{L^t(\Omega; M^{N \times n})} < \varepsilon, \quad \|\text{div}\,(\sigma^{\lambda,\rho} - \sigma)\|_{L^s(\Omega; \mathbb{R}^N)} < \varepsilon, \\[2mm] \rho \in]0, \mu(\lambda, \varepsilon)[, \quad \sigma^{\lambda,\rho} \equiv \sigma_0^\rho + \sum_{k=1}^r \sigma_k^{\lambda,\rho}. \end{cases}$$

For $|\sigma^{\lambda,\rho}(x)|$ we have the estimate $(R \equiv \|\sigma\|_{L^\infty})$

$$|\sigma^{\lambda,\rho}(x)| \leq R\left[\int_\Omega \omega_\rho(x - y)\varphi_0(y)dy + \sum_{k=1}^r \int_{\Omega_\lambda^k} \omega_\rho(x - y)\varphi_k^\lambda(y)dy\right].$$

Decreasing if necessary the numbers $\sigma(\varepsilon)$ and $\mu(\lambda, \varepsilon)$ we can arrange that $[\ldots] \leq 1 + \varepsilon$, in conclusion

$$\|\sigma^{\lambda,\rho}\|_{L^\infty(\Omega)} \leq (1 + \varepsilon)R.$$

Now we take $\varepsilon = \frac{1}{m}$, $\lambda_m = 1 - \frac{1}{2}\delta(\frac{1}{m})$, $\rho_m = \frac{1}{2}\mu(\lambda_m, \frac{1}{m})$ and put $\sigma^m = \sigma^{\lambda_m, \rho_m}$. Then we obtain the statements (1.1)–(1.3) of Lemma A.1.1 together with the inequality

$$\limsup_{m \to \infty} \|\sigma^m\| \le R.$$

On the other hand

$$\|\sigma^m\|_{L^\infty(\Omega)} = \sup_{\|\varkappa\|_{L^1(\Omega)} \le 1} \int_\Omega \sigma^m : \varkappa dx.$$

From this we conclude that

$$\liminf_{m \to +\infty} \|\sigma^m\|_{L^\infty(\Omega)} \ge R \ge \limsup_{m \to \infty} \|\sigma^m\|_{L^\infty(\Omega)},$$

which implies

$$R = \lim_{m \to \infty} \|\sigma^m\|_{L^\infty(\Omega)}.$$

Then the sequence $\tau^m = \alpha_m \sigma^m$, where $\alpha_m = R/\|\sigma^m\|_{L^\infty(\Omega)}$, satisfies all the requirements. Lemma A.1.1 is proved.

$\square$

If we proceed in the same way, we get the following statements:

LEMMA A.1.2 *Let $\sigma \in L^\infty(\Omega; \mathbb{M}^{N \times n})$ and $1 \le t < +\infty$. Then a sequence $\sigma^m \in C^\infty(\overline{\Omega}; \mathbb{M}^{N \times n})$ exists such that (1.1), (1.3), (1.4) hold.*

LEMMA A.1.3 *Let $\sigma \in L^t(\Omega; \mathbb{M}^{n \times n}_{sym})$, $\sigma^D \in L^\infty(\Omega; \mathbb{M}^{n \times n}_{sym})$, $div\, \sigma \in L^s(\Omega; \mathbb{R}^n)$ and $1 \le s \le t < +\infty$. Then a sequence $\sigma^m \in C^\infty(\overline{\Omega}; \mathbb{M}^{n \times n}_{sym})$ exists which has the properties*

(1.5) $\sigma^m \to \sigma$ *in* $L^t(\Omega; \mathbb{M}^{n \times n}_{sym})$;

(1.6) $div\, \sigma^m \to div\, \sigma$ *in* $L^s(\Omega; \mathbb{R}^n)$;

(1.7) $\sigma^{mD} \overset{*}{\rightharpoonup} \sigma^D$ *in* $L^\infty(\Omega; \mathbb{M}^{n \times n}_{sym})$:

(1.8) $\|\sigma^{mD}\|_{L^\infty(\Omega)} = \|\sigma^D\|_{L^\infty(\Omega)}.$

LEMMA A.1.4 *Suppose that $\sigma \in L^t(\Omega; \mathbb{M}^{n \times n}_{sym})$, $\sigma^D \in L^\infty(\Omega; \mathbb{M}^{n \times n}_{sym})$ and $1 \le t < +\infty$. Then a sequence $\sigma^m \in C^\infty(\overline{\Omega}; \mathbb{M}^{n \times n}_{sym})$ exists such that the claims (1.5), (1.7) and (1.8) hold.*

LEMMA A.1.5 *Suppose that* $\sigma \in \Sigma_t = \{\tau : \tau \in L^2(\Omega; M^{n\times n}_{sym}), \tau^D \in L^\infty(\Omega; M^{n\times n}_{sym}),$
div $\tau \in L^t(\Omega; \mathbb{R}^n)\}$ *for some* $2 \le t < +\infty$. *Then* $\sigma \in L^t(\Omega; M^{n\times n}_{sym})$.

Proof.   We have the identity

$$\int_\Omega \sigma : \varepsilon(v)dx = -\int_\Omega v \cdot \text{div } \sigma \, dx \quad \forall v \in C_0^\infty(\Omega; \mathbb{R}^n).$$

From this it follows that

$$(1.9) \quad \frac{1}{n}\int_\Omega \text{tr } \sigma \text{ div } v \, dx = -\int_\Omega (\sigma^D : \varepsilon(v) + v \cdot \text{div } \sigma)dx \quad \forall v \in C_0^\infty(\Omega; \mathbb{R}^n).$$

We denote by $\overset{\circ}{D}_q$ the closure of $C_0^\infty(\Omega; \mathbb{R}^n)$ with respect to the norm $\|\nabla u\|_{L^q(\Omega)}$, by H_q the closure of all solenoidal vector–valued functions from $C_0^\infty(\Omega; \mathbb{R}^n)$ with respect to the same norm and by $\hat{H}_q$ the space of solenoidal vector–valued functions from $\overset{\circ}{D}_q$. We set $q = \frac{t}{t-1} \in]1, 2]$. Since Ω is a bounded Lipschitz domain, we have $H_q = \hat{H}_q$ (see for example [P]).
Next, by $\sigma^D \in L^\infty(\Omega; M^{n\times n}_{sym})$, the linear functional

$$v \mapsto g(v) = \int_\Omega (\sigma^D : \varepsilon(v) + v \cdot \text{div } \sigma)dx$$

is bounded on $\overset{\circ}{D}_q$. It follows from (1.9) that it is equal to zero on $H_q = \hat{H}_q$. Therefore (see for example [P]), there exists $p \in L^t(\Omega)$ such that

$$g(v) = \int_\Omega p \,\text{div } v \, dx.$$

So we have the identity

$$\int_\Omega (\frac{1}{n}\text{tr } \sigma + p)\text{div } v \, dx = 0 \quad \forall v \in \overset{\circ}{D}_2 \, .$$

Since the range of the operator $G : \overset{\circ}{D}_2 \to L^2(\Omega)$, $Gu = \text{div } u$, is the set of all functions from $L^2(\Omega)$ which are orthogonal to 1, we get $\frac{1}{n}\text{tr}\sigma + p = \text{constant}$ (see for example [LS]). Lemma A.1.5 is proved.

$\square$

A.2 Density of smooth functions in spaces of vector–valued functions

Let us consider the space of vector–valued functions $\Omega \to \mathbb{R}^n$ defined as

$$D^{p,q}(\Omega) = \{v = (v_i) : \|v\|_{D^{p,q}(\Omega)} = \|\text{div } v\|_{L^p(\Omega)} + \|v\|_{L^q(\Omega)} + \|\varepsilon^D(v)\|_{L^q(\Omega)} < +\infty\},$$

where $p, q \geq 1$ and Ω denotes a bounded Lipschitz domain. We suppose that

(2.1) $1 \leq q \leq p < +\infty, \quad n \geq q.$

From this it follows that

$$D^{p,q}(\Omega) \subset D^{q,q}(\Omega),$$

and we have

(2.2) $D^{1,1}(\Omega)$ is continuously imbedded into $L^{\frac{n}{n-1}}(\Omega; \mathbb{R}^n)$,

(2.3) $D^{q,q}(\Omega) = W_q^1(\Omega; \mathbb{R}^n)$ if $q > 1$.

The reader can find a proof of statement (2.2) for instance in [MM2]. (2.3) follows from L^p–Korn's inequality (see [MM1], [MM2]), we also refer to Lemma 3.0.1 in chapter 3.

By the imbedding theorem (for Sobolev spaces) we get $(q > 1)$

(2.4) $\begin{cases} \text{the space } D^{q,q}(\Omega) \text{ is imbedded continuously into the} \\ \text{Lebesgue space } L^s(\Omega; \mathbb{R}^n) \text{ for } s \leq \frac{nq}{n-q}, \text{ if } n > q, \\ \text{and for any } s < +\infty, \text{ if } n = q. \end{cases}$

We impose the additional restriction

(2.5) $p \leq \dfrac{nq}{n-q}$ if $n > q$ and $p < +\infty$ if $n = q$.

It follows from (2.4) that if p satisfies conditions (2.1) and (2.5), then we can deduce

(2.6) $\varphi v \in D^{p,q}(\Omega)$ for all $\varphi \in C^1(\overline{\Omega})$ and all $v \in D^{p,q}(\Omega)$.

To prove this it is enough to look at the formula

$$\text{div } (\varphi v) = \varphi \, \text{div } v + \nabla \varphi \cdot v \in L^p(\Omega) \text{ for } \varphi \in C^1(\overline{\Omega}), \ v \in D^{p,q}(\Omega).$$

With standard arguments (see Lemma A.1.1 and also [So1], [So2]) and remembering the fact that (2.5) implies (2.6), we get

LEMMA A.2.1 *Suppose that condition (2.5) holds. Then*

(2.7) $C^\infty(\overline{\Omega}; \mathbb{R}^n)$ *is dense in* $D^{p,q}(\Omega)$.

It is known (see [MM2]) that

(2.8) $D^{p,1}(\Omega)$ *is imbedded continuously in* $L^1(\partial\Omega; \mathbb{R}^n)$,

and, for $q > 1$, by (2.3) and the imbedding theorem for Sobolev spaces

(2.9) $D^{p,q}(\Omega)$ *is imbedded continuously in* $L^q(\partial\Omega; \mathbb{R}^n)$.

Therefore we can define the space

$$\overline{D}_0^{p,q}(\Omega) = \{v \in D^{p,q}(\Omega) : v = 0 \text{ on } \partial\Omega\}.$$

Exploiting the same ideas as in the proof of Lemma A.1.1, it can be proved:

LEMMA A.2.2 *If condition (2.5) holds, then*

(2.10) $\overline{D}_0^{p,q}(\Omega) = D_0^{p,q}(\Omega)$,

where the space $D_0^{p,q}(\Omega)$ *is defined as the closure of* $C_0^\infty(\Omega; \mathbb{R}^n)$ *with respect to the norm of the space* $D^{p,q}(\Omega)$.

Next, let $\partial_1\Omega, \partial_2\Omega \subset \partial\Omega$ be such that

$$\partial_1\Omega \cap \partial_2\Omega = \varnothing, \quad \overline{\partial_1\Omega} \cup \overline{\partial_2\Omega} = \partial\Omega.$$

We introduce the subspace

$$V_0^{p,q}(\Omega) = \{v \in D^{p,q}(\Omega) : v = 0 \text{ on } \partial_1\Omega\}.$$

We are interested under what conditions concerning $\partial_1\Omega$ and $\partial_2\Omega$ the following statement is correct

(2.11) $V_0^{p,q}(\Omega) \cap C^\infty(\overline{\Omega}; \mathbb{R}^n)$ *is dense in* $V_0^{p,q}(\Omega)$.

Lemma A.2.1 and A.2.2 show that (2.11) holds if condition (2.5) and $\partial_1\Omega = \varnothing$ or $\partial_2\Omega = \varnothing$ are satisfied.

There is another interesting case in which (2.11) holds.

LEMMA A.2.3 *Suppose that condition (2.5) is fulfilled and that there are functions φ_1 and φ_2 from $C_0^\infty(\mathbb{R}^n)$ such that $\varphi_1 + \varphi_2 \equiv 1$ in $\overline{\Omega}$, φ_1 vanishes in some neighborhood of $\partial_1\Omega$ and φ_2 vanishes in some neighborhood of $\partial_2\Omega$. Then (2.11) is true.*

The statement of Lemma A.2.3 directly follows from Lemma A.2.1 and Lemma A.2.2.

One may also ask the question whether statement (2.11) is valid if condition (2.5) does not hold. The following lemma shows that statement (2.11) is correct for any p satisfying condition (2.1) provided it is correct for some p satisfying condition (2.5).

LEMMA A.2.4 *Suppose that the following conditions are satisfied:*

$$(2.12) \quad q < n, \quad \frac{nq}{n-q} < p < +\infty,$$

$$(2.13) \quad \begin{array}{l} V_0^{r,q}(\Omega) \cap C^\infty(\overline{\Omega}; \mathbb{R}^n) \text{ is dense in } V_0^{r,q} \text{ for some } r \in [q, \frac{nq}{n-q}], \\ \text{if } q > 1, \text{ and for some } r \in]1, \frac{n}{n-1}], \text{ if } q = 1. \end{array}$$

Then we have statement (2.11).

Proof.   We consider the spaces

$$Y = \Sigma \times L^q(\partial_2\Omega; \mathbb{R}^n), \quad Y^* = \Sigma^* \times L^{q^*}(\partial_2\Omega; \mathbb{R}^n)$$

where

$$\left\{ \begin{array}{l} \Sigma = \{\sigma : \mathrm{tr}\sigma \in L^p(\Omega), \quad \sigma^D \in L^q(\Omega; M^{n\times n}_{\mathrm{sym}})\}, \\[2mm] \Sigma^* = \{\sigma^* : \mathrm{tr}\sigma^* \in L^{p^*}(\Omega), \quad \sigma^{*D} \in L^{q^*}(\Omega; M^{n\times n}_{\mathrm{sym}})\}, \\[2mm] p^* = \frac{p}{p-1}, \quad q^* = \frac{q}{q-1}. \end{array} \right.$$

The duality relation between Y and Y^* takes the form

$$\langle y^*, y \rangle = \int_\Omega \sigma^* : \sigma dx + \int_{\partial_2\Omega} b^* \cdot b dl, \quad y = \{\sigma, b\}, \quad y^* = \{\sigma^*, b^*\}.$$

We introduce the operator $A : D^{p,q}(\Omega) \to Y$,

$$Av = \{\varepsilon(v), -v|_{\partial_2\Omega}\},$$

and note, by the imbeddings (2.4) and (2.9), there are constants c_1 and c_2 such that

$$(2.14) \quad c_1\|v\|_{D^{p,q}(\Omega)} \le \|Av\|_Y \le c_2\|v\|_{D^{p,q}(\Omega)} \quad \forall v \in V_0^{p,q}(\Omega).$$

Let $\ell \in (V_0^{p,q}(\Omega))^*$. We will show that an element $y^* \in Y^*$ exists such that $\ell(v) = \langle y^*, Av \rangle$ for any $v \in V_0^{p,q}(\Omega)$. Indeed, it follows from (2.14) that

$$(2.15) \quad |\ell(v)| \le c_3\|v\|_{D^{p,q}(\Omega)} \le \frac{c_3}{c_1}\|Av\|_Y \quad \forall v \in V_0^{p,q}(\Omega).$$

Let us consider the subspace $Y_0 = A(V_0^{p,q}(\Omega)) \subset Y$ and introduce on Y_0 the linear functional $g(y) = \ell(A^{-1}y)$. By (2.15)

$$|g(y)| \le \frac{c_3}{c_1}\|y\|_Y \quad \forall y \in Y_0.$$

Let G denote a norm preserving extension of g to the whole space Y. Then there exists $y^* \in Y^*$ such that $G(y) = \langle y^*, y \rangle$ for all $y \in Y$ and $g(y) = G(y)$ for $y \in Y_0$. We have

$$g(Av) = \langle y^*, Av \rangle = \ell(v) \quad \forall v \in V_0^{p,q}(\Omega).$$

But it means that

$$\ell(v) = -\int_{\partial_2\Omega} b^* \cdot v\, dl + \int_\Omega \sigma^* : \varepsilon(v)dx \quad \forall v \in V_0^{p,q}(\Omega), \quad y^* = \{\sigma^*, b^*\}.$$

Now we start with the proof of the lemma. We suppose that the claim of the lemma is false, i.e.

$$(2.16) \quad \exists v_0 \in V_0^{p,q}(\Omega) \setminus \overline{V_0^{p,q}(\Omega) \cap C^\infty(\overline{\Omega}; \mathbb{R}^n)}.$$

Then there exists $\ell \in (V_0^{p,q}(\Omega))^*$ such that

$$\begin{cases} \ell(v) = 0 \\ \ell(v_0) = 1 \quad \forall v \in \overline{V_0^{p,q}(\Omega) \cap C^\infty(\overline{\Omega}; \mathbb{R}^n)}. \end{cases}$$

As it was proved above one can find $\sigma^* \in \Sigma^*$ and $b^* \in L^{q^*}(\partial_2\Omega; \mathbb{R}^n)$ with the properties

$$(2.17) \quad \begin{cases} -\int_{\partial_2\Omega} b^* \cdot v_0 dl + \int_\Omega \sigma^* : \varepsilon(v_0)dx = 1, \\ -\int_{\partial_2\Omega} b^* \cdot v\, dl + \int_\Omega \sigma^* : \varepsilon(v)dx = 0 \quad \forall v \in \overline{V_0^{p,q}(\Omega) \cap C^\infty(\overline{\Omega}; \mathbb{R}^n)}. \end{cases}$$

It follows from (2.17) that

$$\frac{1}{n}\int_\Omega tr\sigma^* \mathrm{div}\, v\, dx = -\int_\Omega \sigma^{*D} : \varepsilon(v)dx \quad \forall v \in C_0^\infty(\Omega; \mathbb{R}^n).$$

We put

$$h(v) \equiv -\int_\Omega \sigma^{*D} : \varepsilon(v)dx$$

and note that $h(v) = 0$ for all smooth solenoidal vector fields with compact support in Ω. Since $\sigma^* \in L^{q^*}(\Omega; \mathrm{M}_{\mathrm{sym}}^{n\times n})$,

$$|h(v)| \le \mathrm{constant}\|v\|_{W_q^1(\Omega;\mathbf{R}^n)}.$$

The latter implies (see [P]) the existence of a function $t \in L^{r_0}(\Omega)$, where

$$r_0 = \begin{cases} q^* & \text{if } q > 1 \\ r^* = \frac{r}{r-1} & \text{if } q = 1, \end{cases}$$

such that

$$\int_\Omega t\mathrm{div}\, v\, dx = -\int_\Omega \sigma^{*D} : \varepsilon(v)dx = 0 \quad \forall v \in C_0^\infty(\Omega; \mathbb{R}^n).$$

So we have

$$\int_\Omega (\frac{1}{n}tr\,\sigma^* - t)\mathrm{div}\, v\, dx = 0 \quad \forall v \in C_0^\infty(\Omega; \mathbb{R}^n),$$

and therefore

(2.18) $\dfrac{1}{n}tr\,\sigma^* - t \equiv \text{constant in } \Omega.$

By the conditions of the lemma we know $r^* = \frac{r}{r-1} \le r_0$, and thus we can state that

(2.19) $tr\,\sigma^* \in L^{r^*}(\Omega).$

Since $r < \frac{nq}{n-q}$, $v_0 \in V_0^{r,q}(\Omega)$, and in view of (2.13) a sequence $v_m \in V_0^{r,q}(\Omega) \cap C^\infty(\overline{\Omega}; \mathbb{R}^n) = V_0^{p,q}(\Omega) \cap C^\infty(\overline{\Omega}; \mathbb{R}^n)$ exists such that

(2.20) $v_m \to v_0$ in $D^{r,q}(\Omega).$

It follows from (2.19) and (2.20) that

$$(2.21) \quad \int_\Omega \sigma^* : \varepsilon(v_m)dx \to \int_\Omega \sigma^* : \varepsilon(v)dx.$$

Recalling that $r \geq q$, we also have

$$v_m \to v_0 \text{ in } L^q(\partial_2\Omega; \mathbb{R}^n).$$

From this and (2.17), (2.21) we get the contradiction

$$0 = \int_\Omega \sigma^* : \varepsilon(v_m)dx - \int_{\partial_2\Omega} b^* \cdot v_m dl \to \int_\Omega \sigma^* : \varepsilon(v)dx - \int_{\partial_2\Omega} b^* \cdot v dl = 1.$$

Lemma A.2.4 is proved.

$\square$

A.3   Some properties of the space $BD(\Omega; \mathbb{R}^n)$

We recall that the space $BD(\Omega; \mathbb{R}^n)$ is defined as the set of vector–valued functions with finite norm

$$(3.1) \quad \|v\|_{BD(\Omega;\mathbb{R}^n)} = \|v\|_{L^1(\Omega)} + \int_\Omega |\varepsilon(v)|,$$

where

$$(3.2) \quad \int_\Omega |\varepsilon(v)| = \sup\{-\int_\Omega v \cdot \operatorname{div} \tau \, dx : \tau \in C_0^\infty(\Omega; M_{sym}^{n\times n}), |\tau| \leq 1 \text{ in } \Omega\}.$$

We note that in this case the strain tensor corresponding to the displacement field u generates a bounded positive Radon measure μ which on open subsets $\omega \subset \Omega$ is given by

$$\mu(\omega) = \int_\omega |\varepsilon(v)| = \sup\{-\int_\omega v \cdot \operatorname{div} \tau \, dx : \tau \in C_0^\infty(\omega; M_{sym}^{n\times n}), |\tau| \leq 1 \text{ in } \omega\}.$$

For more information concerning $BD(\Omega; \mathbb{R}^n)$ we refer to the papers [AG1] and [ST2], we are just going to reprove some of its properties.

THEOREM A.3.1 *([AG1], [ST2]) Suppose that Ω is a bounded Lipschitz domain. Then*

(3.3) $\quad\begin{cases} \text{the space } BD(\Omega; \mathbb{R}^n) \text{ is continuously imbedded into the space} \\ L^p(\Omega; \mathbb{R}^n) \text{ for } p \in [1, \frac{n}{n-1}], \text{ moreover, for } p \in [1, \frac{n}{n-1}[\\ \text{this imbedding is compact,} \end{cases}$

(3.4) $\quad\begin{cases} \text{for each } r \in [1, \frac{n}{n-1}] \text{ a constant } C = C(\Omega, r, n) \text{ exists such that} \\ \inf\limits_{v_\times \in V_\times} \left(\int\limits_\Omega |v - v_\times|^r dx \right)^{1/r} \leq C \int\limits_\Omega |\varepsilon(v)| \quad \forall v \in BD(\Omega; \mathbb{R}^n). \end{cases}$

We begin with

LEMMA A.3.1 *Let Ω be a locally star-shaped domain. Then, for any u in $BD(\Omega; \mathbb{R}^n)$, a sequence $\{u^m\}_{m=1}^\infty \subset C_0^\infty(\mathbb{R}^n; \mathbb{R}^n)$ exists such that*

(3.5) $\quad u^m \to u$ in $L^1(\Omega; \mathbb{R}^n)$

(3.6) $\quad \lim\limits_{m \to +\infty} \int\limits_\Omega |\varepsilon(u^m)| dx = \int\limits_\Omega |\varepsilon(u)|.$

Proof.   Local star-shapedness of Ω just means that for any point $x \in \partial\Omega$, there is a neighborhood O_x such that $O_x \cap \Omega$ is star-shaped with respect to some of its point x^*. We have

$$\overline{\Omega} \subset \Omega \cup \left(\bigcup_{x \in \partial\Omega} O_x \right),$$

and since $\overline{\Omega}$ is compact, a finite subcover exists such that

$$\overline{\Omega} \subset \bigcup_{k=0}^r \Omega_k,$$

where $\Omega_0 = \Omega$, $\Omega_k = O_{x_k} \cap \Omega$, and Ω_k is star-shaped with respect to x_k^*, $k = 1, 2, \ldots, r$. Let $\{\varphi_k\}_{k=0}^r \subset C_0^\infty(\mathbb{R}^n)$ be a partition of the unity corresponding to this subcover, i.e.

$$\begin{cases} \text{spt } \varphi_k \subset \Omega_k, \qquad k = 0, 1, 2, \ldots, r, \\ 0 \leq \varphi_k \leq 1 \text{ in } \mathbb{R}^n, \quad k = 0, 1, 2, \ldots, r, \\ \sum\limits_{k=0}^r \varphi_k \equiv 1 \text{ in } \overline{\Omega}. \end{cases}$$

Let λ be a parameter from $]0,1[$. We introduce the sets

$$\Omega_k^\lambda = \{z \in \mathbb{R}^n : x_k^* + \lambda(z - x_k^*) \in \Omega_k\}, \quad k = 1, 2, \ldots, r,$$

and the functions $u_k = \varphi_k u$, $\varkappa_k = u \odot \nabla \varphi_k$, $k = 0, 1, 2, \ldots, r$,

$$\varphi_k^\lambda(x) = \varphi_k(x_k^* + \lambda(x - x_k^*)), \quad x \in \Omega_k^\lambda, \quad k = 1, 2, \ldots, r,$$

$$u_k^\lambda(x) = u_k(x_k^* + \lambda(x - x_k^*)), \quad x \in \Omega_k^\lambda, \quad k = 1, 2, \ldots, r,$$

$$\varkappa_k^\lambda(x) = \varkappa_k(x_k^* + \lambda(x - x_k^*)), \quad x \in \Omega_k^\lambda, \quad k = 1, 2, \ldots, r.$$

For all λ close to 1 we have

$$(3.7) \qquad \Omega \cap \mathrm{spt}\varphi_k^\lambda \subset \Omega_k, \quad k = 1, 2, \ldots, r.$$

Thus we may extend all functions u_k^λ, $\varkappa_k^\lambda$ by zero to Ω and obtain

$$(3.8) \quad \begin{cases} \|u_k^\lambda - u_k\|_{L^1(\Omega_k)} = \|u_k^\lambda - u_k\|_{L^1(\Omega)} \xrightarrow{\lambda \to 1} 0, \\[2mm] \|\varkappa_k^\lambda - \varkappa_k\|_{L^1(\Omega_k)} = \|\varkappa_k^\lambda - \varkappa_k\|_{L^1(\Omega)} \xrightarrow{\lambda \to 1} 0, \quad k = 1, 2, \ldots, r. \end{cases}$$

We further let

$$(3.9) \quad \begin{cases} (u_0)_\rho(x) = \int_\Omega \omega_\rho(x - y)u_0(y)\, dy, \quad (\varkappa_0)_\rho(x) = \int_\Omega \omega_\rho(x - y)\varkappa_0(y)\, dy, \\[2mm] (u_k^\lambda)_\rho(x) = \lambda^{n-1} \int_{\Omega_k^\lambda} \omega_\rho(x - y)u_k^\lambda(y)\, dy, \\[2mm] (\varkappa_k^\lambda)_\rho(x) = \lambda^{n-1} \int_{\Omega_k^\lambda} \omega_\rho(x - y)\varkappa_k^\lambda(y)\, dy, \quad k = 1, 2, \ldots, r. \end{cases}$$

For fixed λ and small enough ρ we have

$$(3.10) \qquad \rho < \mathrm{dist}\,(\partial\Omega_k \cap \Omega,\ \Omega_k \cap \mathrm{spt}\,\varphi_k^\lambda), \quad k = 1, 2, \ldots, r.$$

This means that for λ and ρ satisfying (3.7) and (3.10) the functions $(u_k^\lambda)_\rho$ and $(\varkappa_k^\lambda)_\rho$ vanish near $\partial\Omega_k \cap \Omega$, and moreover, they are equal to zero in $\Omega \setminus \Omega_k$, $k = 1, 2, \ldots, r$. Then, for any fixed λ satisfying (3.7), we get

$$(3.11) \quad \begin{cases} \|(u_k^\lambda)_\rho - \lambda^{n-1}u_k^\lambda\|_{L^1(\Omega)} = \|(u_k^\lambda)_\rho - \lambda^{n-1}u_k^\lambda\|_{L^1(\Omega_k)} \\[2mm] \leq \|(u_k^\lambda)_\rho - \lambda^{n-1}u_k^\lambda\|_{L^1(\Omega_k^\lambda)} \xrightarrow{\rho \to 0} 0, \\[2mm] \|(\varkappa_k^\lambda)_\rho - \lambda^{n-1}\varkappa_k^\lambda\|_{L^1(\Omega)} = \|(\varkappa_k^\lambda)_\rho - \lambda^{n-1}\varkappa_k^\lambda\|_{L^1(\Omega_k)} \\[2mm] \leq \|(\varkappa_k^\lambda)_\rho - \lambda^{n-1}\varkappa_k^\lambda\|_{L^1(\Omega_k^\lambda)} \xrightarrow{\rho \to 0} 0. \end{cases}$$

By coordinate transformation we have

$$
\begin{cases}
(u_k^\lambda)_\rho(x) = \lambda^{n-1} \displaystyle\int\limits_{\Omega_k^\lambda} \omega_\rho(x - y) u_k^\lambda(y) dy \\[4mm]
= \dfrac{1}{\lambda} \displaystyle\int\limits_{\Omega_k} \omega_\rho(x - x_k^* - \dfrac{z - x_k^*}{\lambda}) u_k(z) dz \\[4mm]
= \dfrac{1}{\lambda} \displaystyle\int\limits_{\Omega} \omega_\rho(x - x_k^* - \dfrac{z - x_k^*}{\lambda}) u_k(z) dz, \\[4mm]
(\varkappa_k^\lambda)_\rho(x) = \dfrac{1}{\lambda} \displaystyle\int\limits_{\Omega_k} \omega_\rho(x - x_k^* - \dfrac{z - x_k^*}{\lambda}) \varkappa_k(z) dz \\[4mm]
= \dfrac{1}{\lambda} \displaystyle\int\limits_{\Omega} \omega_\rho(x - x_k^* - \dfrac{z - x_k^*}{\lambda}) \varkappa_k(z) dz.
\end{cases}
\tag{3.12}
$$

We put

$$
\begin{cases}
u^{\lambda,\rho}(x) = (u_0)_\rho(x) + \displaystyle\sum_{k=1}^{r} (u_k^\lambda)_\rho(x), \quad x \in \mathbb{R}^n, \\[4mm]
\varkappa^{\lambda,\rho}(x) = (\varkappa_0)_\rho(x) + \displaystyle\sum_{k=1}^{r} (\varkappa_k^\lambda)_\rho(x), \quad x \in \mathbb{R}^n.
\end{cases}
$$

Obviously $u^{\lambda,\rho} \in C_0^\infty(\mathbb{R}^n; \mathbb{R}^n)$ for all λ and ρ satisfying conditions (3.7), (3.10).

Let us fix an arbitrary number $\delta > 0$. We have

$$
\int\limits_{\Omega} |\varepsilon(u^{\lambda,\rho})| dx = \sup\left\{ \int\limits_{\Omega} \varepsilon(u^{\lambda,\rho}) : \tau \, dx : \tau \in C_0^\infty(\Omega; \mathbb{M}_{\text{sym}}^{n \times n}), \ |\tau| \le 1 \text{ in } \Omega \right\}.
\tag{3.13}
$$

So there exists $\tau \in C_0^\infty(\Omega; \mathbb{M}_{\text{sym}}^{n \times n})$ and $|\tau| \le 1$ in Ω which depends of course on ρ, λ, δ, such that

$$
\begin{cases}
\displaystyle\int\limits_{\Omega} |\varepsilon(u^{\lambda,\rho})| dx < \int\limits_{\Omega} \varepsilon(u^{\lambda,\rho}) : \tau \, dx + \delta = - \int\limits_{\Omega} u^{\lambda,\rho} \cdot \operatorname{div} \tau \, dx + \delta \\[4mm]
= \displaystyle\sum_{k=0}^{r} I_k + \delta,
\end{cases}
\tag{3.14}
$$

$$
\begin{cases}
I_0 = - \displaystyle\int\limits_{\Omega} (u_0)_\rho \cdot \operatorname{div} \tau \, dx, \\[4mm]
I_k = - \displaystyle\int\limits_{\Omega} (u_k^\lambda)_\rho \cdot \operatorname{div} \tau \, dx, \quad k = 1, 2, \ldots, r.
\end{cases}
$$

We have

$$I_0 = -\int_\Omega (u_0)_\rho(x) \cdot \operatorname{div} \tau(x)dx = -\int_\Omega u_0 \cdot (\operatorname{div} \tau)_\rho dx$$

$$= -\int_\Omega u_0 \cdot \operatorname{div} \tau_\rho dx = -\int_\Omega \varphi_0 u \cdot \operatorname{div} \tau_\rho dx,$$

where

$$\tau_\rho(x) = \int_\Omega \omega_\rho(x-y)\tau(y)dy, \quad (\operatorname{div} \tau)_\rho(x) = \int_\Omega \omega_\rho(x-y)\operatorname{div} \tau(y)dy, x \in \mathbb{R}^n.$$

So we get

(3.15) $$I_0 = -\int_\Omega u(y) \cdot \operatorname{div} (\varphi_0\tau_\rho)(y)dy + \int_\Omega \varkappa_0 : \tau_\rho dy.$$

Now let $k \in \{1, 2, \ldots, r\}$. Then

$$I_k = -\frac{1}{\lambda}\int_\Omega \int_\Omega \omega_\rho(x - x_k^* - \frac{z - x_k^*}{\lambda})\varphi_k(z)u(z)dz \cdot \operatorname{div} \tau(x)dx$$

$$= -\frac{1}{\lambda}\int_\Omega \varphi_k(z)u(z)dz \int_\Omega \omega_\rho(x - x_k^* - \frac{z - x_k^*}{\lambda})\operatorname{div} \tau(x)dx$$

$$= \frac{1}{\lambda}\int_\Omega \varphi_k(z)u(z)dz \int_\Omega \left(\tau_{ij}(x)\frac{\partial}{\partial x_j}\omega_\rho(x - x_k^* - \frac{z - x_k^*}{\lambda})\right)dx.$$

On the other hand

$$\frac{\partial}{\partial x_i}\omega_\rho(x - \eta)\big|_{\eta=x_k^*+\frac{x-x_k^*}{\lambda}} = -\frac{\partial}{\partial \eta_i}\omega_\rho(\eta - x)\big|_{\eta=x_k^*+\frac{x-x_k^*}{\lambda}}$$

and therefore

$$\int_\Omega \left(\tau_{ij}(x)\frac{\partial}{\partial x_j}\omega_\rho(x - \eta)\big|_{\eta=x_k^*+\frac{x-x_k^*}{\lambda}}\right)dx$$

$$= -\int_\Omega \left(\frac{\partial\omega_\rho}{\partial \eta_i}(\eta - x)\big|_{\eta=x_k^*+\frac{x-x_k^*}{\lambda}}\tau_{ij}(x)dx\right)$$

$$= -\operatorname{div} \tau_\rho(\eta)\big|_{\eta=x_k^*+\frac{x-x_k^*}{\lambda}}.$$

So we have

$$I_k = -\frac{1}{\lambda}\int_\Omega \varphi_k(z)u(z) \cdot \operatorname{div} \tau_\rho(x_k^* + \frac{z - x_k^*}{\lambda})dz.$$

If we introduce the functions

$$\tau_{(k)}^{\lambda,\rho}(z) = \tau_\rho(x_k^* + \frac{z - x_k^*}{\lambda}), \quad z \in \mathbb{R}^n, \quad k = 1, 2, \ldots, r,$$

then

$$\frac{\partial \tau_{ij(k)}^{\lambda,\rho}}{\partial z_j}(z) = \frac{\partial (\tau_{ij})_\rho(\eta)}{\partial \eta_j}\Big|_{\eta = x_k^* + \frac{z-x_k^*}{\lambda}} \frac{1}{\lambda}.$$

Thus we get

$$(3.16) \quad \begin{cases} I_k = -\int_\Omega \varphi_k(z) u(z) \cdot \text{div } \tau_k^{\lambda,\rho}(z) dz \\ = -\int_\Omega u(z) \cdot \text{div } (\varphi_k \tau_{(k)}^{\lambda,\rho})(z) dz + \int_\Omega (u \odot \nabla \varphi_k)(z) : \tau_k^{\lambda,\rho}(z) dz \\ = -\int_\Omega u(z) \cdot \text{div } (\varphi_k \tau_{(k)}^{\lambda,\rho})(z) dz + \int_\Omega \varkappa_k(z) : \tau_\rho(x_k^* + \frac{z - x_k^*}{\lambda}) dz. \end{cases}$$

From (3.14), (3.15), (3.16) we get the estimate

$$(3.17) \quad \begin{cases} \int_\Omega |\varepsilon(u^{\lambda,\rho})| dx < -\int_\Omega u(z) \cdot \text{div } \left(\varphi_0 \tau_\rho + \sum_{k=1}^r \varphi_k \tau_{(k)}^{\lambda,\rho}\right)(z) dz \\ + \int_\Omega \left(\varkappa_0(z) : \tau_\rho(z) + \sum_{k=1}^r \varkappa_k(z) : \tau_\rho(x_k^* + \frac{z - x_k^*}{\lambda})\right) dz + \delta. \end{cases}$$

We note that

$$|\tau_\rho| \leq 1 \quad \text{in } \mathbb{R}^n$$

and therefore

$$|\tau_{(k)}^{\lambda,\rho}| \leq 1 \quad \text{in } \mathbb{R}^n, \quad k = 1, 2, \ldots, r.$$

Setting

$$\sigma = \varphi_0 \tau_\rho + \sum_{k=1}^r \varphi_k \tau_k^{\lambda,\rho},$$

we get

$$(3.18) \quad |\sigma| \leq 1 \quad \text{in } \mathbb{R}^n.$$

For fixed λ and small enough ρ we have

$$(3.19) \qquad 0 < \rho < \text{dist}(\Omega_k, \partial\Omega_k^\lambda), \quad k = 1, 2, \ldots, r.$$

It is clear that $\tau_{(k)}^{\lambda,\rho} \varphi_k = 0$ in $\Omega \setminus \Omega_k$. If $z \in \partial\Omega \cap \partial\Omega_k$, then $x_k^* + \frac{z - x_k^*}{\lambda} \in \partial\Omega_k^\lambda$, and, by condition (3.19), $\tau_{(k)}^{\lambda,\rho}$ vanishes near $\partial\Omega_k \cap \partial\Omega$. So we can state that

$$\sigma \in C_0^\infty(\Omega; \mathbb{M}_{\text{sym}}^{n \times n}).$$

From this and also from (3.2), (3.17), (3.18) it follows that

$$(3.20) \qquad \begin{cases} \displaystyle\int_\Omega |\varepsilon(u^{\lambda,\rho})| dx \leq \int_\Omega |\varepsilon(u)| + \delta + \\ \displaystyle\int_\Omega \left[\varkappa_0(z) : \tau_\rho(z) + \sum_{k=1}^r \varkappa_k(z) : \tau_\rho(x_k^* + \frac{z - x_k^*}{\lambda}) \right] dz. \end{cases}$$

We wish to estimate the last term on the right hand side of (3.20). We have

$$\int_\Omega \left[\varkappa_0(z) : \tau_\rho(z) + \sum_{k=1}^r \varkappa_k(z) : \tau_\rho(x_k^* + \frac{z - x_k^*}{\lambda}) \right] dz$$

$$= \int_\Omega \varkappa_0 : \tau_\rho dz + \sum_{k=1}^r \int_{\Omega_k} \varkappa_k(z) : \tau_\rho(x_k^* + \frac{z - x_k^*}{\lambda}) dz$$

$$= \int_\Omega \tau : (\varkappa_0)_\rho dz + \sum_{k=1}^r \int_\Omega \tau(x) : \int_{\Omega_k} \omega_\rho(x - x_k^* - \frac{z - x_k^*}{\lambda}) \varkappa_k(z) dz \, dx$$

$$= \int_\Omega \tau : (\varkappa_0)_\rho dz + \sum_{k=1}^n \lambda^n \int_\Omega \tau(x) : \int_{\Omega_k^\lambda} \omega_\rho(x - y) \varkappa_k^\lambda(y) dy \, dx$$

$$= \int_\Omega \tau : (\varkappa_0)_\rho dx + \lambda \sum_{k=1}^r \int_\Omega \tau(x) : (\varkappa_k^\lambda)_\rho(x) dx - \lambda^n \int_\Omega \tau : \sum_{k=0}^r \varkappa_k dx$$

$$\leq \int_\Omega |\varkappa_0 - (\varkappa_0)_\rho| dx + (1 - \lambda^n) \int_\Omega |\varkappa_0| dx$$

$$+ \lambda \sum_{k=1}^r \int_\Omega |(\varkappa_k^\lambda)_\rho - \lambda^{n-1} \varkappa_k^\lambda| dx + \lambda^n \sum_{k=1}^r \int_\Omega |\varkappa_k^\lambda - \varkappa_k| dx.$$

So the final estimate takes the form

$$(3.21) \quad \begin{cases} \displaystyle\int_\Omega |\varepsilon(u^{\lambda,\rho})|\,dx \leq \int_\Omega |\varepsilon(u)| + \delta + \int_\Omega |\varkappa_0 - (\varkappa_0)_\rho|\,dx \\[2mm] \displaystyle +(1-\lambda^n)\int_\Omega |\varkappa_0|\,dx + \sum_{k=1}^{r}\lambda\int_\Omega |(\varkappa_k^\lambda)_\rho - \lambda^{n-1}\varkappa_k^\lambda|\,dx \\[2mm] \displaystyle +\lambda^n \sum_{k=1}^{r}\int_\Omega |\varkappa_k^\lambda - \varkappa_k|\,dx. \end{cases}$$

It remains to estimate the difference

$$(3.22) \quad \begin{cases} \displaystyle\int_\Omega |u^{\lambda,\rho} - u|\,dx = \int_\Omega |(u_0)_\rho - u_0 + \sum_{k=1}^{r}(u_k^\lambda)_\rho - u_k|\,dx \\[2mm] \displaystyle \leq \int_\Omega |(u_0)_\rho - u_0|\,dx + \sum_{k=1}^{r}\left(\int_\Omega |(u_k^\lambda)_\rho - \lambda^{n-1}u_k^\lambda|\,dx \right. \\[2mm] \displaystyle \left. +\lambda^{n-1}\int_\Omega |u_k^\lambda - u_k|\,dx + (1-\lambda^{n-1})\int_\Omega |u_k|\,dx\right). \end{cases}$$

Now we proceed in the following way. Let $\delta = \frac{1}{m}$. There exists $\lambda_m \in \,]0,1[$ such that condition (3.7) holds and

$$(1-\lambda_m^n)\int_\Omega |\varkappa_0|\,dx + \lambda_m^n \sum_{k=1}^{r}\int_\Omega |\varkappa_k^{\lambda_m} - \varkappa_k|\,dx$$

$$+(1-\lambda_m^{n-1})\sum_{k=1}^{r}\int_\Omega |u_k|\,dx + \lambda_m^{n-1}\sum_{k=1}^{r}\int_\Omega |u_k^{\lambda_m} - u_k|\,dx < \frac{1}{m}.$$

Next, for $\lambda = \lambda_m$, we determine ρ_m according to conditions (3.10), (3.19) and such that

$$\int_\Omega |\varkappa_0 - (\varkappa_0)_{\rho_m}|\,dx + \lambda_m \sum_{k=1}^{r}\int_\Omega |(\varkappa_k^{\lambda_m})_{\rho_m} - \lambda_m^{n-1}\varkappa_k^{\lambda_m}|\,dx$$

$$+\int_\Omega |(u_0)_{\rho_m} - u_0|\,dx + \sum_{k=1}^{r}\int_\Omega |(u_k^{\lambda_m})_{\rho_m} - \lambda_m^{n-1}u_k^{\lambda_m}|\,dx < \frac{1}{m}.$$

Setting $u^m = u^{\lambda_m,\rho_m}$ we get from the last two relations the inequalities

$$\int_\Omega |u^m - u|\,dx < \frac{2}{m}, \quad \int_\Omega |\varepsilon(u^m)|\,dx < \int_\Omega |\varepsilon(u)| + \frac{3}{m}.$$

It follows that

$$\lim_{m \to +\infty} \int_\Omega |u^m - u| dx = 0$$

and

$$\limsup_{m \to +\infty} \int_\Omega |\varepsilon(u^m)| dx \le \int_\Omega |\varepsilon(u)|.$$

But since

$$\liminf_{m \to +\infty} \int_\Omega |\varepsilon(u^m)| dx \ge \int_\Omega |\varepsilon(u)|,$$

we have established all statements of the lemma, Lemma A.3.1 is proved.

$\square$

Proof of Theorem A.3.1. The continuity of the imbedding of $BD(\Omega; \mathbb{R}^n)$ into $L^p(\Omega; \mathbb{R}^n)$ for $1 \le p \le \frac{n}{n-1}$ follows from the continuity of the imbedding of $D^{1,1}(\Omega)$ into $L^p(\Omega; \mathbb{R}^n)$ for $1 \le p \le \frac{n}{n-1}$. Indeed, for any $u \in BD(\Omega; \mathbb{R}^n)$, we choose a sequence $\{u^m\}_{m=1}^\infty$ as in Lemma A.3.1. For some constant $C_1 = C_1(\Omega, p, n)$ we have

$$\|u^m\|_{L^p(\Omega)} \le C_1 [\|u^m\|_{L^1(\Omega)} + \|\varepsilon(u^m)\|_{L^1(\Omega)}].$$

Passing to the limit we get

$$\|u\|_{L^p(\Omega)} \le C_1 \|u\|_{BD(\Omega;\mathbb{R}^n)}.$$

Let us now prove the compactness of the imbedding of $BD(\Omega; \mathbb{R}^n)$ into $L^1(\Omega; \mathbb{R}^n)$. Let $\{u^s\}_{s=1}^\infty$ be an arbitrary bounded sequence in $BD(\Omega; \mathbb{R}^n)$. For each u^s we find a sequence $\{u^{s,m}\}_{m=1}^\infty$ as in Lemma A.3.1. For any $s \in \mathbb{N}$ we can determine a number m_s such that

$$(3.23) \qquad \|u^{s,m_s}\|_{D^{1,1}(\Omega)} \le \|u^s\|_{BD(\Omega;\mathbb{R}^n)} + 1,$$

$$(3.24) \qquad \|u^{s,m_s} - u^s\|_{L^1(\Omega;\mathbb{R}^n)} < \frac{1}{s}.$$

The sequence $\{u^{s,m_s}\}_{s=1}^\infty$ is precompact in $L^1(\Omega; \mathbb{R}^n)$ which follows from (3.23) and the compactness of the imbedding of $D^{1,1}(\Omega)$ into $L^1(\Omega; \mathbb{R}^n)$. On the other hand, (3.24) implies precompactness of the sequence u^s in $L^1(\Omega; \mathbb{R}^n)$. For proving compactness of the imbedding of $BD(\Omega; \mathbb{R}^n)$ into $L^r(\Omega; \mathbb{R}^n)$ we use the following facts:

i) Bounded sets in $BD(\Omega; \mathbb{R}^n)$ are bounded in $L^{\frac{n}{n-1}}(\Omega; \mathbb{R}^n)$.

ii) Sets being bounded in $L^{\frac{n}{n-1}}(\Omega; \mathbb{R}^n)$ and precompact in $L^1(\Omega; \mathbb{R}^n)$ are also precompact in $L^r(\Omega; \mathbb{R}^n)$ for any $1 < r < \frac{n}{n-1}$.

So (3.3) is proved and it remains to show (3.4). To do this let us consider the set

$$V^\times(\Omega) = \{u \in BD(\Omega; \mathbb{R}^n) : \int_\Omega u \cdot v \, dx = 0 \quad \forall v \in V_\times\}.$$

We first prove that there exists a constant $C_2(\Omega, n)$ such that

$$(3.25) \qquad \int_\Omega |v| dx \leq C_2 \int_\Omega |\varepsilon(v)| \quad \forall v \in V^\times(\Omega).$$

Suppose that (3.25) is false. Then we can find a sequence $v^m \in V^\times(\Omega)$ such that

$$\int_\Omega |v^m| dx \geq m \int_\Omega |\varepsilon(v^m)|.$$

Let $\tilde{v}^m = v^m / \int_\Omega |v^m| dx$. Then we have

$$\tilde{v}^m \in V^\times(\Omega), \quad \int_\Omega |\tilde{v}^m| dx = 1, \quad \frac{1}{m} \geq \int_\Omega |\varepsilon(\tilde{v}^m)|.$$

So the sequence $\tilde{v}^m$ is bounded in $BD(\Omega; \mathbb{R}^n)$ and we can select a subsequence converging to $v_0 \in V^\times(\Omega)$ strongly in $L^1(\Omega; \mathbb{R}^n)$. Clearly

$$\int_\Omega |v_0| dx = 1, \quad \int_\Omega |\varepsilon(v_0)| = 0.$$

But from the last relation we see that $v_0 \in V_\times$ and thus $v_0 = 0$. This is a contradiction and (3.25) is proved.

Next let $1 \leq r \leq \frac{n}{n-1}$. Then we have

$$\|v\|_{L^r(\Omega;\mathbb{R}^n)} \leq C_1(\Omega, r, n) \left[\int_\Omega |v| dx + \int_\Omega |\varepsilon(v)| \right] \quad \forall v \in V^\times(\Omega),$$

and in view of (3.25) we get that

$$(3.26) \qquad \|v\|_{L^r(\Omega;\mathbb{R}^n)} \leq C_1(C_2 + 1) \int_\Omega |\varepsilon(v)| \quad \forall v \in V^\times(\Omega).$$

We may introduce some orthobasis in $V_\times$ in the following way

$$
\begin{cases}
\displaystyle\int_\Omega e_k \cdot e_j \, dx = 0, & k \neq j, \quad k, j = 1, 2, \ldots, s_n, \\[2mm]
\displaystyle\int_\Omega |e_k|^2 dx = 1, & k = 1, 2, \ldots, s_n. \\[2mm]
e_k \in V_\times, & k = 1, 2, \ldots, s_n.
\end{cases}
$$

For any $v \in BD(\Omega; \mathbb{R}^n)$ we have the decomposition

$$ v = \tilde{v} + v_\times, \quad \tilde{v} \in V^\times(\Omega), \quad v_\times \in V_\times, $$

where

$$ v_\times = \sum_{i=1}^{s_n} e_i \int_\Omega v \cdot e_i \, dx. $$

From (3.26) we get

$$ \inf_{u_\times \in V_\times} \|v - u_\times\|_{L^r(\Omega;\mathbb{R}^n)} \leq \|v - v_\times\|_{L^r(\Omega;\mathbb{R}^n)} $$

$$ \leq C_1(C_2 + 1) \int_\Omega |\varepsilon(v - v_\times)| \leq C_1(C_2 + 1) \int_\Omega |\varepsilon(v)|, $$

and Theorem A.3.1 is proved.

$\square$

A.4   Jensen's inequality

has been an essential tool in the proof of Theorem 2.1.2.

LEMMA A.4.1 *Let the function* $\Phi : \mathbb{R} \to \mathbb{R}$ *be defined as*

$$ (4.1) \qquad \Phi(t) = \begin{cases} \frac{t^2}{2d} & \text{if } |t| \leq d \\[2mm] |t| - \frac{d}{2} & \text{if } |t| > d. \end{cases} $$

Then, for any $u \in BD(\Omega; \mathbb{R}^n)$, *we have the inequality*

$$ (4.2) \qquad \Phi\left(\fint_\Omega |\varepsilon(u)| \right) \leq \fint_\Omega \Phi(|\varepsilon(u)|). $$

We would like to remark that a version of Lemma A.4.1 can be found in the work [DT].

Proof. According to the definition of how to apply a convex function to a measure we have

(4.3)
$$\int_\Omega \Phi(|\varepsilon(u)|) = \sup \Big\{ - \int_\Omega [u \cdot \operatorname{div} \tau + \Phi^*(|\tau|)]dx :$$
$$\tau \in C_0^\infty(\Omega; \mathbb{M}_{\mathrm{sym}}^{n\times n}), \quad |\tau| \le 1 \Big\}.$$

Since

$$\Phi^*(t) = \begin{cases} \frac{d}{2} t^2 & \text{if } |t| \le 1 \\ +\infty & \text{if } |t| > 1 \end{cases}$$

we arrive at the representation

(4.4)
$$\int_\Omega \Phi(|\varepsilon(u)|) = \sup \Big\{ - \int_\Omega [u \cdot \operatorname{div} \tau + \frac{d}{2}|\tau|^2]dx :$$
$$\tau \in C_0^\infty(\Omega; \mathbb{M}_{\mathrm{sym}}^{n\times n}), \quad |\tau| \le 1 \text{ in } \Omega \Big\}.$$

Let the numbers $\rho > 0$, $m \in \mathbb{N}$ satisfy the condition

(4.5) $\qquad \rho < \dfrac{1}{m}.$

We put as usual

$$u_\rho(x) = \int_\Omega w_\rho(x - y)u(y)dy$$

where

$$w_\rho(x) = \frac{1}{\rho^n} w(\frac{|x|}{\rho}), \quad w(t) = c_n \begin{cases} e^{\frac{1}{t^2-1}} & \text{if } |t| < 1 \\ 0 & \text{if } |t| \ge 1, \end{cases}$$

and c_n is chosen so that

$$\int_{\mathbb{R}^n} w_\rho(x)dx = 1.$$

We have

(4.6) $u_\rho \to u$ in $L^1(\Omega; \mathbb{R}^n)$.

Let us introduce the family of sets

$$\Omega_m = \{x \in \Omega : \text{dist}(\partial\Omega, x) > \frac{1}{m}\}$$

and set

(4.7) $\tau = \begin{cases} \frac{1}{d}\varepsilon(u_\rho) & \text{if } |\varepsilon(u_\rho)| \le d \\ \frac{\varepsilon(u_\rho)}{|\varepsilon(u_\rho)|} & \text{if } |\varepsilon(u_\rho)| > d. \end{cases}$

It is clear that $\tau \in L^\infty(\Omega; \mathbb{R}^n)$ and $|\tau| \le 1$ a.e. in Ω. We also introduce the truncation of τ by letting

$$\tau^m = \begin{cases} \tau & \text{if } x \in \Omega_m \\ 0, & x \in \Omega \setminus \Omega_m. \end{cases}$$

Direct calculations show that

(4.8) $\begin{cases} \displaystyle\int_{\Omega_m} \Phi(|\varepsilon(u_\rho)|)dx = \int_{\Omega_m} [\varepsilon(u_\rho) : \tau^m - \Phi^*(|\tau^m|)]dx \\ \displaystyle = \int_{\Omega_m} [\varepsilon(u_\rho) : \tau^m - \frac{d}{2}|\tau^m|^2]dx. \end{cases}$

On the other hand, for each m, a sequence $\{\tau_k^m\}_{k=1}^\infty \in C_0^\infty(\Omega_m; \mathbb{M}_{\text{sym}}^{n\times n})$ exists such that

(4.9) $\begin{cases} |\tau_k^m| \le 1 & \text{in } \Omega_m, \\ \tau_k^m \to \tau^m & \text{in } L^2(\Omega_m; \mathbb{M}_{\text{sym}}^{n\times n}). \end{cases}$

Therefore

$$\int_{\Omega_m} [\varepsilon(u_\rho) : \tau_k^m - \frac{d}{2}|\tau_k^m|^2]dx \overset{k\to+\infty}{\longrightarrow} \int_{\Omega_m} \Phi(|\varepsilon(u_\rho)|)dx.$$

So for any $\delta > 0$ we can find some $\sigma \in C_0^\infty(\Omega_m; \mathbb{M}_{\text{sym}}^{n\times n})$ such that $|\sigma| \le 1$ in Ω and

(4.10) $\displaystyle\int_{\Omega_m} \Phi(|\varepsilon(u_\rho)|)dx < \int_{\Omega_m} [\varepsilon(u_\rho) : \sigma - \frac{d}{2}|\sigma|^2]dx + \delta.$

Next we have

$$
\begin{cases}
\displaystyle\int_{\Omega_m} \varepsilon(u_\rho) : \sigma \, dx = -\int_{\Omega_m} u_\rho \cdot \operatorname{div} \sigma \, dx \\[2mm]
\displaystyle = -\int_{\Omega} u(y) \cdot \int_{\Omega_m} w_\rho(x-y)\operatorname{div}\sigma(x)dx\,dy \\[2mm]
\displaystyle = -\int_{\Omega} u(y) \cdot \int_{\Omega} w_\rho(x-y)\operatorname{div}\sigma(x)dx\,dy = -\int_{\Omega} u(y) \cdot \operatorname{div}\sigma_\rho(y)dy,
\end{cases}
$$

(4.11)

where (as usual)

$$
\sigma_\rho(y) = \int_{\Omega} w_\rho(y-x)\sigma(x)dx = \int_{\Omega_m} w_\rho(y-x)\sigma(x)dx,
$$

and, by condition (4.5),

(4.12) $\quad \sigma_\rho \in C_0^\infty(\Omega; M_{\text{sym}}^{n\times n}), \ |\sigma_\rho| \le 1 \text{ in } \Omega.$

We observe that

$$
\int_{\Omega} |\sigma_\rho|^2 dx \le \int_{\Omega} |\sigma|^2 dx = \int_{\Omega_m} |\sigma|^2 dx.
$$

From this and also from (4.10), (4.11) we derive

$$
\begin{cases}
\displaystyle\int_{\Omega} \Phi(\varepsilon(u_\rho))dx \le -\int_{\Omega}[u \cdot \operatorname{div}\sigma_\rho + \frac{d}{2}|\sigma_\rho|^2]dx + \delta \\[2mm]
\displaystyle \le \int_{\Omega} \Phi(|\varepsilon(u)|) + \delta.
\end{cases}
$$

(4.13)

Next, since the function Φ is convex and differentiable, we have

$$
\Phi(|\varepsilon(u_\rho)(x)|) - \Phi\left(\fint_{\Omega_m} |\varepsilon(u_\rho)|dx\right)
$$
$$
\ge \Phi'\left(\fint_{\Omega_m} |\varepsilon(u_\rho)|dx\right)\left(|\varepsilon(u_\rho)(x)| - \fint_{\Omega_m} |\varepsilon(u_\rho)|dx\right).
$$

Integrating this inequality and using (4.13) gives

$$
|\Omega_m|\Phi\left(\fint_{\Omega_m} |\varepsilon(u_\rho)|dx\right) \le \int_{\Omega} \Phi(|\varepsilon(u)|) + \delta.
$$

Passing to the limit as $\rho \to 0$ and taking into account (4.6), we obtain

$$\liminf_{\rho \to 0} \fint_{\Omega_m} |\varepsilon(u_\rho)| dx \geq \fint_{\Omega_m} |\varepsilon(u)|$$

and

$$|\Omega_m| \Phi\left(\fint_{\Omega_m} |\varepsilon(u)| \right) \leq \int_{\Omega} \Phi(|\varepsilon(u)|) + \delta.$$

Using $\Omega = \bigcup_{m=1}^{\infty} \Omega_m$, $\Omega_m \subset \Omega_{m+1}$, we get

$$\lim_{m \to \infty} \fint_{\Omega_m} |\varepsilon(u)| = \fint_{\Omega} |\varepsilon(u)|,$$

and therefore

$$|\Omega| \Phi\left(\fint_{\Omega} |\varepsilon(u)| \right) \leq \int_{\Omega} \Phi(|\varepsilon(u)|) + \delta.$$

By the arbitrariness of δ we have completed the proof of the lemma.

$\square$

Chapter 3

Quasi-static fluids of generalized Newtonian type

3.0 Preliminaries

Let us briefly review the physical background of the problems under consideration: we discuss the flow of incompressible generalized Newtonian fluids whose behaviour is characterized in terms of differential inequalities. To be precise, let us assume that the fluid occupies a region $\Omega \subset \mathbb{R}^n$ (i.e. a bounded Lipschitz domain), $n \geq 2$; we denote by $v = v(x, t)$, $x \in \Omega$, the velocity field and since the fluid is incompressible, we have

$$(0.1) \qquad \operatorname{div} v = 0.$$

Note that by the mass balance law (0.1) just expresses the fact that the density is a constant function. For a given system $f : \Omega \to \mathbb{R}^n$ of volume forces (which are assumed to be independent of time) the fluid has to obey the equation of motion

$$(0.2) \qquad \operatorname{div} \tau + f = \dot{v}.$$

Here τ is the (apriori unknown) stress tensor (τ_{ij}) and $\dot{v}$ denotes the material derivative, i.e. the quantity $(\nabla v)v + \partial_t v$ with the vector function $(\nabla v)v = (\partial_i v^j v^i)_{1 \leq j \leq n}$. In addition we have to impose

$$(0.3) \qquad \text{various boundary and initial condition for } v \text{ and } \tau.$$

Our main assumption is that the velocity should not depend on t (the time–dependent situation is discussed in the papers [L2–4] of Ladyzhenskaya and in the recent book of Málek, Nečas, Rokyta, Růžička [MNRR]). In this case (0.2) has to be replaced by

(0.2)' $\operatorname{div} \tau + f = (\nabla v)v,$

where now $v = v(x)$, $\tau = \tau(x)$ and (0.3) reduces to boundary conditions for v and τ. Following standard convention (see [DL]) we adress this situation as the stationary or quasi–static case. In the case of slow, steady state motion equation (0.2)' may be replaced by

(0.2)" $\operatorname{div} \tau + f = 0.$

Of course the assumption $(\nabla v)v = 0$ seems to be artificial but, from the mathematical point of view, as we will show later on, sometimes the quasi–static case can be reduced to this situation. In order to characterize the specific fluid under consideration we need a so–called constitutive law which connects the stress tensor τ and the strain velocity $\varepsilon(v) = \frac{1}{2}(\partial_i v^j + \partial_j v^i)$. For various classes of generalized Newtonian fluids and in particular viscoplastic fluids such a constitutive relation can be formulated with the help of a so–called dissipative potential $W : \mathbb{M} \to \mathbb{R}$ which is a given convex function defined on the space of all matrices of order n, a physical discussion is given for example in [AM], [BAH], [DL], [IS], [L1–4], [MM2], [MNRR], [Pr] and [Ser].

We formulate this stress–strain relation as

(0.4) $\tau^D \in \partial W(\varepsilon(v)),$

$\tau^D = \tau - \frac{1}{n}\operatorname{tr} \tau \mathbf{1}$ (stress deviator). Here ∂W is the subdifferential of the convex function W: for $A, B \in \mathbb{M}$ we have $A \in \partial W(B)$ if and only if $W(T) \geq W(B) + A : (T - B)$ for all $T \in \mathbb{M}$. If the dissipative potential is a differentiable function, then (0.4) is just another formulation of the equation $\tau^D = \frac{\partial W}{\partial E}(\varepsilon(v))$. A typical example for a nonsmooth W occurs in the case of a classical Bingham fluid: with positive constants μ and k_* we have (see [DL], [MM2])

$$W(E) = \mu|E|^2 + \sqrt{2}k_*|E|, \ E \in \mathbb{M},$$

and from (0.4) we get the equation

$$\varepsilon(v) = \begin{cases} \frac{1}{2\mu}\left(1 - \frac{\sqrt{2}k_*}{|\tau^D|}\right)\tau^D & \text{if } |\tau^D| \geq \sqrt{2}k_* \\ \\ 0 & \text{if } |\tau^D| < \sqrt{2}k_* \end{cases}$$

which means that for small stresses the fluid behaves like a rigid body (i.e. $\varepsilon(v) = 0$). The problem under investigation can be formulated as follows:

$$\begin{cases} \text{to find a pair } (v, \tau) \text{ of functions such that } (0.1), (0.2)' \text{ (or } (0.2)''), (0.3) \\ \\ \text{and } (0.4) \text{ are satisfied.} \end{cases}$$

For simplicity let us assume that we have fixed Dirichlet boundary data for the unknown velocity field, i.e. v has to be found in the class $\mathbb{K} = \{w : \Omega \to \mathbb{R}^n | \text{div } w = 0, w = v_0 \text{ on } \partial\Omega\}$ with given function v_0. Let us further assume that (v, τ) denotes a pair of smooth solutions under the hypothesis $(0.2)''$. Then it is easy to show that the velocity field v minimizes the variational integral

$$J(v) = \int_\Omega \big(W(\varepsilon(v)) - f \cdot v\big) \, dx$$

among functions in $\mathbb{K}$, and it is therefore reasonable to study the variational problem

(0.5) $\qquad J \to \min$ in $\mathbb{K}$

separately. As far as the existence of (weak) solutions is concerned we may argue along standard lines: the dissipative function W is convex with appropriate growth rates so that theorems on lower semicontinuous variational integrals together with Korn type inequalities show solvability of (0.5) provided the class $\mathbb{K}$ is imbedded in a suitable Sobolev space. In order to come back to the original problem we then have to analyze the regularity properties of solutions to (0.5), for example we ask if minimizers are of class C^1. This will be done in the next section. Unfortunately, apart from the two–dimensional case $n = 2$, we only obtain partial C^1–regularity which means that minimizers are smooth outside a (closed) singular set with small measure, and there is no hope to improve these results: in the Bingham case for example the dissipative potential is irregular at $0 \in \mathbb{M}$ which means that C^1–regularity of a minimizer v can only be expected on the set $[\varepsilon(v) \neq 0]$. But even if the potential W is smooth but non–quadratic, the experience in regularity theory (see [Gi]) suggests that singularities may occur in dimensions $n \geq 3$.

The next results concern local boundedness of the strain velocity: for various classes of generalized Newtonian fluids including the Bingham model we prove in section 2 that the tensor $\varepsilon(v)$ is in the space $L^\infty_{\text{loc}}(\Omega; \mathbb{M})$ which implies Hölder continuity of v for any exponent $0 < \alpha < 1$. Moreover, we will show that the classical derivative $\nabla u(x)$ exists for almost all $x \in \Omega$.

Up to now all our discussions are limited to the variational setting which means that we start from the assumption $(\nabla v)v = 0$ which leads us to the variational problem (0.5). It is worth remarking that the results described above are valid in any dimension n, moreover, as a byproduct of our investigation we obtain theorems for certain nonlinear Stokes systems. In fact, if the dissipative function W is of class C^1, then local J–minimizers v with respect to the constraint div $v = 0$ are solutions of

$$(0.6) \qquad \int_\Omega \frac{\partial W}{\partial E}(\varepsilon(v)) : \varepsilon(\varphi)\, dx = \int_\Omega f \cdot \varphi\, dx, \ \varphi \in C_0^\infty(\Omega; \mathbb{R}^n), \ \mathrm{div}\,\varphi = 0,$$

which for $W(E) = \frac{1}{2}|E|^2$, $E \in \mathbb{M}$, reduces to the classical Stokes system studied for example in [L1]: if the dissipative potential is quadratic, then

$$\int_\Omega \frac{\partial W}{\partial E}(\varepsilon(v)) : \varepsilon(\varphi)\, dx = \frac{1}{2}\int_\Omega \nabla v : \nabla \varphi\, dx$$

for solenoidal test vectors φ and (0.6) turns into the set of equations

$$\begin{cases} \mathrm{div}\,v = 0 \\[2mm] -\frac{1}{2}\Delta v = f + \nabla p \end{cases}$$

with a suitable pressure function p.

Let us now discuss the exact equation of motion (0.2)'. By the same reasoning as described above we can show: if (v, τ) are smooth solutions of (0.1), (0.2)' and (0.4) and if for simplicity we just impose Dirichlet boundary conditions on v, then v is a solution of the problem

$$(0.7) \qquad \begin{cases} \mathrm{div}\,v = 0, \\[2mm] \displaystyle\int_\Omega \frac{\partial W}{\partial E}(\varepsilon(v)) : \varepsilon(\varphi) + (\nabla v)v \cdot \varphi\, dx = \int_\Omega f \cdot \varphi\, dx \\[2mm] \text{for all } \varphi \in C_0^\infty(\Omega; \mathbb{R}^n), \ \mathrm{div}\,\varphi = 0; \\[2mm] v = v_0 \text{ on } \partial\Omega \end{cases}$$

provided W has continuous first derivatives. In the general case of nonsmooth W (0.7) has to be replaced by a variational inequality which for the Bingham model (w.l.o.g. $\mu = 1/2$, $\sqrt{2}k_* = 1$) reads

$$(0.8) \quad \begin{cases} \text{to find } v \in \mathbb{K} \text{ such that} \\[2mm] \displaystyle\int_\Omega \varepsilon(v) : (\varepsilon(w) - \varepsilon(v)) + |\varepsilon(w)| - |\varepsilon(v)| + (\nabla v)v \cdot (w - v)\, dx \\[2mm] \geq \displaystyle\int_\Omega f \cdot (w - v)\, dx \\[2mm] \text{holds for all } w \in \mathbb{K}. \end{cases}$$

Equation (0.7) is just an extension of the time independent Navier–Stokes system (see [L1] or [Ga2])

$$\begin{cases} \operatorname{div} v = 0 \,,\; v = v_0 \text{ on } \partial\Omega, \\[2mm] -\tfrac{1}{2}\Delta v + (\nabla v)v = f + \nabla p \end{cases}$$

which is easily derived from (0.7) by letting $W(E) = \tfrac{1}{2}|E|^2$, $E \in \mathbb{M}$. In contrast to the variational case the existence of weak solutions for (0.7) or (0.8) in suitable energy spaces is no longer obvious, in fact, we are confronted with the same difficulties as for the stationary nonlinear Navier–Stokes system. Following the lines of [L1] and [DL] we get at least the existence of generalized solutions for homogeneous boundary data in dimensions $n = 2$ and 3 using elementary arguments. But since we concentrate on regularity properties, we will just assume that we already have some local weak solution of (0.7) and (0.8). Fortunately most of the regularity results obtained in the variational setting extend to solutions of (0.7) and (0.8): let us define the artificial volume load $\tilde{f} = f - (\nabla v)v$, where v solves (0.7) or (0.8) in the weak sense. Then v is local minimizer of

$$w \mapsto \int_\Omega \left(W(\varepsilon(w)) - \tilde{f} \cdot w \right) dx$$

subject to the constraint $\operatorname{div} w = 0$. For proving regularity in the variational setting it is sufficient to know that the volume forces f belong to some Morrey space which means that now we have to analyze the nonlinearity $(\nabla v)v$. In dimensions $n = 2$ and 3 we will show some growth properties of $\int_{B_r(x_0)} |\nabla v|^2 dx$, $B_r(x_0) \subset \Omega$, from which we derive that $\tilde{f}$ is in the "right" Morrey class. For details we refer to section 4.

Next we give some important examples of dissipative potentials W whose physical motivation can be found in [AM], [BAH], [DL], [IS], [MM2], [Ka], [Ki] and [Pr].

a) fluids of Bingham type: we let

$$W(E) = \frac{1}{m}|E|^m + \kappa|E|, E \in \mathbb{M},$$

with constants $m > 1$, $\kappa \geq 0$. As subcases this model includes

  i) Newton fluids: $m = 2$, $\kappa = 0$
     In this case (0.5) is reduced to the Stokes system, whereas (0.7) is the
     stationary Navier–Stokes system.
  ii) classical Bingham fluids: $m = 2$, $\kappa > 0$
  iii) Norton fluids: $m > 2$, $\kappa = 0$ (see [N])

b) power law models: for nonnegative real numbers μ_0, μ_∞ such that
   $\mu_0 + \mu_\infty > 0$ we let

$$W(E) = \mu_\infty |E|^2 + \mu_0 \left\{ \begin{array}{ll} (\delta + |E|^2)^{p/2} & \delta > 0, \, p \geq 1 \\ \text{or} & \\ |E|^p , & p > 1 \end{array} \right\}.$$

c) Powell–Eyring model: according to [PE] we define

$$W(E) = \mu_\infty |E|^2 + \mu_1 \int_0^{|E|} \operatorname{ar sinh} t \, dt, \ E \in \mathbb{M},$$

with constants $\mu_\infty, \mu_1 > 0$.

d) Prandtl–Eyring model: passing to the limit $\mu_\infty \to 0$ in the previous case
   we get the dissipative potential

$$W(E) = \mu_1 \int_0^{|E|} \operatorname{ar sinh} t \, dt, \ E \in \mathbb{M},$$

which can be found in the work [E].

As the reader might guess from the definition, the last model requires separate
methods, for example, we have to define appropriate function spaces. For this
reason Prandtl–Eyring fluids are postponed to chapter 4. We wish to remark
that a condensed version of chapter 3 can be found in the paper [Fu6] whereas
a more detailed version was presented in [Fu4].

We finish this section by recalling some auxiliary results we shall use later.

Let $G \subset \mathbb{R}^n$ denote a smooth bounded region and suppose that $p \in]1, \infty[$ is a
given real number. We use the symbol c to denote various positive constants

depending on n, p and the domain G.

The first lemma is an extension of the classical Korn's inequality (see [Ko 1,2]; for a proof and further references see also [Fi], [Fri], [Str], [Z]).

LEMMA 3.0.1 *Suppose* $u \in L^p(G; \mathbb{R}^n)$ *and* $\varepsilon(u) = \frac{1}{2}(\nabla u + \nabla u^T) \in L^p(G; \mathbb{M})$. *Then* u *belongs to the Sobolev-space* $W_p^1(G; \mathbb{R}^n)$ *and we have the estimate*

$$\|u\|_{W_p^1} \leq c\{\|u\|_{L^p} + \|\varepsilon(u)\|_{L^p}\}.$$

For $v \in \overset{\circ}{W}_p^1(G; \mathbb{R}^n)$ *the inequality*

$$\|v\|_{W_p^1} \leq c\|\varepsilon(v)\|_{L^p}$$

holds.

We refer the reader to [MM1] (see also [IS], [Fu2]).

Sometimes we make use of the following interpolation inequality

LEMMA 3.0.2 *For any function* $u \in W_p^1(G; \mathbb{R}^n)$ *with* $p \geq 2$ *we have* $\|u\|_{L^p} \leq c\{\|u\|_{L^2} + \|\varepsilon(u)\|_{L^p}\}$.

This result can easily be obtained by contradiction using Lemma 3.0.1.

Let $V_{\times}$ denote the space of rigid motions, i.e. $u \in V_{\times}$ has the form $u(x) = a + Bx$ with $a \in \mathbb{R}^n$ and $B \in \mathbb{M}$ skew-symmetric.

LEMMA 3.0.3 *i) A function* $u \in W_p^1(G; \mathbb{R}^n)$ *belongs to the space* $V_{\times}$ *if and only if* $\varepsilon(u) = 0$.

ii) For all $u \in W_p^1(G; \mathbb{R}^n)$ *we have* $\inf_{r \in V_{\times}} \|u - r\|_{L^p} \leq c\|\varepsilon(u)\|_{L^p}$.

iii) The infimum in ii) is achieved, i.e. for each $u \in W_p^1(G; \mathbb{R}^n)$ *we find* $r \in V_{\times}$ *such that* $\|u - r\|_{L^p} \leq c\|\varepsilon(u)\|_{L^p}$.

In order to construct suitable comparison functions we make essential use of

LEMMA 3.0.4 *Suppose* $f \in L^p(G)$ *with* $\fint_G f \, dx = 0$. *Then there exists a function* $u \in \overset{\circ}{W}{}^1_p(G; \mathbb{R}^n)$ *satisfying*

$$div\ u = f \quad and \quad \begin{cases} \|\nabla u\|_{L^p} \le c\|f\|_{L^p} \\ \|\nabla u\|_{L^2} \le c\|f\|_{L^2}. \end{cases}$$

Proof.   cf. [LS], [P], [Ga1], III, Theorem 3.1, [Fu2], proof of Lemma 1.

Next we collect various Campanato–type estimates for the linear Stokes system which are rather well–known but difficult to trace in the literature.

LEMMA 3.0.5 *Let* A *denote a symmetric strictly positive bilinear form on the space of all symmetric* $(n \times n)$*–matrices. Then there is a constant* $c_* = c_*(n, A)$ *independent of* G *such that for* $v \in W^1_2(G; \mathbb{R}^n)$ *satisfying*

$$\begin{cases} div\ v = 0 \\ \\ \displaystyle\int_G A(\varepsilon(v), \varepsilon(\varphi)) \, dx = 0, \ \varphi \in \overset{\circ}{W}{}^1_2(G; \mathbb{R}^n), \ div\ \varphi = 0 \end{cases}$$

we have $v \in C^\infty(G; \mathbb{R}^n)$ *and further (here* $B_R = B_R(x_0) \subset G$*):*

i) $\displaystyle\int_{B_{R/2}} |\varepsilon(v)|^2 \, dx \le c_* R^{-2} \int_{B_R} |v - u_\times|^2 \, dx, \quad u_\times \in V_\times;$

ii) $\displaystyle\int_{B_{R/2}} |\nabla v|^2 \, dx \le c_* R^{-2} \int_{B_R} |v - \xi|^2 \, dx, \quad \xi \in \mathbb{R}^n;$

iii) $\displaystyle\int_{B_r} |\nabla v|^2 \, dx \le c_* (\frac{r}{R})^n \int_{B_R} |\nabla v|^2 \, dx, \quad B_r \subset B_R;$

iv) $\displaystyle\int_{B_r} |v - (v)_r|^2 \, dx \le c_* (\frac{r}{R})^{n+2} \int_{B_R} |v - (v)_R|^2 \, dx;$

v) $\displaystyle\int_{B_r} |\varepsilon(v) - (\varepsilon(v))_r|^2 \, dx \le c_* (\frac{r}{R})^{n+2} \int_{B_R} |\varepsilon(v) - (\varepsilon(v))_R|^2 \, dx \ and$

vi) $\displaystyle\int_{B_r} |\nabla v - (\nabla v)_r|^2 \, dx \le c_* (\frac{r}{R})^{n+2} \int_{B_R} |\nabla v - (\nabla v)_R|^2 \, dx.$

Proof: i) (cf. [GM], Thm.1.1, for a similar argument). Let η denote a cut–off function satisfying $\eta \equiv 1$ on $B_{R/2}$, spt $\eta \subset B_R$, $|\nabla\eta| \leq 2/R$, $0 \leq \eta \leq 1$, and define $\varphi_1 := \eta^2(v - u_\times)$. Then

$$\operatorname{div} \varphi_1 = 2\eta \, \nabla\eta \cdot (v - u_\times)$$

and according to Lemma 3.0.4 we can find $w \in \overset{\circ}{W}{}^1_2(B_R; \mathbb{R}^n)$ satisfying

$$\operatorname{div} w = 2\eta \, \nabla\eta \cdot (v - u_\times),$$

$$\int_{B_R} |\nabla w|^2 \, dx \leq c_* \int_{B_R} |\nabla\eta \cdot (v - u_\times)|^2 \, dx.$$

$\varphi := \varphi_1 - w$ is in $\overset{\circ}{W}{}^1_2(B_R; \mathbb{R}^n)$ with div $\varphi = 0$, hence

$$\int_{B_R} A(\varepsilon(v), \varepsilon(\varphi)) \, dx = 0.$$

Observing $\varepsilon(\varphi) = \varepsilon(\varphi_1) - \varepsilon(w) = \eta^2\varepsilon(v) + \eta\{\nabla\eta \otimes (v - u_\times) + [\nabla\eta \otimes (v - u_\times)]^T\} - \varepsilon(w)$ we arrive at

$$\int_{B_R} \eta^2 A(\varepsilon(v), \varepsilon(v)) \, dx \leq c_* \left\{ \int_{B_R} \eta|\varepsilon(v)||\nabla\eta||v - u_\times| \, dx + \int_{B_R} |\varepsilon(v)||\varepsilon(w)| \, dx \right\}.$$

From Young's inequality we deduce for small positive δ

$$\int_{B_{R/2}} |\varepsilon(v)|^2 \, dx \leq \delta \int_{B_R} |\varepsilon(v)|^2 \, dx + c_*(\delta) R^{-2} \int_{B_R} |v - u_\times|^2 \, dx.$$

With obvious modifications (replacing $R/2$ by some radius s and R by t taken from $]R/2, R[$) the result follows from [GM], Lemma 0.5.

To see ii) we combine i) and Lemma 3.0.1 (the dependence on the radius of the constants occuring in Korn's inequality is easily checked by scaling).

In the next step one shows $v \in W^k_{2,\mathrm{loc}}(G; \mathbb{R}^n)$ for all $k \in \mathbb{N}$, thus $v \in C^\infty(G; \mathbb{R}^n)$ and any partial derivative of arbitrary order is a solution of the constant–coefficient problem. It is then easy to proceed along the lines of [Gi] using i), ii) from above for the derivatives of v by the way proving iii)–vi).

For the readers convenience we show $v \in W^2_{2,\mathrm{loc}}(G; \mathbb{R}^n)$. The main ideas are taken from [GM], proof of Theorem 1.3. Choose a ball $B_R \subset G$ and define η as in i). Let

$$\psi_1 := \tau_{-h}(\eta^2 \tau_h v)$$

where $\tau_h w(x) := \frac{1}{h}[w(x + he_\gamma) - w(x)]$ for some fixed direction e_γ, $\gamma = 1, \ldots, n$. We then define $\psi := \psi_1 - w$ with $w \in \mathring{W}_2^1(B_R; \mathbb{R}^n)$ such that div $w =$ div ψ_1 and

$$\|\nabla w\|_{L^2(B_R)} \leq c_* \|\text{div } \psi_1\|_{L^2(B_R)}.$$

(Of course we assume $|h|$ to be sufficiently small so that spt $\psi_1 \subset B_R$.)

We obtain

$$\int_{B_R} A\big(\varepsilon(v), \varepsilon[\tau_{-h}(\eta^2 \tau_h v)]\big) \, dx = \int_{B_R} A\big(\varepsilon(v), \varepsilon(w)\big) \, dx.$$

The first integral equals

$$\int_{B_R} A\big(\varepsilon(\tau_h v), \varepsilon(\eta^2 \tau_h v)\big) \, dx \geq c_* \int_{B_R} \eta^2 |\tau_h \varepsilon(v)|^2 \, dx - c_* \int_{B_R} |\varepsilon(\tau_h v)| |\eta| |\nabla \eta| |\tau_h v| \, dx$$

and with Young's inequality we deduce

$$\int_{B_R} \eta^2 |\varepsilon(\tau_h v)|^2 \, dx \leq c_* \Big\{ \int_{B_R} |\nabla \eta|^2 |\tau_h v|^2 \, dx + \int_{B_R} |\varepsilon(v)| |\varepsilon(w)| \, dx \Big\}.$$

We next observe

$$\int_{B_R} |\varepsilon(w)|^2 \, dx \leq \int_{B_R} |\nabla w|^2 \, dx \leq c_* \int_{B_R} |\tau_{-h}(\text{div}[\eta^2 \tau_h v])|^2 \, dx$$

$$= c_* \int_{B_R} |\tau_{-h}(2\eta \nabla \eta \cdot \tau_h v)|^2 \, dx$$

$$\leq c_* \int_{B_R} |\tau_{-h}(\eta \nabla \eta)|^2 |\tau_h v|^2 \, dx + c_* \int_{B_R} \eta^2 |\nabla \eta|^2 |\tau_{-h}(\tau_h v)|^2 \, dx;$$

here we have used the fact that v is divergence–free. From Hölder's inequality we have

$$\int_{B_R} |\varepsilon(v)| |\varepsilon(w)| \, dx \leq c(\delta) R^{-2} \int_{B_R} |\varepsilon(v)|^2 \, dx + \delta R^2 \int_{B_R} |\varepsilon(w)|^2 \, dx.$$

Since

$$\int_{B_{R-|h|}} |\tau_h w|^2 \, dx \leq \int_{B_R} |\nabla w|^2 \, dx$$

we end up with

$$
\int_{B_{R/2}} |\varepsilon(\tau_h v)|^2 dx \leq \delta \int_{B_R} |\tau_h \nabla v|^2 dx + c(\delta) R^{-2} \int_{B_R} |\nabla v|^2 dx.
$$

Finally we apply Korn's inequality to the function $\tau_h v$ yielding

$$
\int_{B_{R/2}} |\nabla(\tau_h v)|^2 dx \leq c_* \Big\{ \int_{B_{R/2}} R^{-2} |\tau_h v|^2 dx + \int_{B_{R/2}} |\varepsilon(\tau_h v)|^2 dx \Big\}.
$$

Combining the last two inequalities we have

$$
\int_{B_{R/2}} |\nabla(\tau_h v)|^2 dx \leq \delta \int_{B_R} |\nabla(\tau_h v)|^2 dx + c(\delta) R^{-2} \int_{B_R} |\nabla v|^2 dx,
$$

and the claim follows from Lemma 0.5 in [GM].

Let us indicate how to obtain v). First of all, for any weak solution u to the above system, we have inequality vi), i.e.

$$
\int_{B_r} |\nabla u - (\nabla u)_r|^2 dx \leq c_* \Big(\frac{r}{R}\Big)^{n+2} \int_{B_R} |\nabla u - (\nabla u)_R|^2 dx.
$$

This is a consequence of iv), we just have to replace the function by some arbitrary derivative. iv) (and in the same way iii)) in turn can be deduced from i), ii) and the $W^k_{2,loc}$-regularity along the lines of [GM], Proposition 1.9.

Assume now $r \leq R/2$ and let $\ell(x) = (\nabla v)_r(x)$. Then

$$
\int_{B_r} |\varepsilon(v) - (\varepsilon(v))_r|^2 dx = \int_{B_r} |\varepsilon(v - \ell)|^2 dx
$$

$$
\leq \int_{B_r} |\nabla(v - \ell)|^2 dx = \int_{B_r} |\nabla v - (\nabla v)_r|^2 dx
$$

$$
\leq_{(vi)} c_* \int_{B_{R/2}} |\nabla v - (\nabla v)_{R/2}|^2 dx \Big(\frac{r}{R}\Big)^{n+2}
$$

$$
\leq c_* \Big(\frac{r}{R}\Big)^{n+2} \int_{B_{R/2}} |\nabla v - A|^2 dx
$$

for any $n \times n$ matrix A. Let $w(x) = v(x) - (\varepsilon(v))_R x$ and choose a rigid motion $\gamma(x)$ such that

$$\int_{B_R} |w - \gamma|^2 dx \le c_* R^2 \int_{B_R} |\varepsilon(w)|^2 dx.$$

Let us also set $u = w - \gamma$, $A = (\varepsilon(v))_R + \nabla\gamma$. This gives

$$\int_{B_{R/2}} |\nabla v - A|^2 dx = \int_{B_{R/2}} |\nabla u|^2 dx$$

$$\le c_* R^{-2} \int_{B_R} |u|^2 dx,$$

the last inequality follows from ii) applied to u. Putting together the estimates we find that

$$\int_{B_{R/2}} |\nabla v - A|^2 dx \le c_* \int_{B_R} |\varepsilon(w)|^2 dx,$$

hence

$$\int_{B_r} |\varepsilon(v) - (\varepsilon(v))_r|^2 dx \le c_* \left(\frac{r}{R}\right)^{n+2} \int_{B_R} |\varepsilon(v) - (\varepsilon(v))_R|^2 dx$$

and v) is established.

$\square$

3.1 Partial C^1 regularity in the variational setting

In [Se11,12] partial regularity results for the variational problem (0.5) in case of a Bingham fluid, i.e. $W(E) = \mu|E|^2 + \sqrt{2}k_*|E|$, were obtained. Later on (see [FGR]) this result was extended to general fluids of Bingham type. We have the following

MAIN THEOREM 3.1.1 *Suppose we are given a real number $p \geq 2$, a function $v_0 \in W_p^1(\Omega; \mathbb{R}^n)$ such that $\operatorname{div} v_0 = 0$ and consider a volume load f_* in $L^p(\Omega; \mathbb{R}^n)$ which belongs to some Morrey space $L_{loc}^{p',n-p'+p'\lambda}(\Omega; \mathbb{R}^n)$, $0 < \lambda < 1$, $p' = p/p - 1$. Then the variational problem*

$$\begin{cases} I(u) = \displaystyle\int_\Omega \left(|\varepsilon(u)|^p + |\varepsilon(u)| - f_* \cdot u\right) dx \to \min \ in \\ \mathbb{K} = \{v \in W_p^1(\Omega; \mathbb{R}^n) : \operatorname{div} v = 0, \ v = v_0 \ on \ \partial\Omega\} \end{cases}$$

admits a unique solution u. Define the sets

$$\Omega^* = \{x \in \Omega : \quad x \ is \ a \ Lebesgue \ point \ for \ \varepsilon(u) \ and$$

$$\lim_{r\downarrow 0} \fint_{B_r(x)} |\varepsilon(u) - (\varepsilon(u))_{x,r}|^p dz = 0\},$$

$$\Omega_+^* = \{x \in \Omega^* : \quad \varepsilon(u)(x) \neq 0\}.$$

Then we have the following statements:

a) Ω_+^* is open and $\varepsilon(u) \in C^{0,\alpha}(\Omega_+^*; \mathbb{M})$ for any $\alpha < \lambda$.

b) $\nabla u \in C^{0,\alpha}(\Omega_+^*; \mathbb{M})$ for any $\alpha < \lambda$.

Moreover, $\varepsilon(u) = 0$ a.e. on $\Omega - \Omega_+^*$.

The proof of the theorem (taken from the paper [FGR]) is organized in several steps: for technical simplicity we may assume that the volume forces f_* vanish. We first consider a more regular functional of the form

$$J(u) = \int_\Omega |\varepsilon(u)|^p + f(\varepsilon(u)) \, dx$$

with f smooth, convex but growing of order less than p. Let u denote a J-minimizer in the class $\mathbb{K}$; existence and uniqueness of such u (and indeed for minimizers of the original problem) follow by standard methods; see e.g. [Gi]. We show that $\varepsilon(u)$ is Hölder–continuous near a point $x_0 \in \Omega$ provided x_0 is a Lebesgue–point for $\varepsilon(u)$, $\varepsilon(u)(x_0) \neq 0$ and

$$\lim_{r \downarrow 0} \fint_{B_r(x_0)} |\varepsilon(u) - (\varepsilon(u))_{x_o,r}|^p \, dz = 0.$$

The main ingredient here is a blow–up Lemma which for general $p \geq 2$ is more complicated than in the quadratic case. Next we replace f by a sequence $f_\delta(Q) := \sqrt{\delta^2 + |Q|^2}$ with corresponding J_δ–minimizers u_δ. It is not hard to show that $u_\delta \to u$ as $\delta \downarrow 0$ strongly in W_p^1 where u is the solution from the Main Theorem. Moreover, by the first step, we have uniform partial regularity for $\varepsilon(u_\delta)$ which gives part a) of the Main Theorem by letting δ tend to 0.

To be precise, let

$$J(u) = \int_\Omega |\varepsilon(u)|^p + f(\varepsilon(u)) \, dx$$

with $f : \mathbb{M} \to [0, \infty[$ of class C^2 and convex such that the growth condition

$$(1.1) \qquad |D^2 f(Q)| \leq a + b(1 + |Q|^2)^{\frac{q-2}{2}}$$

holds for constants a, b and q such that $a, b > 0$, $1 \leq q < p$. A typical example is $f(Q) = (\delta^2 + |Q|^2)^{q/2}$, $\delta \neq 0$. We have the following result analogous to the Main Theorem (we use the notation of that theorem):

THEOREM 3.1.2 *The functional J admits a unique minimizer u in the class* $\mathbb{K}$. *This minimizer satisfies:*

a) Ω_+^* *is open, and* $\varepsilon(u) \in C^{0,\alpha}(\Omega_+^*)$ *for all* α *such that* $0 < \alpha < 1$.

b) $\nabla u \in C^{0,\alpha}(\Omega_+^*)$ *for all* α *such that* $0 < \alpha < 1$.

The key element in the proof of Theorem 3.1.2 is the growth estimate of Lemma 3.1.1 (cf.[EG], section 2, or [FR]). The proof of Theorem 3.1.2 given Lemma 3.1.1 is standard; the arguments are a simplification of those used in the proof of the Main Theorem given after Lemma 3.1.5.

LEMMA 3.1.1 (Blow–up lemma) *Let A be a non–zero symmetric matrix with norm 2σ, and let u be a local minimizer for J in $\mathbb{K}$. Then there exists a constant c_0 depending only on A and C^2 bounds on f in the σ–neighborhood of*

A such that for every $t \in]0,1[$ there exists ε depending only on t, f and A such that the conditions

(1.2) $|(\varepsilon(u))_{x_0,R} - A| \leq \sigma$,

$E(u, B_R(x_0)) :=$

(1.3)
$$\fint_{B_R(x_0)} |\varepsilon(u) - (\varepsilon(u))_{x_0,R}|^p \, dx + \fint_{B_R(x_0)} |\varepsilon(u) - (\varepsilon(u))_{x_0,R}|^2 \, dx < \varepsilon^2$$

for some ball $B_R(x_0) \subset \Omega$ together imply the estimate

(1.4) $E(u, B_{tR}(x_0)) \leq c_0 t^2 E(u, B_R(x_0))$.

Proof.   If the proposition were false we could find, for any $c_0 > 0$ and $t \in]0,1[$, a sequence of balls $\{B_{R_k}(x_k)\} \subset \Omega$ and a sequence $\{v_k\}$ of local minimizers for J in $\mathbb{K}$ such that the estimates

(1.5) $|(\varepsilon(v_k))_{x_k,R_k} - A| \leq \sigma$,

(1.6) $E(v_k, B_{R_k}(x_k)) := \varepsilon_k^2 \searrow 0$ as $k \to \infty$

and

(1.7) $E(v_k, B_{tR_k}(x_k)) > c_0 t^2 \varepsilon_k^2$

were all valid.

Set $A_k := (\varepsilon(v_k))_{x_k,R_k}$, and define $g_k(z) = v_k(x_k + R_k z) - R_k A_k z$, for $z \in B_1$. It is immediate that $g_k \in W_p^1(B_1; \mathbb{R}^n)$, so by Lemma 3.0.3 (iii) there exists $p_k \in V_\times$ such that

(1.8)
$$\int_{B_1} |g_k(z) - p_k(z)|^2 \, dz \leq c_1 \int_{B_1} |\varepsilon(g_k)|^2 \, dz$$

for c_1 depending only on n and p. Define now

$$u_k(z) := \frac{1}{R_k \varepsilon_k} (g_k(z) - p_k(z)).$$

We see by (1.8) (using also (1.6))

(1.9)
$$\begin{cases} \displaystyle\int_{B_1} |u_k|^2 \, dz & \leq \ \frac{c_1^2}{R_k^2 \varepsilon_k^2} \displaystyle\int_{B_1} |\varepsilon(g_k)|^2 \, dz \\[2mm] & = \ \frac{c_1^2}{R_k^n \varepsilon_k^2} \displaystyle\int_{B_{R_k}} |\varepsilon(v_k)(y) - (\varepsilon(v_k))_{x_k,R_k}|^2 \, dy \\[2mm] & \leq \ c_1^2 \omega_n, \end{cases}$$

where ω_n denotes the volume of the unit ball. Further we have

(1.10) $\displaystyle\fint_{B_1} |\varepsilon(u_k)|^2 \, dz \leq 1,$

(1.11) $\displaystyle\fint_{B_1} |\varepsilon(u_k)|^p \, dz \leq \varepsilon_k^{2-p}.$

We thus have the existence of $v \in W_2^1(B_1; \mathbb{R}^n)$ and $A_0 \in \mathbb{M}$ such that, after passing to an appropriate subsequence,

(1.12) $u_k \rightharpoonup v$ in $W_2^1(B_1; \mathbb{R}^n)$

(1.13) $u_k \to v$ in $L^2(B_1; \mathbb{R}^n)$

(1.14) $\varepsilon_k^{1-2/p} \varepsilon(u_k) \rightharpoonup 0$ in $L^p(B_1; \mathbb{M})$

and

(1.15) $A_k \to A_0;$

here (1.12) follows from (1.9), (1.10) and Korn's inequality, (1.13) follows from (1.12) and (1.15) follows from (1.5). To see (1.14) we note from (1.11) there exists h such that $\varepsilon^{1-2/p}\varepsilon(u_k) \rightharpoonup h$ in L^p; for any $\psi \in C_0^\infty(B; \mathbb{M})$ we thus have

$$\left| \int_{B_1} h : \psi \, dz \right| = \lim_{k \to \infty} \left| \int_{B_1} \varepsilon_k^{1-2/p} \varepsilon(u_k) : \right.$$

$$\left. \psi \, dz \right| \leq \limsup_{k \to \infty} \varepsilon_k^{1-2/p} \|\varepsilon(u_k)\|_{L^2(B_1)} \|\psi\|_{L^2(B_1)} = 0$$

by (1.6) and (1.10).

Assume now the conclusion of Lemma 3.1.3, viz. that the weak convergence in (1.12) and (1.14) can be improved to strong convergence (after passing to appropriate subsequences). Assume further Lemma 3.1.2, the blow–up equation, and set $c_0 := 2c_*$, where c_* is the constant in Lemma 3.0.5 corresponding to the bilinear form associated with equation (1.18); note that c_* depends only on A and C^2 bounds on f in the σ–neighborhood of A. Lemma 3.0.5 (v) then says

(1.16) $\displaystyle\fint_{B_t} |\varepsilon(v) - (\varepsilon(v))_t|^2 \, dz \leq \frac{c_0}{2} t^2 \fint_{B_1} |\varepsilon(v) - (\varepsilon(v))_1|^2 \, dz.$

On the other hand, rewriting (1.7) in terms of the rescaled map u_k yields

$$E(u_k, B_t) > c_0 t^2,$$

which by Lemma 3.1.3 then implies

$$(1.17) \quad \fint_{B_t} |\varepsilon(v) - (\varepsilon(v))_t|^2 \, dz \geq c_0 t^2.$$

We have from (1.12) and (1.10)

$$\fint_{B_1} |\varepsilon(v)|^2 \, dz \leq \liminf_{k \to \infty} \fint_{B_1} |\varepsilon(u_k)|^2 \, dz \leq 1$$

which, combined with (1.17), yields

$$\fint_{B_t} |\varepsilon(v) - (\varepsilon(v))_t|^2 \, dz \geq c_0 t^2 \fint_{B_1} |\varepsilon(v) - (\varepsilon(v))_1|^2 \, dz;$$

comparing this to (1.16) gives the desired contradiction.

$$\square$$

We next establish the auxiliary results discussed above, viz. the blow–up equation and the desired strong convergences.

LEMMA 3.1.2 (Blow–up equation) *With notation as above, the limit–map v satisfies the equation*

$$\int_{B_1} p|A_0|^{p-2} \big[\varepsilon(v) + (p-2)|A_0|^{-2}(A_0 : \varepsilon(v))A_0\big] : \varepsilon(\varphi) \, dz$$

$$(1.18)$$

$$= -\int_{B_1} \frac{\partial^2 f}{\partial Q^2}(A_0)(\varepsilon(v), \varepsilon(\varphi)) \, dz$$

for all $\varphi \in C_0^1(B_1; \mathbb{R}^n)$ such that $\operatorname{div} \varphi = 0$.

Proof. From the Euler–Lagrange equation for v_k we have

$$\int_{B_{R_k}(x_k)} p|\varepsilon(v_k)|^{p-2} \varepsilon(v_k) : \varepsilon(\psi) \, dy = -\int_{B_{R_k}(x_k)} \frac{\partial f}{\partial Q}(\varepsilon(v_k)) : \varepsilon(\psi) \, dy$$

for all $\psi \in C_0^1(B_{R_k}(x_k); \mathbb{R}^n)$ such that div $\psi = 0$. Rewriting this in terms of the rescaled map u_k yields

(1.19)

$$\int_{B_1} p|\varepsilon_k\varepsilon(u_k) + A_k|^{p-2}(\varepsilon_k\varepsilon(u_k) + A_k) : \varepsilon(\varphi) \, dz$$

$$= -\int_{B_1} \frac{\partial f}{\partial Q}(\varepsilon_k\varepsilon(u_k) + A_k) : \varepsilon(\varphi) \, dz$$

for all $\varphi \in C_0^1(B_1; \mathbb{R}^n)$ such that div $\varphi = 0$; for the remainder of the proof, we fix such a φ. Noting $\int_{B_1} |A_k|^{p-2}A_k : \varepsilon(\varphi) \, dz = 0$, we subtract this quantity from the left–hand side of (1.19), and similarly subtract $0 = \int_{B_1} \frac{\partial f}{\partial Q}(A_k) : \varepsilon(\varphi) \, dz$ from the right–hand side; multiplying through by ε_k^{-1} then yields $0 = I_k + I\!I_k$,

where $$I_k := p\,\varepsilon_k^{-1}\int_{B_1}\left[|\varepsilon_k\varepsilon(u_k) + A_k|^{p-2}(\varepsilon_k\varepsilon(u_k) + A_k) - |A_k|^{p-2}A_k\right] : \varepsilon(\varphi) \, dz$$

and $$I\!I_k := \varepsilon_k^{-1}\int_{B_1}\left[\frac{\partial f}{\partial Q}(\varepsilon_k\varepsilon(u_k) + A_k) - \frac{\partial f}{\partial Q}(A_k)\right] : \varepsilon(\varphi) \, dz.$$

We rewrite

$$I_k := p\varepsilon_k^{-1}\int_{B_1}\int_0^1 \frac{d}{ds}\left[|A_k + s\varepsilon_k\varepsilon(u_k)|^{p-2}(A_k + s\varepsilon_k\varepsilon(u_k)) - |A_k|^{p-2}A_k\right] ds : \varepsilon(\varphi) \, dz$$

$$= I_k^1 + I_k^2,$$

where

$$I_k^1 := p\int_{B_1}\int_0^1 |A_k + s\varepsilon_k\varepsilon(u_k)|^{p-2} ds\, \varepsilon(u_k) : \varepsilon(\varphi) \, dz,$$

$$I_k^2 := p(p-2)\int_{B_1}\int_0^1 \left[|A_k + s\varepsilon_k\varepsilon(u_k)|^{p-4}(A_k + s\varepsilon_k\varepsilon(u_k)) : \varepsilon(u_k)\right](A_k + s\varepsilon_k\varepsilon(u_k)) \, ds : \varepsilon(\varphi) \, dz.$$

Similarly we have

$$\mathbf{II}_k = \varepsilon_k^{-1} \int_{B_1} \int_0^1 \frac{d}{ds} \left[\frac{\partial f}{\partial Q}(A_k + s\varepsilon_k \varepsilon(u_k)) - \frac{\partial f}{\partial Q}(A_k) \right] ds : \varepsilon(\varphi) \, dz$$

$$= \int_{B_1} \int_0^1 \frac{\partial^2 f}{\partial Q^2}(A_k + s\varepsilon_k \varepsilon(u_k))(\varepsilon(u_k), \varepsilon(\varphi)) \, ds dz.$$

The desired equation (1.18) follows directly, once we can establish the following asymptotic behaviour (possibly after passing to a subsequence):

$$(1.20) \qquad \lim_{k \to \infty} I_k^1 = p|A_0|^{p-2} \int_{B_1} \varepsilon(v) : \varepsilon(\varphi) \, dz;$$

$$(1.21) \qquad \lim_{k \to \infty} I_k^2 = p(p-2)|A_0|^{p-4} \int_{B_1} (A_0 : \varepsilon(v)) A_0 : \varepsilon(\varphi) \, dz; \text{ and}$$

$$(1.22) \qquad \lim_{k \to \infty} \mathbf{II}_k = \int_{B_1} \frac{\partial^2 f}{\partial Q^2}(A_0)(\varepsilon(v), \varepsilon(\varphi)) \, dz.$$

From (1.10) and (1.15), equation (1.20) is immediate once we establish

$$(1.23) \qquad \lim_{k \to \infty} I_k^\alpha = 0,$$

where

$$I_k^\alpha := \int_{B_1} \left[\int_0^1 |A_k + s\varepsilon_k \varepsilon(u_k)|^{p-2} \, ds - |A_k|^{p-2} \right] \varepsilon(u_k) : \varepsilon(\varphi) \, dz.$$

To see this, we begin by noting, from (1.11)

$$(1.24) \qquad \varepsilon_k \varepsilon(u_k) \to 0 \text{ in } L^p, \text{ with pointwise convergence a.e. in } B_1.$$

Given $\kappa > 0$, we can thus apply Egorov's theorem to find $M = M(\kappa) \subset B_1$ with $|M| < \kappa$ such that $\varepsilon_k \varepsilon(u_k) \to 0$ uniformly on $B_1 \setminus M$. We then estimate

$$|I_k^\alpha| \leq \|\varepsilon(\varphi)\|_\infty \sup_{B_1 \setminus M} \left| \int_0^1 |A_k + s\varepsilon_k \varepsilon(u_k)|^{p-2} \, ds - |A_k|^{p-2} \right| \int_{B_1 \setminus M} |\varepsilon(u_k)| dz$$

$$+ \|\varepsilon(\varphi)\|_\infty \int_M \left| \int_0^1 |A_k + s\varepsilon_k \varepsilon(u_k)|^{p-2} \, ds - |A_k|^{p-2} \right| |\varepsilon(u_k)| \, dz.$$

The first term tends to zero as $k \to \infty$ by (1.12) and the uniform convergence of $\varepsilon_k \varepsilon(u_k)$ to zero on $B_1 \setminus M$. To estimate the second term, we note by (1.5)

$$\left|1 + \frac{\varepsilon_k |\varepsilon(u_k)|}{|A_k|}\right|^{p-2} \le c_1 \left(1 + \varepsilon_k^{p-2}|\varepsilon(u_k)|^{p-2}\right)$$

for $c_1 := 2^{p-2} \max\{\sigma^{2-p}, 1\}$.

We thus have

$$\int_M \left|\int_0^1 |A_k + s\varepsilon_k\varepsilon(u_k)|^{p-2}\, ds - |A_k|^{p-2}\right| |\varepsilon(u_k)|\, dz$$

$$\le |A_k|^{p-2} \int_M \left[(1 + \frac{\varepsilon_k|\varepsilon(u_k)|}{|A_k|})^{p-2} + 1\right]|\varepsilon(u_k)|\, dz$$

$$\le 3^{p-2}\sigma^{p-2}(c_1 + 1)\left[\int_M |\varepsilon(u_k)|dz + \varepsilon_k^{p-2} \int_M |\varepsilon(u_k)|^{p-1}\, dz\right]$$

$$\le 3^{p-2}\sigma^{p-2}(c_1 + 1)\left[\sqrt{\kappa} + \kappa^{1/p}\varepsilon_k^{1-2/p}\right],$$

the last line following from Hölder's inequality, (1.10) and (1.11). By (1.6) and the arbitrariness of κ, this establishes (1.23), and hence (1.20).

In a completely analogous manner we derive (1.21) and (1.22); we use the structure condition (1.1) in showing (1.22). By the remarks above, this completes the derivation of (1.18).

$$\square$$

It remains to show

LEMMA 3.1.3 (Strong convergence) *The convergence properties stated in (1.12), (1.14) can be improved to strong convergence, i.e.*

$$\varepsilon(u_k) \to \varepsilon(v) \qquad in\ L^2_{loc}(B_1; \mathbb{M})$$

$$\varepsilon_k^{1-2/p}\varepsilon(u_k) \to 0 \quad in\ L^p_{loc}(B_1; \mathbb{M}).$$

Proof.   As in [EG] (see also [FR]) we define the scaled energy density

$$f_k(Q) = \varepsilon_k^{-2}\left(|A_k + \varepsilon_k Q|^p - |A_k|^p - p|A_k|^{p-2}A_k : \varepsilon_k Q\right)$$

$$+\varepsilon_k^{-2}\left(f(A_k + \varepsilon_k Q) - f(A_k) - \tfrac{\partial f}{\partial Q}(A_k) : \varepsilon_k Q\right), \quad Q \in \mathbb{M}.$$

For $0 < r \le 1$ we let

$$I_k^r(w, B_r) := \int_{B_r} f_k(\varepsilon(w)) dz$$

for any $w \in W_p^1(B_r; \mathbb{R}^n)$ with div $w = 0$. In view of the original problem it is obvious that u_k is a minimizer of $I_k^r(\cdot, B_r)$, i.e.

$$(1.25) \qquad \int_{B_r} f_k(\varepsilon(u_k)) dz \le \int_{B_r} f_k(\varepsilon(\tilde{u})) dz$$

whenever $\tilde{u} \in W_p^1(B_1; \mathbb{R}^n)$, spt $(u_k - \tilde{u}) \subset\subset B_r$, div $\tilde{u} = 0$.

We claim

$$(1.26) \qquad 0 \le f_k(Q) \le c_1 \left(|Q|^2 + \varepsilon_k^{p-2} |Q|^p \right) \quad \forall Q \in M$$

for some positive constant c_1 independent of k but depending on the same parameters as c_0 from Lemma 3.1.1. This follows from

$$\varepsilon_k^{-2} \left(|A_k + \varepsilon_k Q|^p - |A_k|^p - p\varepsilon_k |A_k|^{p-2} A_k : Q \right)$$

$$= p \int_0^1 \left\{ |A_k + s\varepsilon_k Q|^{p-2} Q : Q + (p-2)|A_k + s\varepsilon_k Q|^{p-4} \right.$$

$$\left. \times \left[Q : (A_k + s\varepsilon_k Q) \right]^2 \right\} (1 - s)\, ds$$

$$(1.27) \qquad = p\, \varepsilon_k^{p-2} \int_0^1 \left\{ \left| \frac{A_k}{\varepsilon_k} + sQ \right|^{p-2} Q : Q + (p-2) \left| \frac{A_k}{\varepsilon_k} + sQ \right|^{p-4} \right.$$

$$\left. \times \left[Q : \left(\tfrac{A_k}{\varepsilon_k} + sQ \right) \right]^2 \right\} (1 - s)\, ds$$

$$\le c_2\, \varepsilon_k^{p-2} |Q|^2 \int_0^1 \left| \frac{A_k}{\varepsilon_k} + sQ \right|^{p-2} ds$$

$$\le c_3\, \varepsilon_k^{p-2} |Q|^2 \left(\left| \tfrac{A_k}{\varepsilon_k} \right|^{p-2} + |Q|^{p-2} \right)$$

$$\le c_4 \left(|Q|^2 + \varepsilon_k^{p-2} |Q|^p \right)$$

and (using the growth condition (1.1) imposed on f)

$$\varepsilon_k^{-2} \ \left(f(A_k + \varepsilon_k Q) - f(A_k) - \tfrac{\partial f}{\partial Q}(A_k) : \varepsilon_k Q \right)$$

$$\leq \int_0^1 \left| \frac{\partial^2 f}{\partial Q^2}(A_k + s\varepsilon_k Q) \right| |Q|^2 (1-s) \, ds$$

$$\leq \int_0^1 |Q|^2 \left(a + b(1 + |A_k + s\varepsilon_k Q|^2)^{\frac{q-2}{2}} \right) (1-s) \, ds$$

$$\leq c_5 |Q|^2 \left(1 + \varepsilon_k^{p-2} |Q|^{p-2} \right)$$

Following Evans–Gariepy [EG] we define the measures

$$\mu_k(Z) := \int_Z |\varepsilon(u_k)|^2 + \varepsilon_k^{p-2} |\varepsilon(u_k)|^p \, dx, \quad Z \subset B_1(0).$$

According to (1.12) and (1.14) the sequence μ_k is uniformly bounded, and thus we have the existence of a Radon measure μ on B_1 such that, after passing to an appropriate subsequence, $\mu_k \rightharpoonup \mu$.

We choose $0 < r < 1$ such that $\mu(\partial B_r) = 0$. Let $0 < s < r$. We define

$$\varphi := \begin{cases} v & \text{on } B_s(0) \\ \Psi + u_k + \eta(v - u_k) & \text{on } B_r(0) - B_s(0) \end{cases}$$

where $\eta \in C_0^1(B_r)$ is a cut–off function chosen to satisfy $0 \leq \eta \leq 1$, $\eta = 1$ on B_s, $|\nabla \eta| \leq \frac{c_6}{r-s}$, and $\Psi \in \overset{\circ}{W}{}^1_p(B_r - B_s; \mathbb{R}^n)$ is a solution of

$$\text{div } \Psi = -\text{div } (u_k + \eta[v - u_k]).$$

According to Lemma 3.0.4 Ψ satisfies ($t = 2, p$)

$$(1.28) \qquad \int_{B_r \backslash B_s} |\varepsilon(\Psi)|^t dx \leq \int_{B_r \backslash B_s} |\nabla \Psi|^t dx \leq c_7(r,s) \int_{B_r \backslash B_s} |\nabla \eta(v - u_k)|^t dx.$$

Combining (1.25), (1.26) and (1.28) we deduce

(1.29)
$$
\begin{cases}
I_k^r(u_k) - I_k^r(v) \le I_k^r\big(\Psi + u_k + \eta(v - u_k)\big) - I_k^r(v) \\[2ex]
\le \displaystyle\int_{B_r \setminus B_s} f_k\Big(\varepsilon\big((\Psi + u_k + \eta(v - u_k))\big)\Big)\, dx \\[2ex]
\le c_8 \displaystyle\int_{B_r \setminus B_s} \Big\{|\varepsilon\big(\Psi + u_k + \eta(v - u_k))|^2 + \varepsilon_k^{p-2} \\[1ex]
\qquad \times |\varepsilon\big(\Psi + u_k + \eta(v - u_k))|^p\Big\}\, dx \\[2ex]
\le c_9 \displaystyle\int_{B_r \setminus B_s} \Big\{|\varepsilon(\Psi)|^2 + |\varepsilon(u_k)|^2 + |\nabla\eta|^2|v - u_k|^2 + |\varepsilon(v)|^2 \\[1ex]
\qquad + \varepsilon_k^{p-2}\big(|\varepsilon(\Psi)|^p + |\varepsilon(u_k)|^p + |\nabla\eta|^p|v - u_k|^p + |\varepsilon(v)|^p\big)\Big\}\, dx \\[2ex]
\le c_{10} \displaystyle\int_{B_r \setminus B_s} \Big\{|\varepsilon(u_k)|^2 + \varepsilon_k^{p-2}|\varepsilon(u_k)|^p + |\varepsilon(v)|^2 + \varepsilon_k^{p-2}|\varepsilon(v)|^p\Big\}\, dx \\[2ex]
\qquad + c_{11}(r, s) \displaystyle\int_{B_r \setminus B_s} \Big\{|v - u_k|^2 + \varepsilon_k^{p-2}|v - u_k|^p\Big\}\, dx.
\end{cases}
$$

Now we consider the last integral. We already know

$$
\int_{B_r \setminus B_s} |v - u_k|^2 \longrightarrow 0, \text{ as } k \to \infty.
$$

On the other hand we have by Lemma 3.0.2

$$
\|\varepsilon_k^{1-2/p}(v - u_k)\|_{L^p(B_r \setminus B_s)} \le c_{12}(r, s)\Big\{\|\varepsilon_k^{1-2/p}(v - u_k)\|_{L^2(B_r \setminus B_s)}
$$
$$
+ \|\varepsilon\big(\varepsilon_k^{1-2/p}(v - u_k)\big)\|_{L^p(B_r \setminus B_s)}\Big\}.
$$

Both terms on the right hand side are uniformly bounded. Applying Korn's inequality, Lemma 3.0.1, we obtain

$$
\|\varepsilon_k^{1-2/p}(v - u_k)\|_{W_p^1(B_r \setminus B_s)}
$$

$$
\le c_{13}(r, s)\Big\{\|\varepsilon_k^{1-2/p}(v - u_k)\|_{L^2(B_r \setminus B_s)} + \|\varepsilon_k^{1-2/p}\varepsilon(v - u_k)\|_{L^p(B_r \setminus B_s)}\Big\}.
$$

Since $\varepsilon_k^{1-2/p}(v - u_k) \to 0$ in $L^2(B_r \setminus B_s)$ we end up with

$\varepsilon_k^{1-2/p}(v - u_k) \to 0$ strongly in $L^p(B_r \setminus B_s)$.

We now pass to the limit $k \to \infty$ in (1.29) with the result

$$\limsup_{k\to\infty} \left(I_k^r(u_k) - I_k^r(v) \right) \leq c_{14} \left\{ \mu(\overline{B_r \setminus B_s}) + \int_{B_r\setminus B_s} |\varepsilon(v)|^2 \, dx \right\}.$$

Taking the limit $s \nearrow r$ we arrive at

(1.30) $\qquad \displaystyle\limsup_{k\to\infty} \left(I_k^r(u_k) - I_k^r(v) \right) \leq 0$ for almost all $r \in (0,1)$.

We next claim

(1.31)
$$\limsup_{k\to\infty} \left(I_k^r(u_k) - I_k^r(v) \right)$$
$$\geq c_{15} \limsup_{k\to\infty} \int_{B_r} \left(|\varepsilon(u_k) - \varepsilon(v)|^2 + \varepsilon_k^{p-2}|\varepsilon(u_k) - \varepsilon(v)|^p \right) dx$$

for some positive constant c_{15}. Inequality (1.31) follows from

$$I_k^r(u_k) - I_k^r(v) = \int_{B_r} \left(f_k(\varepsilon(u_k)) - f_k(\varepsilon(v)) \right) dx$$

$$= \varepsilon_k^{-2} \int_{B_r} \left(|A_k + \varepsilon_k\varepsilon(u_k)|^p - |A_k + \varepsilon_k\varepsilon(v)|^p - p|A_k|^{p-2}A_k : \right.$$

$$\left. \varepsilon_k(\varepsilon(u_k) - \varepsilon(v)) \right) dx$$

$$+ \varepsilon_k^{-2} \int_{B_r} \left(f(A_k + \varepsilon_k\varepsilon(u_k)) - f(A_k + \varepsilon_k\varepsilon(v)) - \frac{\partial f}{\partial Q}(A_k) : \right.$$

$$\left. \varepsilon_k(\varepsilon(u_k) - \varepsilon(v)) \right) dx$$

$$= \varepsilon_k^{-2} \int_{B_r} \left(|A_k + \varepsilon_k\varepsilon(u_k)|^p - |A_k + \varepsilon_k\varepsilon(v)|^p - p|A_k + \varepsilon_k\varepsilon(v)|^{p-2} \right.$$

$$\left. \times (A_k + \varepsilon_k\varepsilon(v)) : \varepsilon_k(\varepsilon(u_k) - \varepsilon(v)) \right) dx +$$

$$+\varepsilon_k^{-2} \int_{B_r} \Big(p(|A_k + \varepsilon_k\varepsilon(v)|^{p-2}(A_k + \varepsilon_k\varepsilon(v)) - |A_k|^{p-2}A_k) :$$

$$\varepsilon_k(\varepsilon(u_k) - \varepsilon(v))\Big)\,dx$$

$$+\varepsilon_k^{-2} \int_B \Big(f(A_k + \varepsilon_k\varepsilon(u_k)) - f(A_k + \varepsilon_k\varepsilon(v)) - \frac{\partial f}{\partial Q}(A_k) :$$

$$\varepsilon_k(\varepsilon(u_k) - \varepsilon(v))\Big)\,dx$$

$$=: \quad I_k + II_k + III_k,$$

(1.32)

$$
\left\{
\begin{aligned}
I_k \;&= \varepsilon_k^{-2} \int_{B_r} \int_0^1 \frac{d^2}{ds^2}|A_k + \varepsilon_k\varepsilon(v) + s\varepsilon_k(\varepsilon(u_k) - \varepsilon(v))|^p\,(1-s)\,ds\,dx \\[2mm]
&\geq p \int_{B_r} \int_0^1 |A_k + \varepsilon_k\varepsilon(v + s)\varepsilon_k(\varepsilon(u_k) - \varepsilon(v))|^{p-2}\,(1-s)\,ds \\[2mm]
&\quad\times |\varepsilon(u_k) - \varepsilon(v)|^2\,dx \\[2mm]
&\geq c_{16} \int_{B_r} \big[|A_k + \varepsilon_k\varepsilon(v)|^{p-2} + \varepsilon_k^{p-2} \\[2mm]
&\quad\times |\varepsilon(u_k) - \varepsilon(v)|^{p-2}\big]|\varepsilon(u_k) - \varepsilon(v)|^2\,dx \\[2mm]
&\geq c_{17} \int_{B_r} \big(|\varepsilon(u_k) - \varepsilon(v)|^2 + \varepsilon_k^{p-2}|\varepsilon(u_k) - \varepsilon(v)|^p\big)\,dx
\end{aligned}
\right.
$$

For the last estimate we use the local boundedness of $\varepsilon(v)$ so that $c_{18} \geq |A_k + \varepsilon_k\varepsilon(v)|^{p-2} \geq c_{19}$ for k large enough. Analogous calculations to those used in estimating the term I_k^α in Lemma 3.1.2 (cf. 1.23) show

(1.33) $\qquad \lim_{k\to\infty} II_k = 0$

Finally by convexity of f

$$
\begin{aligned}
\mathrm{III}_k \; &\geq \varepsilon_k^{-2} \int\limits_{B_r} \left(\frac{\partial f}{\partial Q}(A_k + \varepsilon_k \varepsilon(v)) - \frac{\partial f}{\partial Q}(A_k) \right) : \varepsilon_k(\varepsilon(u_k) - \varepsilon(v))\, dx \\
&= \int\limits_{B_r} \int\limits_0^1 \frac{\partial^2 f}{\partial Q^2}(A_k + s\varepsilon_k\varepsilon(v))\big(\varepsilon(v), \varepsilon(u_k) - \varepsilon(v)\big)\, ds\, dx,
\end{aligned}
$$

and the last term vanishes as $k \to \infty$ (recall $\varepsilon(v) \in L^\infty(B_r)$). Combining this with (1.32) and (1.33), inequality (1.31) follows and the claim of Lemma 3.1.3 is then a consequence of (1.30).

$\square$

In order to complete the proof of the Main Theorem we need to establish that, for the family of functionals $\{J_\delta\}_{\delta \in (0,1]}$, with $J_\delta(v) := \int_\Omega |\varepsilon(v)|^p + f_\delta(\varepsilon(v))dx$, where $f_\delta(Q):= (\delta^2 + |Q|^2)^{1/2}$, the quantity ε occuring in Lemma 3.1.1 can be chosen uniformly in δ. Precisely we have

LEMMA 3.1.4 *Let A be a non–zero symmetric matrix with norm 2σ, and let u be a local minimizer for J_δ in $\mathbb{K}$. Then there exists a constant c_0 depending only on A such that for every $t \in\ ]0,1[$ there exists ε depending on t and A such that the conditions*

(1.34) $|(\varepsilon(u))_{x_0,R} - A| \leq \sigma$

(1.35) $E\big(u, B_R(x_0)\big) < \varepsilon^2$

for some ball $B_R(x_0) \subset \Omega$ together imply the estimate

(1.36) $E\big(u, B_{tR}(x_0)\big) \leq c_0 t^2 E\big(u, B_R(x_0)\big).$

Proof.   If the proposition were false we could find, for any $c_0 > 0$ and $t \in]0,1[$, a sequence $\{\delta_k\} \subset]0,1]$ and corresponding local J_δ–minimizers $\{v_k\}$ and balls $\{B_{R_k}(x_k)\} \subset \Omega$ such that conditions (1.5)–(1.7) held. If $\delta_0 := \liminf_{k \to \infty} \delta_k > 0$ we have an immediate contradiction (set $\varepsilon = \min_{\delta \in [\delta',1]} \varepsilon(t, f_\delta, A)$, where $\delta' = \inf_k \delta_k$, in (1.35) (here $\varepsilon(t, f_\delta, A)$ is given by Lemma 3.1.1); since $\delta' > 0$ by assumption, $\varepsilon > 0$). Hence, after passing to an appropriate subsequence, we may assume $\delta_k \to 0$.

As in the proof of Lemma 3.1.1 we define the rescaled functions $\{u_k\}$ and deduce the convergences given in (1.12)–(1.15). The remainder of the proof of Lemma 3.1.4 is directly analogous to that of Lemma 3.1.1, given that we can establish the analogues of Lemma 3.1.2 (the blow–up equation) and Lemma 3.1.3 (desired strong convergences). In order to establish the blow–up equation, we again argue as in Lemma 3.1.2; cf. [Se12], Lemma 1. We need to discuss the integral $I\!I_k$ with f replaced by f_{δ_k}, i.e.

$$
\begin{aligned}
I\!I_k \;=\;& \varepsilon_k^{-1} \int_{B_1} \big[(\delta_k^2 + |\varepsilon_k \varepsilon(u_k) + A_k|^2)^{-1/2} (\varepsilon_k \varepsilon(u_k) + A_k) \\
& - (\delta_k^2 + |A_k|^2)^{-1/2} A_k \big] : \varepsilon(\varphi)\, dx \\[2mm]
=\;& I_k^1 + I_k^2,
\end{aligned}
$$

where

$$
I_k^1 := \varepsilon_k^{-1} \int_{B_1} \frac{\varepsilon_k \varepsilon(u_k)}{\alpha_k} : \varepsilon(\varphi)\, dz
$$

and

$$
I_k^2 := \varepsilon_k^{-1} \int_{B_1} \Big(\frac{1}{\beta_k} - \frac{1}{\alpha_k} \Big) (\varepsilon_k \varepsilon(u_k) + A_k) : \varepsilon(\varphi)\, dz
$$

for

$$
\alpha_k := (\delta_k^2 + |A_k|^2)^{1/2}, \quad \beta_k := (\delta_k^2 + |\varepsilon_k \varepsilon(u_k) + A_k|^2)^{1/2}.
$$

It is immediate that $I_k^1 \to \frac{1}{|A_0|} \int_{B_1} \varepsilon(v) : \varepsilon(\varphi)\, dz$. Further, we see

$$
\begin{aligned}
I_k^2 \;=\;& \varepsilon_k^{-1} \int_{B_1} \frac{(\alpha_k^2 - \beta_k^2)}{\alpha_k \beta_k (\alpha_k + \beta_k)} (\varepsilon_k \varepsilon(u_k) + A_k) : \varepsilon(\varphi)\, dz \\[2mm]
=\;& \varepsilon_k^{-1} \int_{B_1} (-2\varepsilon(u_k) : A_k - \varepsilon_k^2 |\varepsilon(u_k)|^2) \frac{(\varepsilon_k \varepsilon(u_k) + A_k)}{\alpha_k \beta_k (\alpha_k + \beta_k)} : \varepsilon(\varphi)\, dz.
\end{aligned}
$$

Since

$$
\left| \frac{\varepsilon_k \varepsilon(u_k) + A_k}{\alpha_k \beta_k (\alpha_k + \beta_k)} \right| \le \alpha_k^{-2} \to |A_0|^{-2} < \sigma^{-2} \text{ as } k \to \infty
$$

and $\dfrac{\varepsilon_k \varepsilon(u_k) + A_k}{\alpha_k \beta_k (\alpha_k + \beta_k)} \to \dfrac{A_0}{2|A_0|^3}$ pointwise a.e. by (1.24), we have

$$\lim_{k \to \infty} I\!\!I_k = \int_{B_1} \left[\frac{\varepsilon(v)}{|A_0|} - (\varepsilon(v) : A_0) \frac{A_0}{|A_0|^3} \right] : \varepsilon(\varphi) \, dz;$$

thus the analogue of (1.18) is

(1.37)
$$\begin{cases} \displaystyle\int_{B_1} p|A_0|^{p-2} \big[\varepsilon(v) + (p-2)|A_0|^{-2}(A_0 : \varepsilon(v))A_0 \big] : \varepsilon(\varphi) \, dz \\[4mm] \displaystyle = \int_{B_1} \left[(A_0 : \varepsilon(v)) \frac{A_0}{|A_0|^3} - \frac{\varepsilon(v)}{|A_0|} \right] : \varepsilon(\varphi) \, dz. \end{cases}$$

By the above remarks this completes the proof of Lemma 3.1.4 modulo the strong convergences to be shown next.

□

We note that analogous results hold for other choices of the f_δ, for example $f_\delta(Q) := (\delta^2 + |Q|^2)^{q/2}$, $1 < q < p$.

In order to obtain the appropriate version of Lemma 3.1.3 we let

(1.38) $f_{\delta_k}(Q) := \varepsilon_k^{-2} \big(|A_k + \varepsilon_k Q|^p - |A_k|^p - p \varepsilon_k |A_k|^{p-2} A_k : Q \big) + \varepsilon_k^{-2} \tilde{f}_{\delta_k}(Q)$

where $\tilde{f}_{\delta_k}(Q) := \sqrt{\delta_k^2 + |A_k + \varepsilon_k Q|^2} - \sqrt{\delta_k^2 + |A_k|^2} - \varepsilon_k Q : \dfrac{A_k}{\sqrt{\delta_k^2 + |A_k|^2}}$,

notice $\mathrm{Lip}\,(\tilde{f}_{\delta_k}) \le 1$.

We now claim that estimate (1.26) ist still valid where c_1 remains independent of δ_k. To see this we have to establish (1.26) for the function $\tilde{f}_{\delta_k}$.

Let us first consider the case $\varepsilon_k|Q| \le \frac{\sigma}{2}$, where $\sigma = \frac{|A|}{2}$ as in Lemma 3.1.4. Then

$$
(1.39) \quad
\begin{cases}
\varepsilon_k^{-2} \tilde{f}_{\delta_k}(Q) &= \varepsilon_k^{-2} \displaystyle\int_0^1 \frac{\partial^2 \tilde{f}_{\delta_k}}{\partial Q^2}(A_k + s\varepsilon_k Q)(\varepsilon_k Q, \varepsilon_k Q)(1-s)\, ds \\[4mm]
&\le c_2|Q|^2 \displaystyle\int_0^1 \left(\delta_k^2 + |\varepsilon_k s Q + A_k|^2\right)^{-1/2} ds \\[4mm]
&\le c_2|Q|^2 \dfrac{1}{|A_k| - \varepsilon_k|Q|} \\[4mm]
&\le c_3|Q|^2 \dfrac{1}{\sigma} \\[4mm]
&\le c_4(\sigma)|Q|^2.
\end{cases}
$$

Now let $\varepsilon_k|Q| \ge \dfrac{\sigma}{2}$: we have

$$
(1.40) \quad
\begin{cases}
\varepsilon_k^{-2} \tilde{f}_{\delta_k}(Q) &\le \varepsilon_k^{-2}\left(2\,\mathrm{Lip}\,(\tilde{f}_{\delta_k})\varepsilon_k|Q|\right) \\[4mm]
&\le 2\dfrac{|Q|}{\varepsilon_k} \\[4mm]
&\le c_5(\sigma)|Q|^2
\end{cases}
$$

Estimates (1.39) and (1.40) together with (1.27) give (1.26) for a suitable constant c_1.

The integral $I\!I\!I_k$ from the proof of Lemma 3.1.3 now reads

$$
(1.41) \quad
\begin{cases}
I\!I\!I_k := \varepsilon_k^{-2} \displaystyle\int_{B_r} \left(\tilde{f}_{\delta_k}(A_k + \varepsilon_k \varepsilon(u_k)) - \tilde{f}_{\delta_k}(A_k + \varepsilon_k \varepsilon(v)) - \frac{\partial \tilde{f}_{\delta_k}}{\partial Q}(A_k) : \right. \\[4mm]
\left. \varepsilon_k(\varepsilon(u_k) - \varepsilon(v)) \right) dx \\[4mm]
\ge \varepsilon_k^{-2} \displaystyle\int_{B_r} \left(\frac{\partial \tilde{f}_{\delta_k}}{\partial Q}(A_k + \varepsilon_k \varepsilon(v)) - \frac{\partial \tilde{f}_{\delta_k}}{\partial Q}(A_k) \right) : \varepsilon_k(\varepsilon(u_k) - \varepsilon(v))\, dx.
\end{cases}
$$

Since $\varepsilon(v) \in L^\infty(B_r)$ we have $|A_k + \varepsilon_k\varepsilon(v)| \ge \dfrac{\sigma}{2}$ in B_r for k large enough. This gives

$$
(1.42) \quad \varepsilon_k^{-1}\left[\frac{\partial \tilde{f}_{\delta_k}}{\partial Q}(A_k + \varepsilon_k\varepsilon(v)) - \frac{\partial \tilde{f}_{\delta_k}}{\partial Q}(A_k)\right] \xrightarrow[k \to \infty]{} \frac{\partial^2 f_0}{\partial Q^2}(A_0)(\varepsilon(v), \cdot)
$$

uniformly in B_r, where $f_0(Q) = |Q|$.

On the other hand we already know $\varepsilon(u_k) \rightharpoonup \varepsilon(v)$ in $L^2(B_r)$ so that (1.41) and (1.42) imply:

$$\lim_{k \to \infty} \mathrm{I\!I\!I}_k \geq 0.$$

Let us now suppose that $u \in W_p^1(\Omega; \mathbb{R}^n)$ is the minimizer of $I(v) = \int_\Omega |\varepsilon(v)|^p + |\varepsilon(v)|\, dx$ in the class $\mathbb{K}$ defined by Dirichlet–boundary conditions and the requirement div $v \equiv 0$. We set $(\delta > 0)$

$$f_\delta(Q) := (\delta^2 + |Q|^2)^{1/2}, \ Q \in \mathbb{M},$$

which is a smooth convex function $\mathbb{M} \to [0, \infty[$ satisfying the growth condition (1.1) for $q = 1$.

LEMMA 3.1.5 *Let u_δ denote the minimizer of $v \mapsto \int_\Omega |\varepsilon(v)|^p + f_\delta(\varepsilon(v))\, dx$ in the class $\mathbb{K}$. Then $u_\delta \to u$ strongly in $W_p^1(\Omega, \mathbb{R}^n)$ as $\delta \downarrow 0$.*

Proof.   From Korn's inequality and

$$\int_\Omega |\varepsilon(u_\delta)|^p\, dx \ \leq I_\delta(u_\delta) \leq I_\delta(u_0) \leq I_1(u_0)$$

we infer

$$\sup_{\delta > 0} \|u_\delta\|_{W_p^1} < \infty$$

so that $u_\delta \rightharpoonup \tilde{u}$ weakly in W_p^1 as $\delta \searrow 0$ for some element $\tilde{u} \in W_p^1$ at least for a subsequence.

Using

$$I(u_\delta) \leq I_\delta(u_\delta) \leq I_\delta(u)$$

we get

(1.43) $I(\tilde{u}) \leq \liminf_{\delta \searrow 0} I(u_\delta) \leq \liminf_{\delta \searrow 0} I_\delta(u) = I(u).$

Since I is strictly convex, we conclude $\tilde{u} = u$ and we have

$$u_\delta \rightharpoonup u \text{ in } W_p^1 \text{ as } \delta \searrow 0$$

(not only for a subsequence).

We now claim

(1.44) $$\limsup_{\delta \searrow 0} \|\varepsilon(u_\delta)\|_p \le \|\varepsilon(u)\|_p$$

which completes the proof using uniform convexity of L^p:

from (1.43) we have

$$I(u) \le \liminf_{\delta \searrow 0} I(u_\delta) \le \liminf_{\delta \searrow 0} I_\delta(u_\delta) \le \limsup_{\delta \searrow 0} I_\delta(u_\delta) \le \limsup_{\delta \searrow 0} I_\delta(u)$$

that is

(1.45) $$\lim_{\delta \searrow 0} I_\delta(u_\delta) = I(u).$$

Now

(1.46)
$$
\begin{cases}
\displaystyle \int_\Omega |\varepsilon(u_\delta)|^p \, dx = I_\delta(u_\delta) - I(u) + \int_\Omega \left(|\varepsilon(u)|^p + |\varepsilon(u)| \right) dx \\[2mm]
\displaystyle \qquad - \int_\Omega \sqrt{\delta^2 + |\varepsilon(u_\delta)|^2} \, dx \\[2mm]
\displaystyle \le I_\delta(u_\delta) - I(u) + \int_\Omega |\varepsilon(u)|^p \, dx + \int_\Omega |\varepsilon(u)| \, dx - \int_\Omega |\varepsilon(u_\delta)| \, dx.
\end{cases}
$$

Notice (by lower semicontinuity)

(1.47) $$\limsup_{\delta \searrow 0} \left\{ \int_\Omega |\varepsilon(u)| dx - \int_\Omega |\varepsilon(u_\delta)| dx \right\} \le 0$$

and (1.44) is a consequence of (1.45), (1.46) and (1.47).

$$\square$$

We define $\Omega_+^* := \Big\{ x \in \Omega \colon x$ is a Lebesgue point for $\varepsilon(u)$,

$$|\varepsilon(u)(x)| > 0 \text{ and } \lim_{r \downarrow 0} E\big(u, B_r(x)\big) = 0 \Big\}.$$

Choose $x_0 \in \Omega_+^*$ and let $A := \varepsilon(u)(x_0) \ne 0$, $\sigma := \frac{1}{2}|A|$. We fix a radius R such that

(1.48) $|(\varepsilon(u))_{x_0,R} - A| < \dfrac{\sigma}{2}, \ E(u, B_R(x_0)) < \tilde{\varepsilon}(t)$

with $\tilde{\varepsilon}(t) < \varepsilon^2(t)$ $(\varepsilon(t)$ defined in Lemma 3.1.4$)$ to be specified later. Here t is chosen such that $c_0 t^2 \leq \frac{1}{2}$. For δ small enough (recall Lemma 3.1.5) we deduce from (1.48)

(1.49)$_\delta$ $|(\varepsilon(u_\delta))_{x_0,R} - A| < \frac{\sigma}{2}, \ E(u_\delta, B_R(x_0)) < \tilde{\varepsilon}(t);$

hence (via Lemma 3.1.4)

(1.50)$_\delta$ $E(u_\delta, B_{tR}(x_0)) \leq \frac{1}{2} E(u_\delta, B_R(x_0)).$

Combining (1.49)$_\delta$, (1.50)$_\delta$ we get $E(u_\delta, B_{tR}(x_0)) < \frac{1}{2}\tilde{\varepsilon}(t)$ and

$$|(\varepsilon(u_\delta))_{x_0,tR} - A| \leq \frac{\sigma}{2} + t^{-n} E(u_\delta, B_R(x_0))^{1/2} < \sigma$$

provided $t^{-n}\tilde{\varepsilon}(t)^{1/2} < \frac{\sigma}{2}$. In this case Lemma 3.1.4 gives

$$E(u_\delta, B_{t^2 R}(x_0)) \leq \frac{1}{4} E(u_\delta, B_R(x_0)).$$

In order to iterate this argument we observe

$$|(\varepsilon(u_\delta))_{t^{k+1}R} - A| \leq |(\varepsilon(u_\delta))_{x_0,R} - A| + \sum_{l=0}^{k} t^{-n} E(u_\delta, B_{t^l R}(x_0))^{1/2}.$$

Then it is easy to check that for $\tilde{\varepsilon}(t)$ satisfying

$$t^{-n} \sum_{l=0}^{\infty} 2^{-l/2} \sqrt{\tilde{\varepsilon}(t)} < \frac{\sigma}{2}$$

we end up with

$$E(u_\delta, B_{t^k R}(x_0)) \leq 2^{-k} E(u_\delta, B_R(x_0))$$

being valid for all k and small enough δ. Now from standard arguments it follows that there exists $\gamma \in (0,1)$ such that

(1.51) $\varepsilon(u_\delta) \in C^{0,\gamma}(B_{R/2}(x_0))$

uniformly with respect to δ. (Obviously a slight refinement of the above calculations gives (1.51) with γ replaced by an arbitrary exponent α.) From Lemma 3.1.5 we immediately deduce $\varepsilon(u) \in C^{0,\gamma}\big(B_{R/2}(x_0)\big)$ which proves the first part of the Main Theorem. To conclude the proof we fix a compact subregion G of Ω_+^* and observe the equation

$$(1.52) \qquad \int_G \lambda \varepsilon(u) : \varepsilon(\varphi)\, dx = \int_G -F : \varepsilon(\varphi)\, dx$$

for all $\varphi \in C_0^1(G; \mathbb{R}^n)$, div $\varphi = 0$. Here we have set

$$\lambda = p|\varepsilon(u)|^{p-2}, \quad F = \frac{\varepsilon(u)}{|\varepsilon(u)|}.$$

Suppose $B_R(0) \subset G$ and let w denote the minimizer of $\int_{B_R(0)} |\varepsilon(w)|^2\, dx$ for boundary values u and subject to the constraint div $w = 0$.

Inserting $\varphi = u - w$ into (1.52) and observing that

$$|\lambda(x) - \lambda(0)| \le cR^\alpha, \quad |F(x) - F(0)| \le cR^\alpha$$

for $x \in B_R(0)$ we deduce $\big($note $\varepsilon(u) \in L^\infty(G)\big)$

$$\int_{B_R(0)} |\varepsilon(u) - \varepsilon(w)|^2\, dx \le c\, R^{n+2\alpha};$$

on account of Korn's inequality this yields

$$\int_{B_R(0)} |\nabla u - \nabla w|^2\, dx \le c\, R^{n+2\alpha}.$$

Finally we make use of the Campanato estimate for w to get for $r < R$

$$\int_{B_r(0)} |\nabla u - (\nabla u)_r|^2\, dx \le c\left\{ \left(\frac{r}{R}\right)^{n+2} \int_{B_R(0)} |\nabla u - (\nabla u)_R|^2\, dx + R^{n+2\alpha} \right\};$$

hence $\nabla u \in C^{0,\alpha}(G)$, and the proof of the Main Theorem is complete.

$\square$

In contrast to the Bingham fluid model we obtain partial regularity up to a closed set of vanishing Lebesgue measure in Ω for the Powell–Eyring model and certain types of power law fluids.

Let us first discuss the power model

$$(1.53) \quad W(E) = \mu_\infty |E|^2 + \mu_0 \left\{ \begin{array}{ll} (\delta + |E|^2)^{p/2}, & \delta > 0,\ p \geq 1 \\ \\ \text{or} \\ \\ |E|^p, & p > 1 \end{array} \right\}$$

with constants $\mu_\infty,\ \mu_0 \geq 0$, $\mu_\infty + \mu_0 > 0$. We define $m > 1$ in such a way (depending on the various choices for μ_∞, μ_0 and p) that

$$(1.54) \quad J(u) = \int_\Omega \big(W(\varepsilon(u)) - g \cdot u\big)\, dx \to \min$$

is uniquely solvable in $\mathbb{K} = \{v \in W^1_m(\Omega; \mathbb{R}^n) : \operatorname{div} v = 0,\ v = v_0 \text{ on } \partial\Omega\}$. For technical simplicity we just let $g = 0$, the next results (with obvious restrictions on the exponent α) remain valid for volume forces in some Morrey space.

THEOREM 3.1.3 *Let $u \in \mathbb{K}$ denote the solution of (1.54) with W defined in (1.53). Suppose further that $p \geq 2$.*

   a) *If $\mu_\infty > 0$, then there exists an open subset Ω_0 of Ω such that $|\Omega - \Omega_0| = 0$ and $u \in C^{1,\alpha}(\Omega_0; \mathbb{R}^n)$ for any $\alpha < 1$.*

   b) *In case $\mu_\infty = 0$ we have:*

   1) *Let $W(E) = \mu_0(\delta + |E|^2)^{p/2}$. Then the result of a) holds.*

   2) *Let $W(E) = \mu_0|E|^p$. Then there is an open subset Ω_+^* of $\Omega_+ = \{x \in \Omega : x \text{ is a Lebesgue point of } \varepsilon(u) \text{ and } \varepsilon(u)(x) \neq 0\}$ such that $|\Omega_+ - \Omega_+^*| = 0$ and $u \in C^{1,\alpha}(\Omega_+^*; \mathbb{R}^n)$.*

THEOREM 3.1.4 *Let $u \in \mathbb{K}$ denote the solution of (1.54) with W from (1.53), let $\mu_\infty > 0$ and $1 \leq p < 2$.*

   a) *Suppose that $W(E) = \mu_\infty|E|^2 + \mu_0(\delta + |E|^2)^{p/2}$. Then we have the partial regularity result from Theorem 3.1.3 a).*

   b) *If $W(E) = \mu_\infty|E|^2 + \mu_0|E|^p$, then the statement from Theorem 3.1.3 b) 2) holds.*

REMARK 3.1.1 We conjecture that for any dissipative potential

$$W(E) = \mu_\infty |E|^2 + \mu_0 \left\{ \begin{array}{ll} (\delta + |E|^2)^{p/2}, & \delta > 0, \ p \geq 1 \\ & \text{or} \\ |E|^p, & p > 1 \end{array} \right\}$$

(local) minimizers are of class C^1 without interior singular points.

REMARK 3.1.2 In order to get better regularity results it will be necessary to study the smoothness properties of local minimizers of $\int_\Omega |\varepsilon(v)|^m dx$ for growth rates $m > 1$ and with respect to the constraint div $v = 0$.

Next we consider the variational problem (1.54) for the dissipative potential

$$W(E) = \mu_\infty |E|^2 + \mu_0 \int_0^{|E|} \text{ar sinh } t \, dt$$

assuming that $g \in L^{2,n-2+2\lambda}(\Omega; \mathbb{R}^n)$ for some $0 < \lambda < 1$.

THEOREM 3.1.5 *With notation introduced above let* $u \in \{v \in W_2^1(\Omega; \mathbb{R}^n) : \text{div } v = 0, \ v = v_0 \text{ on } \partial\Omega\}$ *denote the unique J–minimizer. Then there is an open set* Ω_0 *such that* $|\Omega - \Omega_0| = 0$ *and* $u \in C^{1,\alpha}(\Omega_0; \mathbb{R}^n), \ \alpha < \lambda$.

Proof of Theorem 3.1.3, 3.1.4, 3.1.5: Most of the results follow from the proof of Theorem 3.1.2. In the situation of Theorem 3.1.5 we apply Theorem 3.1.2 with $p = 2$ and $f(E) = \mu_0 \int_0^{|E|} \text{ar sinh } t \, dt$. It is then easy to check that the blow–up lemma 3.1.1 is true without assuming that the matrix A is different from zero. We therefore deduce Hölder continuity of $\varepsilon(u)$ near a point $x_0 \in \Omega$ if and only if x_0 is a Lebesgue point for $\varepsilon(u)$ and $\lim_{r \to 0} \fint_{B_r(x_0)} |\varepsilon(u) - (\varepsilon(u))_{x_0,r}|^2 dx = 0$. Exactly the same argument (we now let $f(E) = \mu_0(\delta + |E|^2)^{p/2}$ in Theorem 3.1.2) proves Theorem 3.1.4 a). For obtaining the results of Theorem 3.1.3 a) and b) 1) we again go through the proof of Theorem 3.1.2 taking care of the fact that now the dissipative potentials W under consideration are nondegenerate, hence the requirement $A \neq 0$ of Lemma 3.1.1 is superfluous, which proves the results. The statement of Theorem 3.1.3 b) 2) is obvious. In the situation of Theorem 3.1.4 b) we would like to apply Theorem 3.1.2 again with the choice $f(E) = |E|^p$. Since f is not of class C^2 we have to consider the approximation $(\delta + |E|^2)^{p/2}, \ \delta > 0$, and to prove uniform partial regularity with respect to δ. But this is just another version of the Main Theorem 3.1.1 which does not require additional arguments.

□

REMARK 3.1.3 In section 3 we will show how to improve the above results in the twodimensional case.

REMARK 3.1.4 It should be clear without further comments that all theorems of this section extend to local minimizers.

3.2   Local boundedness of the strain velocity

In this section we discuss dissipative potentials of the type
$W(E) = \frac{1}{2}|E|^2 + S(E)$ with $S : \mathbb{M} \to \mathbb{R}$ defined according to

$$(2.1) \qquad S(E) = \begin{cases} |E| & \text{or} \\ \int_0^{|E|} \operatorname{ar\,sinh} t \, dt & \text{or} \\ |E|^p, \ 1 < p < 2, & \text{or} \\ (\delta + |E|^2)^{p/2}, \ 1 \le p < 2, \ \ \delta > 0, \end{cases} , E \in \mathbb{M},$$

in particular, the next results apply to the classical Bingham fluid, the Powell–Eyring model and certain power laws. The reader should note that the case $W(E) = \frac{1}{m}|E|^m + S(E)$, $m > 2$, is still open.

In order to formulate our results not only for the specific fluid models mentioned above we assume

$$(2.2) \qquad W(E) = \frac{1}{2}|E|^2 + f_0(|E|)$$

with $f_0 : [0, \infty[\to \mathbb{R}$ convex and nondecreasing. For $v_0 \in W_2^1(\Omega; \mathbb{R}^n)$ with div $v_0 = 0$ we let as usual $\mathbb{K} = \{u \in W_2^1(\Omega; \mathbb{R}^n) : \operatorname{div} u = 0, \ u = v_0 \text{ on } \partial\Omega\}$ and consider the variational integral

$$J : \mathbb{K} \to \mathbb{R}, \ J(u) = \int_\Omega W(\varepsilon(u)) \, dx;$$

the case of non–zero volume forces requires some minor modifications.

THEOREM 3.2.1 *With W defined according to (2.2) let $u \in \mathbb{K}$ denote the unique J–minimizer in $\mathbb{K}$. Assume further that*

$$(2.3) \qquad \lim_{t \to \infty} t^{-2} f_0(t) = 0.$$

Then u is Hölder continuous on Ω with any exponent < 1. Moreover, the classical derivative $\nabla u(x)$ exists for Lebesgue almost all x in Ω.

REMARK 3.2.1 Condition (2.3) holds for $f_0(|E|) = S(E)$ with S defined according to (2.1).

REMARK 3.2.2 We do not claim that the set of points x for which $\nabla u(x)$ exists is open.

REMARK 3.2.3 The statement of Theorem 3.2.1 is true for any continuous function f_0 satisfying (2.3) and for any local J–minimizer in $\mathbb{K}$. But in this more general setting the existence of minimizers is not guaranteed, we may then look at the relaxed variational problem and according to [Fu1], [FS1] the statement of Theorem 3.2.1 extends to minimizers of the relaxed problem.

Of course (2.3) excludes quadratic growth at infinity but it is possible to replace (2.3) by the following assumptions:

$$(2.4) \qquad f_0 \text{ is of class } C^1 \text{ on some interval } [a, \infty[, \ a > 0;$$

$$(2.5) \qquad \ell = \lim_{t \to \infty} \frac{1}{t} f_0'(t) \text{ exists in } [0, \infty[.$$

THEOREM 3.2.2 *Define W according to (2.2) and let (2.4), (2.5) hold for f_0. Then the conclusion of Theorem 2.1 is still true.*

REMARK 3.2.4 Conditions (2.4), (2.5) are easily checked for $f_0(|E|) = S(E)$ with S from (2.1).

Next we state the main result of this section.

THEOREM 3.2.3 *Let $u \in \mathbb{K}$ denote the unique J–minimizer for the dissipative potential W from (2.2). Let (2.4) hold and assume further:*

$$(2.6) \qquad \sup_{t \geq a} \frac{1}{t} f_0'(t) < \infty$$

(2.7) *there exists $L \geq a$ such that* $\displaystyle\sup_{s \geq L} \frac{s}{|s - t|} \left| \frac{f_0'(s)}{s} - \frac{f_0'(t)}{t} \right| \to 0$ *as $t \to \infty$.*

Then $\varepsilon(u)$ is a locally bounded function.

REMARK 3.2.5 From $\varepsilon(u) \in L^\infty_{\text{loc}}(\Omega; \mathbb{M})$ we deduce with Korn's inequality that, for any $p < \infty$, we have $u \in W^1_{p,\text{loc}}(\Omega; \mathbb{R}^n)$. This provides an alternative proof of the Hölder results from Theorem 3.2.1 and 3.2.2.

REMARK 3.2.6 In the papers [Fu3] and [FS1] we proved that the nonlinearities S occuring in (2.1) satisfy all the assumptions from Theorem 3.2.3, in particular, we get local boundedness of the strain velocity for the classical Bingham fluid.

REMARK 3.2.7 The statement of Theorem 3.2.3 extends to the following setting: with W as in the theorem we let $W' = W + \vartheta$ where $\vartheta : \mathbb{M} \to \mathbb{R}$ is just continuous with compact support. It was then shown in [FS1] that the quasiconvex envelope QW' of W' equals W outside a large ball in $\mathbb{M}$, moreover, we have $\varepsilon(u) \in L^{\infty}_{\text{loc}}(\Omega; \mathbb{M})$, where $u \in \mathbb{K}$ now denotes a minimizer of the relaxed problem $\int_{\Omega} QW'(\varepsilon(u))\, dx \to \min$ in $\mathbb{K}$.

We now will give the proofs of the various regularity results.

For Theorem 3.2.1 we follow the paper [Fu1] replacing the exponent p occuring there by 2 and taking $\varepsilon(u)$ in place of ∇u. Then Hölder continuity of u is proved as in [Fu1], Theorem 1. With the same changes differentiability almost everywhere of minimizers is a consequence of [Fu1], Theorem 3. We leave the details to the reader.

Proof of Theorem 3.2.2: We just show differentiability almost everywhere. We may assume that f_0 is of class C^1 on $[0, \infty[$ satisfying in addition $0 \leq \frac{1}{t} f_0'(t) \leq A_1$ on $[0, \infty[$ for some $A_1 \geq 0$: if this is not the case then we may either proceed as in the proof of Theorem 3.2.3 by splitting the integrals involving f_0' or we may argue as in [FS1] by writing $W = \frac{1}{2}| \cdot |^2 + \tilde{f}_0(| \cdot |) + \vartheta$ where $\tilde{f}_0 : [0, \infty[\to \mathbb{R}$ is a convex, nondecreasing C^1 function equal to f_0 on $[a, \infty[$ and ϑ denotes a function with compact support. Exactly this situation was studied in [FS1], Theorem 1.3. The main step towards the desired claim is

LEMMA 3.2.1 *Let* $G = \{x \in \Omega : \sup\limits_{r > 0} \fint_{B_r(x)} |\nabla u|^2 dz < \infty\}$. *Then, for* $x \in G$, *there exist positive constants* K_x *and* $R_x \leq \frac{1}{2}$ *dist* $(x, \partial\Omega)$ *with the following property: if, for some* $R \leq R_x$, *we have* $\text{osc}_{B_R(x)} u \geq K_x R$, *then* $\text{osc}_{B_{R/2}(x)} u \leq \frac{1}{2} \text{osc}_{B_R(x)} u$.

Having proved Lemma 3.2.1 we may argue as in [Fu1] to get the result of Theorem 3.2.2.

Proof of Lemma 3.2.1: Assuming that the statement is wrong we find a point $x \in G$ (w.l.o.g. $x = 0$) and sequences $K_k \to \infty$, $R_k \to 0$ such that

$$(2.8) \qquad \omega_k := \text{osc}_{B_{R_k}} u \geq K_k R_k, \quad \text{osc}_{B_{R_k}/2} u \geq \frac{1}{2}\omega_k.$$

Let us define $u_k(z) = \frac{1}{\omega_k}\big(u(R_k z) - (u)_{R_k}\big)$, $z \in B_1$, $(u)_{R_k} = \fint_{B_{R_k}} u\, dx$. From (2.8) we deduce

$$(2.9) \qquad \text{osc}_{B_1} u_k = 1, \quad \text{osc}_{B_{1/2}} u_k > \frac{1}{2}, \quad (u_k)_1 = 0.$$

Clearly div $u_k = 0$ and u_k locally minimizes the energy

$$J_k(v) = \int\limits_{B_1} \frac{1}{2}|\varepsilon(v)|^2 + f_k(\varepsilon(v))\, dx$$

with respect to the constraint div $v = 0$. Here we have abbreviated

$$f_k(E) = R_k^2 \omega_k^{-2} f_0\left(\frac{\omega_k}{R_k}|E|\right).$$

From our modified assumptions concerning f_0 we deduce $f_0(t) \leq \frac{A_1}{2}t^2 + f_0(0)$. Using $R_k\omega_k^{-1} \to 0$ as $k \to \infty$ we see that the integrand of the energy J_k satisfies the hypotheses of [Gi], chapter V, Theorem 3.1, hence

$$(2.10) \qquad \int\limits_{B_{R/2}(x)} |\varepsilon(u_k)|^2 dz \leq c_1 \left[R^{-2} \int\limits_{B_R(x)} |u_k - (u_k)_R|^2 dz + R^n \right]$$

for arbitrary balls $B_R(x) \subset B_1$. Here and in what follows $c_1, c_2 \ldots$ denote various positive constants independent of k. (Remark: in order to obtain the Caccioppoli inequality (2.10) for the symmetric derivative and also under the constraint div $u = 0$ one has to modify Giaquinta's argument along the lines of [GM], see also proof of Lemma 3.0.5i)).

Combining (2.9) and (2.10) we get boundedness of the sequence $\{\varepsilon(u_k)\}$ in the space $L^2_{\mathrm{loc}}(B_1; \mathbb{M})$ and by Korn's inequality $\{u_k\}$ is bounded in $W^1_{2,\mathrm{loc}}(B_1; \mathbb{R}^n)$. After passing to a subsequence we may therefore assume

$$(2.11) \qquad \begin{cases} u_k \rightharpoonup \tilde{u} & \text{weakly in } W^1_{2,\mathrm{loc}}(B_1; \mathbb{R}^n), \\[2mm] u_k \to \tilde{u} & \text{strongly in } L^2_{\mathrm{loc}}(B_1; \mathbb{R}^n) \text{ and a.e.} \\[2mm] u_k \rightharpoonup \tilde{u} & \text{weakly in } L^2_{\mathrm{loc}}(B_1; \mathbb{R}^n) \end{cases}$$

for some function $\tilde{u} \in W^1_{2,\mathrm{loc}}(B_1; \mathbb{R}^n)$. In a next step we prove a uniform local Hölder condition for the functions u_k. To this purpose fix a ball $B_R(x) \subset B_1(0)$ and let v_k denote the solution of the Stokes problem

$$(2.12) \qquad \begin{cases} \int\limits_{B_R(x)} \varepsilon(v_k) : \varepsilon(\varphi)\, dz = 0 \text{ for } \varphi \in \overset{\circ}{W}{}^1_2(B_R(x); \mathbb{R}^n),\ \text{div } \varphi = 0, \\[2mm] \text{div } v_k = 0 \quad \text{a.e.}, \\[2mm] v_k = u_k \quad \text{on } \partial B_R(x). \end{cases}$$

For v_k we have the estimate (see Lemma 3.0.5 iii))

$$(2.13) \qquad \int\limits_{B_r(x)} |\nabla v_k|^2 \, dz \le c_2 \left(\frac{r}{R}\right)^n \int\limits_{B_R(x)} |\nabla v_k|^2 \, dz, \ 0 < r < R,$$

with c_2 only depending on the dimension n. (2.13) clearly implies

$$(2.14) \qquad \int\limits_{B_r(x)} |\nabla u_k|^2 \, dz \le c_3 \left\{ \left(\frac{r}{R}\right)^n \int\limits_{B_R(x)} |\nabla u_k|^2 \, dz + \int\limits_{B_R(x)} |\nabla u_k - \nabla v_k|^2 \, dz \right\}.$$

It therefore remains to discuss $\int\limits_{B_R(x)} |\nabla u_k - \nabla v_k|^2 \, dz$ or equivalently (observe Korn's inequality) the term $\int\limits_{B_R(x)} |\varepsilon(u_k) - \varepsilon(v_k)|^2 \, dz$. From the minimum property of u_k we deduce

$$(2.15) \qquad \begin{cases} \int\limits_{B_R(x)} \varepsilon(u_k) : \varepsilon(\varphi) + D f_k(\varepsilon(u_k)) : \varepsilon(\varphi) \, dz = 0 \text{ for all} \\[2mm] \varphi \in \overset{\circ}{W}{}^1_2(B_R(x); \mathbb{R}^n), \ \mathrm{div}\, \varphi = 0, \end{cases}$$

with $D f_k(\varepsilon(u_k)) = R_k \omega_k^{-1} f_0'(\omega_k R_k^{-1} |\varepsilon(u_k)|) |\varepsilon(u_k)|^{-1} \varepsilon(u_k)$. Let $l := \lim\limits_{y \to \infty} \frac{1}{y} f_0'(y)$ and $s_k := \omega_k R_k^{-1} |\varepsilon(u_k)|$. Then

$$\int\limits_{B_R(x)} (1 + l) |\varepsilon(u_k) - \varepsilon(v_k)|^2 \, dz$$

$$\overset{(2.12)}{=} \int\limits_{B_R(x)} (1 + l) \, \varepsilon(u_k) : (\varepsilon(u_k) - \varepsilon(v_k)) \, dz$$

$$= \int\limits_{B_R(x)} \left(1 + \frac{1}{s_k} f_0'(s_k)\right) \varepsilon(u_k) : (\varepsilon(u_k) - \varepsilon(v_k)) \, dz$$

$$+ \int\limits_{B_R(x)} \left(l - \frac{1}{s_k} f_0'(s_k)\right) \varepsilon(u_k) : (\varepsilon(u_k) - \varepsilon(v_k)) \, dz$$

$$\overset{(2.15)}{\le} \int\limits_{B_R(x)} |l - \frac{1}{s_k} f_0'(s_k)| \, |\varepsilon(u_k) - \varepsilon(v_k)| \, |\varepsilon(u_k)| \, dz,$$

hence

$$(2.16) \quad \int_{B_R(x)} |\varepsilon(u_k) - \varepsilon(v_k)|^2 \, dz \leq c_4 \int_{B_R(x)} |l - \frac{1}{s_k} f_0'(s_k)| \, |\varepsilon(u_k)| \, |\varepsilon(u_k) - \varepsilon(v_k)| \, dz.$$

Given a large positive number κ we select $y_0 > 0$ such that

$$(2.17) \quad |l - \frac{1}{y} f_0'(y)| \leq 1/\kappa \quad \text{for all } y \geq y_0.$$

Therefore, if y is some number $\geq \kappa$, we can find $k_0 = k_0(\kappa)$ such that

$$(2.17)' \quad |l - R_k \omega_k^{-1} y^{-1} f_0'(\omega_k R_k^{-1} y)| \leq 1/\kappa$$

holds for all $k \geq k_0$. Note that k_0 does not depend on y.
Let us introduce the sets $\Omega_k := \{z \in B_R(x) : |\varepsilon(u_k)(z)| \leq \kappa\}$ and $\Omega_k' = B_R(x) - \Omega_k$. Then

$$\int_{\Omega_k} |l - \frac{1}{s_k} f_0'(s_k)| \, |\varepsilon(u_k)| \, |\varepsilon(u_k) - \varepsilon(v_k)| \, dz$$

$$\leq 2 A_1 \kappa \int_{B_R(x)} |\varepsilon(u_k) - \varepsilon(v_k)| \, dz$$

and

$$\int_{\Omega_k'} |l - \frac{1}{s_k} f_0'(s_k)| \, |\varepsilon(u_k)| \, |\varepsilon(u_k) - \varepsilon(v_k)| \, dz \leq (2.17)'$$

$$\leq \frac{1}{\kappa} \int_{B_R(x)} |\varepsilon(u_k)| \, |\varepsilon(u_k) - \varepsilon(v_k)| \, dz.$$

Applying Young's inequality we derive from (2.16)

$$(2.18) \quad \int_{B_R(x)} |\varepsilon(u_k) - \varepsilon(v_k)|^2 \, dz \leq c_5 \{ R^n (1 + \kappa) + \frac{1}{\kappa} \int_{B_R(x)} |\varepsilon(u_k)|^2 \, dz \}.$$

Inserting (2.18) into (2.14) we end up with

$$(2.19) \quad \int_{B_r(x)} |\nabla u_k|^2 \, dz \leq c_6 \{ [(\frac{r}{R})^n + \frac{1}{\kappa}] \int_{B_R(x)} |\nabla u_k|^2 \, dz + (1 + \kappa) R^n \}$$

being valid for all $r < R$ and all $k \geq k_0(\kappa)$. Going through the proof of Lemma 2.1 in [Gi], chapter III, we deduce from (2.19) the growth estimate (κ sufficiently large, $k \geq k_0(\kappa)$)

$$\int_{B_r(x)} |\nabla u_k|^2 \, dz \leq c_7 \left(\frac{r}{R}\right)^{n-1} \int_{B_R(x)} |\nabla u_k|^2 \, dz$$

which implies $u_k \in C^{0,1/2}(B_1; \mathbb{R}^n)$ with local Hölder constant independent of k, especially $u_k \to \tilde{u}$ locally uniformly on B_1 which gives (recall (2.9)) $\mathrm{osc}_{B_{1/2}} \tilde{u} \geq 1/2$. On the other hand we have for $t < 1$

$$\int_{B_t} |\nabla u_k|^2 \, dz = R_k^{2-n} \omega_k^{-2} \int_{B_{tR_k}} |\nabla u|^2 \, dx \leq R_k^2 \omega_k^{-2} \fint_{B_{R_k}} |\nabla u|^2 \, dx \to 0$$

as $k \to \infty$ so that $\nabla \tilde{u} = 0$ which is the desired contradiction. $\qquad\square$

Proof of Theorem 3.2.3: We first derive a variational inequality for the minimizer u. Consider $\varphi \in C_0^1(\Omega; \mathbb{R}^n)$, $\mathrm{div}\, \varphi = 0$, and $\delta > 0$. For $t > 0$ we have

$$\int_\Omega W(\varepsilon(u)) \, dx \leq \int_\Omega W(\varepsilon(u) + t\varepsilon(\varphi)) \, dx$$

which implies $([|\varepsilon(u)| > a + \delta] = \{x \in \Omega : |\varepsilon(u)(x)| > a + \delta\},$ etc. $)$

$$\frac{1}{t} \int_\Omega \frac{1}{2} |\varepsilon(u) + t\varepsilon(\varphi)|^2 - \frac{1}{2} |\varepsilon(u)|^2 \, dx +$$

$$+ \frac{1}{t} \int_{[|\varepsilon(u)| > a + \delta]} f_0(|\varepsilon(u) + t\varepsilon(\varphi)|) - f_0(|\varepsilon(u)|) \, dx$$

$$\geq \frac{1}{t} \int_{[|\varepsilon(u)| \leq a + \delta]} f_0(|\varepsilon(u)|) - f_0(|\varepsilon(u) + t\varepsilon(\varphi)|) \, dx.$$

Let us further assume that $t < \delta / \|\nabla \varphi\|_\infty$. Then, on the set $[|\varepsilon(u)| > a + \delta]$, we have

$$\frac{1}{t} \big(f_0(|\varepsilon(u) + t\varepsilon(\varphi)|) - f_0(|\varepsilon(u)|) \big) \longrightarrow$$

$$f_0'(|\varepsilon(u)|) \frac{1}{|\varepsilon(u)|} \varepsilon(u) : \varepsilon(\varphi) \qquad \text{as } t \downarrow 0,$$

on $[|\varepsilon(u)| \le a + \delta]$ we observe (in case $\delta < a/2$)

$$\frac{1}{t}\left(f_0(|\varepsilon(u)|) - f_0(|\varepsilon(u) + t\varepsilon(\varphi)|)\right) \ge -M|\varepsilon(\varphi)|,$$

where M denotes the Lipschitz constant for f_0 on $[0, 2a]$. After passing to the limit $t \downarrow 0$ we thus get

$$\int_{\Omega} \varepsilon(u) : \varepsilon(\varphi)\, dx + \int_{[|\varepsilon(u)|>a+\delta]} f_0'(|\varepsilon(u)|)\frac{1}{|\varepsilon(u)|}\varepsilon(u) : \varepsilon(\varphi)\, dx$$

$$\ge \int_{[|\varepsilon(u)|\le a+\delta]} (-M)|\varepsilon(\varphi)|\, dx.$$

Next we pass to the limit $\delta \downarrow 0$, replace φ by $-\varphi$ and use an approximation argument to derive

$$(2.20) \quad \begin{cases} \displaystyle\int_{\Omega} \varepsilon(u) : \varepsilon(\varphi)\, dx + \int_{[|\varepsilon(u)|>a]} f_0'(|\varepsilon(u)|)\frac{1}{|\varepsilon(u)|}\varepsilon(u) : \varepsilon(\varphi)\, dx \\[4mm] \displaystyle \le M \int_{[|\varepsilon(u)|\le a]} |\varepsilon(\varphi)|dx, \quad \varphi \in \overset{\circ}{W}{}_2^1(\Omega; \mathbb{R}^n), \ \operatorname{div} \varphi = 0. \end{cases}$$

We fix a ball $B_R(x_0)$ and denote by v the unique solution of the Stokes problem

$$(2.21) \quad \begin{cases} \displaystyle\int_{B_R(x_0)} \varepsilon(v) : \varepsilon(\varphi)\, dx = 0 \text{ for all } \varphi \in \overset{\circ}{W}{}_2^1(B_R(x_0); \mathbb{R}^n), \ \operatorname{div} \varphi = 0, \\[2mm] v = u \text{ on } \partial B_R(x_0), \ \operatorname{div} v = 0 \text{ a.e. on } B_R(x_0). \end{cases}$$

According to Lemma 3.0.5 v satisfies

$$(2.22) \quad \int_{B_r(x_0)} |\varepsilon(v) - (\varepsilon(v))_{x_0,r}|^2 dx \le c_1 \left(\frac{r}{R}\right)^{n+2} \int_{B_R(x_0)} |\varepsilon(v) - (\varepsilon(v))_{x_0,R}|^2 dx,$$

$0 \le r \le R$, for some constant $c_1 = c_1(n)$. From (2.22) we deduce

$$(2.23) \quad \begin{cases} \displaystyle\int_{B_r(x_0)} |\varepsilon(u) - (\varepsilon(u))_{x_0,r}|^2 dx \ \le c_2 \Big\{ \int_{B_R(x_0)} |\varepsilon(u) - \varepsilon(v)|^2 dx \\[4mm] \displaystyle \hspace{3cm} + \left(\frac{r}{R}\right)^{n+2} \int_{B_R(x_0)} |\varepsilon(u) - (\varepsilon(u))_{x_0,R}|^2 dx \Big\}. \end{cases}$$

In order to estimate the first integral on the right-hand side of (2.23) we introduce the quantities $s = |\varepsilon(u)|$, $t = \left(\fint_{B_R(x_0)} |\varepsilon(u)|^2 dx \right)^{1/2}$; we further observe that (2.20) remains valid with a replaced by the number L defined in (2.7), provided M now denotes the Lipschitz constant for f_0 on $[0, 2L]$. Let us also assume that

$$(2.24) \qquad t > 2L.$$

From (2.20) and (2.21) we then deduce by letting $\varphi = u - v$:

$$\left(1 + \frac{1}{t} f_0'(t)\right) \int_{B_R(x_0)} |\varepsilon(u) - \varepsilon(v)|^2 dx$$

$$= \left(1 + \frac{1}{t} f_0'(t)\right) \int_{B_R(x_0)} \varepsilon(u) : (\varepsilon(u) - \varepsilon(v))\, dx$$

$$= \int_{B_R(x_0)} \varepsilon(u) : (\varepsilon(u) - \varepsilon(v))\, dx$$

$$+ \int_{B_R(x_0) \cap [s > L]} f_0'(s)\frac{1}{s}\varepsilon(u) : (\varepsilon(u) - \varepsilon(v))\, dx$$

$$+ \int_{B_R(x_0)} \frac{1}{t} f_0'(t)\varepsilon(u) : (\varepsilon(u) - \varepsilon(v))\, dx$$

$$- \int_{B_R(x_0) \cap [s > L]} f_0'(s)\frac{1}{s}\varepsilon(u) : (\varepsilon(u) - \varepsilon(v))\, dx$$

$$\leq M \int_{B_R(x_0) \cap [s \leq L]} |\varepsilon(u) - \varepsilon(v))|\, dx$$

$$+ \int_{B_R(x_0) \cap [s > L]} \left(\frac{1}{t} f_0'(t) - \frac{1}{s} f_0'(s)\right)\varepsilon(u) : (\varepsilon(u) - \varepsilon(v))\, dx$$

$$+ \int_{B_R(x_0) \cap [s \leq L]} \frac{1}{t} f_0'(t)\varepsilon(u) : (\varepsilon(u) - \varepsilon(v))\, dx \leq (2.6)$$

$$\leq (M + L \sup_{y \geq 2L} \{\frac{1}{y} f_0'(y)\}) \int_{B_R(x_0) \cap [s \leq L]} |\varepsilon(u) - \varepsilon(v)| \, dx$$

$$+ \int_{B_R(x_0) \cap [s > L]} s |\frac{1}{t} f_0'(t) - \frac{1}{s} f_0'(s)| \, |\varepsilon(u) - \varepsilon(v)| \, dx$$

$$\leq c_3 \int_{B_R(x_0) \cap [s \leq L]} |\varepsilon(u) - \varepsilon(v)| \, dx + \alpha(t) \int_{B_R(x_0) \cap [s > L]} |s - t| \, |\varepsilon(u) - \varepsilon(v)| \, dx$$

where $\alpha(t) = \sup_{s \geq L} \left\{ \frac{s}{|s-t|} |\frac{f_0'(s)}{s} - \frac{f_0'(t)}{t}| \right\}$.

With Young's inequality we get

$$\int_{B_R(x_0)} |\varepsilon(u) - \varepsilon(v)|^2 \, dx \leq c_4 \{|B_R(x_0) \cap [s \leq L]|$$

$$+ \alpha(t)^2 \int_{B_R(x_0)} |s - t|^2 \, dx\}$$

$$\leq c_5 \{|B_R(x_0) \cap [s \leq L]| + \alpha(t)^2 \int_{B_R(x_0)} |\varepsilon(u) - (\varepsilon(u))_{x_0,R}|^2 \, dx\}.$$

For the measure of $B_R(x_0) \cap [s \leq L]$ we observe (recall (2.24))

$$|B_R(x_0) \cap [s \leq L]| \leq \frac{2}{t} \int_{B_R(x_0) \cap [s \leq L]} |s - t| \, dx$$

$$\leq c_6 \frac{1}{t} |B_R(x_0) \cap [s \leq L]|^{1/2} \left(\int_{B_R(x_0)} |\varepsilon(u) - (\varepsilon(u))_{x_0,R}|^2 \, dx \right)^{1/2},$$

hence

$$\int_{B_R(x_0)} |\varepsilon(u) - \varepsilon(v)|^2 \, dx \leq c_7 \beta(t) \int_{B_R(x_0)} |\varepsilon(u) - (\varepsilon(u)))_{x_0,R}|^2 \, dx,$$

$$\beta(t) = \frac{1}{t^2} + \alpha(t)^2.$$

Inserting this into (2.23) we finally have shown

(2.25)
$$\int_{B_r(x_0)} |\varepsilon(u) - (\varepsilon(u))_{x_0,r}|^2 \, dx \leq c_8 \{(\frac{r}{R})^{n+2} + \beta(t)\}$$

$$\int_{B_R(x_0)} |\varepsilon(u) - (\varepsilon(u))_{x_0,R}|^2 \, dx.$$

Note that (2.25) is valid for any $0 < r < R < \text{dist}(x_0, \partial\Omega)$, $x_0 \in \Omega$, provided (2.24) holds. It is now more or less standard to derive local boundedness of $\varepsilon(u)$ from (2.25) (see [Se14]). For completeness we give the arguments: let

$$\Psi(x_0, R) = \left(\fint_{B_R(x_0)} |\varepsilon(u) - (\varepsilon(u))_{x_0,R}|^2 dx \right)^{1/2}.$$

(2.25) implies

$$(2.26) \qquad \Psi(x_0, r) \leq c_9 \left\{ \frac{r}{R} + \gamma(t)\left(\frac{r}{R}\right)^{-n/2} \right\} \Psi(x_0, R)$$

for any $r \in]0, R]$ with $\gamma(s) \to 0$ as $s \to \infty$ $\left(\gamma(s) = \sqrt{\beta(s)}\right)$, and (2.26) is valid under the assumption that $\left(\fint_{B_{2R}(x_0)} |\varepsilon(u)|^2 dx \right)^{1/2} > 2L$. We first select $\tau \in]0, 1[$ according to

$$(2.27) \qquad 2c_9\sqrt{\tau} < 1.$$

The number $\tilde{L}$ is defined through the condition

$$(2.28) \qquad \tau^{-n/2}\gamma(s) < \tau \text{ for all } s \geq \tilde{L}.$$

Let $M = \max\{2L, \tilde{L}\} + 1$. We then consider a point $x_0 \in \Omega$ for which $\lim_{\rho \downarrow 0} \fint_{B_\rho(x_0)} |\varepsilon(u)|^2 dx$ exists (which is true for a.a. $x_0 \in \Omega$).

Two cases can occur:

Case 1. $\displaystyle\lim_{\rho \downarrow 0} \fint_{B_\rho(x_0)} |\varepsilon(u)|^2 dx \leq M^2$

Case 2. $\displaystyle\lim_{\rho \downarrow 0} \fint_{B_\rho(x_0)} |\varepsilon(u)|^2 dx > M^2.$

In case 2 we let

$$R_1 = \sup \left\{ r \in]0, \text{dist}(x_0, \partial\Omega)[: \right.$$

$$\left. \fint_{B_\rho(x_0)} |\varepsilon(u)|^2 dx > M^2 \text{ for all } 0 < \rho < r \right\}.$$

Then

$$(2.29) \qquad \left(\fint_{B_r(x_0)} |\varepsilon(u)|^2 dx \right)^{1/2} > 2L$$

and (observe (2.28))

$$(2.30) \quad \tau^{-n/2}\gamma\left(\left(\fint_{B_r(x_0)} |\varepsilon(u)|^2 dx\right)^{1/2}\right) < \tau$$

for all $0 < r \leq R_1$. By (2.26), (2.27), (2.29) and (2.30) we find

$$\Psi(x_0, \tau R) \leq \sqrt{\tau}\Psi(x_0, R), \ 0 < R \leq R_1,$$

which gives after iteration (starting at $R = R_1$)

$$(2.31) \quad \Psi(x_0, \tau^k R_1) \leq \tau^{k/2}\Psi(x_0, R_1), \ k \in \mathbb{N}.$$

From (2.31) we deduce

$$\left|\left(\fint_{B_{\tau^k R_1}(x_0)} |\varepsilon(u)|^2 dx\right)^{1/2} - \left(\fint_{B_{R_1}(x_0)} |\varepsilon(u)|^2 dx\right)^{1/2}\right|$$

$$\leq \sum_{i=0}^{k-1}\left|\left(\fint_{B_{\tau^{i+1}R_1}(x_0)} |\varepsilon(u)|^2 dx\right)^{1/2} - \left(\fint_{B_{\tau^i R_1}(x_0)} |\varepsilon(u)|^2 dx\right)^{1/2}\right|$$

$$\leq \sum_{i=0}^{k-1}(1+\tau^{-n})^{1/2}\Psi(x_0, \tau^i R_1)$$

$$\leq \sum_{i=0}^{k-1}(1+\tau^{-n})^{1/2}\tau^{i/2}\Psi(x_0, R_1) \leq c_{10}(\tau)\left(\fint_{B_{R_1}(x_0)} |\varepsilon(u)|^2 dx\right)^{1/2},$$

$$c_{10}(\tau) = \frac{1}{1-\sqrt{\tau}}(1+\tau^{-n})^{1/2},$$

and therefore

$$\lim_{\rho \downarrow 0}\left(\fint_{B_\rho(x_0)} |\varepsilon(u)|^2 dx\right)^{1/2}$$

$$\leq (1+c_{10}(\tau))\left(\fint_{B_{R_1}(x_0)} |\varepsilon(u)|^2 dx\right)^{1/2}.$$

In case $R_1 = \text{dist}\,(x_0, \partial\Omega)$ we get

$$\lim_{\rho \downarrow 0}\left(\fint_{B_\rho(x_0)} |\varepsilon(u)|^2 dx\right)^{1/2}$$

$$\leq (1+c_{10}(\tau))|B_1(0)|^{-1/2}\,\text{dist}\,(x_0, \partial\Omega)^{-n/2}\left(\int_\Omega |\varepsilon(u)|^2 dx\right)^{1/2},$$

for $R_1 < \text{dist}\,(x_0, \partial\Omega)$ we just use

$$\left(\fint_{B_{R_1}(x_0)} |\varepsilon(u)|^2 dx\right)^{1/2} \leq M.$$

Collecting our results we have finally shown

$$|\varepsilon(u)(x_0)| \leq (1 + c_{10}(\tau))$$

$$\times \max\left\{M, |B_1(0)|^{-1/2}\,\text{dist}\,(x_0, \partial\Omega)^{-n/2}\left(\int_\Omega |\varepsilon(u)|^2\,dx\right)^{1/2}\right\}$$

for almost all $x_0 \in \Omega$, this completes the proof of Theorem 3.2.3.

$\square$

REMARK 3.2.8 For sufficiently regular boundary data v_0 and $\partial\Omega$ of class C^2 we expect global L^∞–bounds for the tensor $\varepsilon(u)$.

REMARK 3.2.9 We proved Theorem 3.2.3 for the case of zero volume forces f. The statement and also the proof remain valid if we consider volume forces f in the space $L^2(\Omega; \mathbb{R}^n) \cap L^{2,n-2+2\mu}_{\text{loc}}(\Omega; \mathbb{R}^n)$ for some $0 < \mu < 1$.

REMARK 3.2.10 As in section 1 all our results continue to hold for local minimizers of the energies under consideration.

3.3   The two–dimensional case

This section continues the study of the smoothness properties of the strain velocity which we started in section 1. We will show that for two–dimensional
problems the tensor $\varepsilon(v)$ is continuous on Ω without singular points provided v
is the minimizer of the energy functional associated to the Bingham, the Powell–
Eyring and certain power law fluid models.

Let us first assume that W is of the form

$$(3.1) \qquad W(E) = \mu_\infty |E|^2 + \mu_0 \left\{ \begin{array}{c} (\delta + |E|^2)^{p/2} \\ \text{or} \\ \displaystyle\int_0^{|E|} \text{ar} \sinh t \, dt \end{array} \right\} , E \in \mathbb{M},$$

with constants μ_0, μ_∞, $\delta > 0$, $1 \le p < 2$. For volume forces g in the space
$L^2(\Omega; \mathbb{R}^n)$ we consider the variational problem (1.54) in the class $\mathbb{K} = \{v \in W_2^1(\Omega; \mathbb{R}^n) : \operatorname{div} v = 0, v = v_0 \text{ on } \partial\Omega\}$.

THEOREM 3.3.1 *Let $v \in \mathbb{K}$ denote the solution of (1.54) with W defined as
in (3.1). Then v has second generalized derivatives, i.e. $v \in W_{2,\mathrm{loc}}^2(\Omega; \mathbb{R}^n)$.*

THEOREM 3.3.2 *With notation as in Theorem 3.3.1 let us in addition as-
sume that $g \in L^{2,n-2+2\mu}(\Omega; \mathbb{R}^n)$ for some $0 < \mu < 1$. Then there is an open
subset Ω_0 of Ω such that $v \in C^{1,\alpha}(\Omega_0; \mathbb{R}^n)$ for any $\alpha < \mu$ and $\mathcal{H}^{n-2}(\Omega - \Omega_0) = 0$.
Here $\mathcal{H}^{n-2}$ denotes the $(n-2)$–dimensional Hausdorff measure. In particular,
if $n = 2$, then we have no singular points, i.e. $\Omega_0 = \Omega$.*

REMARK 3.3.1 Our first result holds for any dimension $n \ge 2$.

Proof of Theorem 3.3.1: Instead of modifying the arguments used for the proof
of Lemma 3.0.5 we give a variant which can be found in the paper [FS2]. W.l.o.g.
we may assume that $\partial\Omega$ is of class C^2 (otherwise we replace Ω by a smooth
subdomain). Let $u \in W_2^1(\Omega; \mathbb{R}^n)$ denote the unique solution of the linear Stokes
problem

$$(3.2) \qquad \left\{ \begin{array}{ll} \displaystyle\int_\Omega \varepsilon(u) : \varepsilon(\varphi) \, dx = \int_\Omega g \cdot \varphi \, dx & \text{for all } \varphi \in \overset{\circ}{W}_2^1(\Omega; \mathbb{R}^n), \ \operatorname{div}\varphi = 0, \\[2ex] \operatorname{div} u = 0 \text{ on } \Omega & \text{and} \\[2ex] u|_{\partial\Omega} = 0. \end{array} \right.$$

From the work [L1] we deduce that u is in the space $W_2^2(\Omega; \mathbb{R}^n)$ together with the estimate

$$(3.3) \qquad \int_\Omega |\nabla u|^2 + |\nabla^2 u|^2 dx \leq c_1(\Omega) \int_\Omega |g|^2 dx.$$

Using (3.3) as well as the minimum property of v we find

$$(3.4) \qquad \int_\Omega \left(\frac{\partial W}{\partial \varepsilon}(\varepsilon(v)) : \varepsilon(\varphi) - \varepsilon(u) : \varepsilon(\varphi) \right) dx = 0$$

for any $\varphi \in \overset{\circ}{W}_2^1(\Omega; \mathbb{R}^n)$, $\operatorname{div} \varphi = 0$. According to [L1] and [LS] there exists a pressure function $p \in L^2(\Omega)$ such that (3.4) can be written as

$$(3.5) \qquad \int_\Omega \left(\frac{\partial W}{\partial \varepsilon}(\varepsilon(v)) : \varepsilon(\varphi) - \varepsilon(u) : \varepsilon(\varphi) \right) dx = \int_\Omega p \operatorname{div} \varphi \, dx$$

being valid now for any φ in the space $\overset{\circ}{W}_2^1(\Omega; \mathbb{R}^n)$. Let $\eta \in C_0^2(\Omega)$ and consider $h \in \mathbb{R}^n$ such that $|h| < \operatorname{dist}(\operatorname{spt}\eta, \partial\Omega)$. We further let $\Delta g(x) = g(x + h) - g(x)$ for functions g. (3.5) implies

$$\int_\Omega \left(\frac{\partial W}{\partial \varepsilon}(\varepsilon(v)(x + h)) - \frac{\partial W}{\partial \varepsilon}(\varepsilon(v)(x)) \right) : \varepsilon(\eta^2 \Delta_h v) \, dx$$

$$= \int_\Omega \varepsilon(\Delta_h u) : \varepsilon(\eta^2 \Delta_h v) \, dx + \int_\Omega \Delta_h p \operatorname{div}(\eta^2 \Delta_h v) \, dx,$$

or equivalently

$$(3.6) \quad \begin{cases} \displaystyle \int_\Omega \eta^2 \left[\frac{\partial W}{\partial \varepsilon}(\varepsilon(v)(x + h)) - \frac{\partial W}{\partial \varepsilon}(\varepsilon(v)(x)) \right] : \varepsilon(\Delta_h v) \, dx \\[2ex] \displaystyle = -2 \int_\Omega \eta \left[\frac{\partial W}{\partial \varepsilon}(\varepsilon(v)(x + h)) - \frac{\partial W}{\partial \varepsilon}(\varepsilon(v)(x)) \right] : (\Delta_h v \odot \nabla\eta) \, dx \\[2ex] \displaystyle + \int_\Omega \eta^2 \varepsilon(\Delta_h u) : \varepsilon(\Delta_h v) \, dx + 2 \int_\Omega \eta \varepsilon(\Delta_h u) : (\Delta_h v \odot \nabla\eta) \, dx \\[2ex] \displaystyle + \int_\Omega \Delta_h p \nabla\eta^2 \cdot \Delta_h v \, dx. \end{cases}$$

Here we recall the notation $a \odot b = \frac{1}{2}(a_i b_j + a_j b_i)$ for $a, b \in \mathbb{R}^n$. From definition (3.1) we deduce

$$\mu_\infty |X|^2 \leq D^2 W(Y)(X, X) \leq \Lambda |X|^2$$

for any $X, Y \in \mathbb{M}$ with a suitable constant $\Lambda > 0$, and we may use this inequality to get a lower bound for the left–hand side of (3.6):

(3.7)
$$\begin{cases}
\displaystyle \int_\Omega \eta^2 |\varepsilon(\Delta_h v)|^2 \, dx \leq c_2 \Big\{ \Big(\int_\Omega \eta^2 |\varepsilon(\Delta_h v)|^2 dx \Big)^{1/2} \\[2mm]
\displaystyle \times \Big(\int_\Omega \eta^2 |\varepsilon(\Delta_h u)|^2 dx + \int_\Omega |\Delta_h v|^2 |\nabla \eta|^2 dx \Big)^{1/2} \\[2mm]
\displaystyle + \int_\Omega \eta |\varepsilon(\Delta_h u)| \, |\Delta_h v| \, |\nabla \varphi| \, dx + \int_\Omega \Delta_h p \nabla \eta^2 \cdot \Delta_h v \, dx \Big\}.
\end{cases}$$

For the last integral on the right–hand side of (3.7) we observe

$$\int_\Omega \Delta_h p \nabla \eta^2 \cdot \Delta_h v \, dx = 2 \int_\Omega \Delta_h p \nabla \eta \cdot \eta \Delta_h v \, dx$$

$$\leq 2 \|\Delta_h p \nabla \eta\|^* \, \|\nabla(\eta \Delta_h v)\|_{L^2(\Omega)} \leq 4 \|\Delta_h p \nabla \eta\|^* \, \|\varepsilon(\eta \Delta_h v)\|_{L^2(\Omega)},$$

where

$$\|\Delta_h p \nabla \eta\|^* = \sup \Big\{ \int_\Omega \Delta_h p \nabla \eta \cdot w \, dx : w \in \overset{\circ}{W}{}^1_2(\Omega; \mathbb{R}^n), \|\nabla w\|_{L^2(\Omega)} \leq 1 \Big\}$$

$$\leq c_3 |h| \Big(\int_\Omega p^2 \, dx \Big)^{1/2}.$$

Inserting this into (3.7) and using Young's inequality we find that

$$\int_\Omega \eta^2 |\varepsilon(\Delta_h v)|^2 dx \leq c_4 \int_\Omega \Big(|\varepsilon(\Delta_h u)|^2 + |\Delta_h v|^2 + |h|^2 p^2 \Big) \, dx$$

where all integrals have to be calculated w.r.t. spt η. Recalling (3.3) as well as standard estimates for the pressure p we get the final inequality

$$\int_\Omega \eta^2 |\nabla(\varepsilon(v))|^2 dx \leq c_5 \int_\Omega |g|^2 dx,$$

and the proof of Theorem 3.3.1 is complete.

$\square$

REMARK 3.3.2 With similar arguments (for details see [Se11], [Se12]) we can show the following: suppose that $g \in L^{2,n-2+2\mu}(\Omega; \mathbb{R}^n)$. Then there is a constant $c > 0$ with the property

$$\int_{B_{R/2}(x_0)} |\nabla\varepsilon(v)|^2 dx \le c\Big\{R^{n-2+2\mu} + R^{-2} \int_{B_R(x_0)} |\varepsilon(v) - (\varepsilon(v))_{x_0,R}|^2 dx\Big\}$$

for any $B_R(x_0) \subset \Omega$. This estimate can be used to give a "direct" proof of Theorem 3.3.2 following standard arguments used in the theory of nonlinear elliptic systems (see [Gi]).

Proof of Theorem 3.3.2: From the proofs of Theorem 3.1.4 a) and Theorem 3.1.5 we immediately deduce that v is of class $C^{1,\alpha}$, $\alpha < \mu$, in a neighborhood of $x_0 \in \Omega$ if and only if

i) $\sup_{r>0} |(\varepsilon(v))_{x_0,r}| < \infty$ and ii) $\lim_{r\downarrow 0} \fint_{B_r(x_0)} |\varepsilon(v) - (\varepsilon(v))_{x_0,r}|^2 dx < \infty.$

According to Theorem 3.2.3, i) is true for every $x_0 \in \Omega$. By Poincaré's inequality (and Theorem 3.3.1) we have

$$\fint_{B_r(x_0)} |\varepsilon(v) - (\varepsilon(v))_{x_0,r}|^2 dx \le c\,r^2 \fint_{B_r(x_0)} |\nabla^2 v|^2 dx,$$

and

$$r^2 \fint_{B_r(x_0)} |\nabla^2 v|^2 dx \to 0 \text{ as } r \downarrow 0 \text{ for } \mathcal{H}^{n-2} - \text{almost all } x_0 \in \Omega. \qquad \square$$

We next discuss the Bingham case and give the proof of a theorem which was obtained in [Se13].

THEOREM 3.3.3 *Suppose that $n = 2$ and define*
$I(u) = \int_\Omega (\mu|\varepsilon(u)|^2 + \sqrt{2}k_*|\varepsilon(u)| - g \cdot u) \, dx$ *on the class $\mathbb{K} = \{v \in W_2^1(\Omega; \mathbb{R}^2) :$ div $v = 0$, $v = v_0$ on $\partial\Omega\}$, μ and k_* denoting positive constants. Suppose further that $g \in L^{2,2\lambda}(\Omega; \mathbb{R}^2)$ for some $\lambda > 0$. Then, for the unique I-minimizer $v \in \mathbb{K}$, we have $\varepsilon(v) \in C^0(\Omega; \mathbb{M})$.*

REMARK 3.3.3 We do not claim that $\varepsilon(v)$ belongs to some Hölder space.

REMARK 3.3.4 We note that Theorems 3.3.1, 3.3.2 and 3.3.3 hold under the assumption that v is only locally minimizing.

The proof of Theorem 3.3.3 is divided in several steps: first we show an analogue
of Theorem 3.3.1 for minimizers of the energy I introduced above.

THEOREM 3.3.4 *Suppose that $n \geq 2$ and $g \in L^2(\Omega; \mathbb{R}^n)$. Then the unique
I-minimizer in the class $\mathbb{K}$ is in the space $W_{2,\text{loc}}^2(\Omega; \mathbb{R}^n)$.*

Proof of Theorem 3.3.4: We argue by approximation. Let $F : \mathbb{M} \to [0, \infty[$
denote a convex Lipschitz function of class C^2 and define

$$J(u) = \int_\Omega \left(\frac{1}{2}|\varepsilon(u)|^2 + F(\varepsilon(u)) - g \cdot u\right) dx, \quad u \in \mathbb{K}.$$

Let $v \in \mathbb{K}$ denote the unique J-minimizer. From the proof of Theorem 3.3.1
it should be clear that $v \in W_{2,\text{loc}}^2(\Omega; \mathbb{R}^n)$ but now we need an estimate for the
second generalized derivatives of v which does not involve an upper bound for
$D^2 F$: in the concrete approximation below this quantity can not be controlled.
We write $W = \frac{1}{2}|\cdot|^2 + F$ and define u as in (3.2) assuming that $\partial\Omega$ is smooth
enough. Then as before

$$(3.8) \qquad \int_\Omega \frac{\partial W}{\partial E}(\varepsilon(v)) : \varepsilon(\varphi)\, dx = \int_\Omega \varepsilon(u) : \varepsilon(\varphi)\, dx + \int_\Omega p \operatorname{div} \varphi\, dx$$

for any $\varphi \in \overset{\circ}{W}_2^1(\Omega; \mathbb{R}^n)$. Here $p \in L^2(\Omega)$ is a pressure function such that $\int_\Omega p\, dx = 0$. It is easy to show that

$$(3.9) \qquad \int_\Omega p^2\, dx \leq c_1 \int_\Omega \left(|\varepsilon(v)|^2 + |\varepsilon(u)|^2\right) dx$$

where c_1 depends also on F but only the Lipschitz constant of F enters. We
replace φ in (3.8) by $\partial_\gamma \varphi$ for some $\gamma = 1, \ldots, n$ and get

$$(3.10) \qquad \int_\Omega \partial_\gamma\left\{\frac{\partial W}{\partial E}(\varepsilon(v))\right\} : \varepsilon(\varphi)\, dx = \int_\Omega \varepsilon(\partial_\gamma u) : \varepsilon(\varphi)\, dx + \int_\Omega \partial_\gamma p \operatorname{div} \varphi\, dx.$$

Consider now $\eta \in C_0^2(\Omega)$, $\eta \geq 0$, $\xi \in V_\times$, and insert $\varphi = \eta^3 [\partial_\gamma v - \xi]$ in equation

(3.10). We have

(3.11)
$$
\begin{cases}
\displaystyle\int_\Omega \eta^3 \partial_\gamma \{\frac{\partial W}{\partial E}(\varepsilon(v))\} : \varepsilon(\partial_\gamma v)\, dx \\[2mm]
\displaystyle = \int_\Omega \eta^3 D^2 W(\varepsilon(v))(\varepsilon(\partial_\gamma v),\, \varepsilon(\partial_\gamma v))\, dx \\[2mm]
\displaystyle = \int_\Omega \eta^3 |\varepsilon(\partial_\gamma v)|^2 dx + \int_\Omega \eta^3 \partial^2 F(\varepsilon(v))(\varepsilon(\partial_\gamma v),\, \varepsilon(\partial_\gamma v))\, dx \\[2mm]
\displaystyle \geq \int_\Omega \eta^3 |\varepsilon(\partial_\gamma v)|^2 dx
\end{cases}
$$

by convexity and smoothness of F. On the other hand

$$\int_\Omega \eta^3 \partial_\gamma \{\frac{\partial W}{\partial E}(\varepsilon(v))\} : \varepsilon(\partial_\gamma v)\, dx$$

$$= \int_\Omega \partial_\gamma \{\frac{\partial W}{\partial E}(\varepsilon(v))\} : \varepsilon\big(\eta^3 [\partial_\gamma v - \xi]\big)\, dx$$

$$- \int_\Omega \partial_\gamma \{\frac{\partial W}{\partial E}(\varepsilon(v))\} : \big(\nabla \eta^3 \odot [\partial_\gamma v - \xi]\big)\, dx = (3.10)$$

$$= \int_\Omega \varepsilon(\partial_\gamma u) : \varepsilon\big(\eta^3 [\partial_\gamma v - \xi]\big)\, dx + \int_\Omega \partial_\gamma p\, \operatorname{div}\, \big[\eta^3 (\partial_\gamma v - \xi)\big]\, dx$$

$$- \int_\Omega \partial_\gamma \{\frac{\partial W}{\partial E}(\varepsilon(v))\} : \big(\nabla \eta^3 \odot [\partial_\gamma v - \xi]\big)\, dx$$

$$= (I) + (II) - (III),$$

and we discuss these integrals separately. Recalling (3.3) we use Young's inequality with arbitrary $\delta > 0$ to see (w.l.o.g. $0 \leq \eta \leq 1$)

(3.12)
$$
\begin{cases}
\displaystyle (I) \;\leq\; c_2(\delta) \int_\Omega |\nabla^2 u|^2 dx + \delta \int_\Omega |\varepsilon\big(\eta^3 [\partial_\gamma v - \xi]\big)|^2 dx \\[2mm]
\displaystyle \qquad \leq\; c_3(\delta)\{ \int_\Omega |g|^2 dx + \int_\Omega |\nabla \eta|^2 |\partial_\gamma v - \xi|^2 dx \} \\[2mm]
\displaystyle \qquad\quad + \delta \int_\Omega \eta^3 |\varepsilon(\partial_\gamma v)|^2 dx.
\end{cases}
$$

Next we have

$$
\begin{aligned}
(\boldsymbol{II}) \;=\; & \int_\Omega \partial_\gamma p\, \nabla \eta^3 \cdot (\partial_\gamma v - \xi)\, dx = - \int_\Omega p\, \partial_\gamma \big[\nabla \eta^3 \cdot (\partial_\gamma v - \xi) \big]\, dx \\
\leq\; & \int_\Omega |p|\, |\partial_\gamma \nabla \eta^3|\, |\partial_\gamma v - \xi|\, dx \\
& + \int_\Omega |p|\, |\nabla \eta^3|\, |\partial_\gamma (\partial_\gamma v - \xi)|\, dx.
\end{aligned}
$$

Let us assume that η is supported on some ball $B_R(x_0)$ and $|\nabla^k \eta| \leq c_4 R^{-k}$, $k = 1, 2$. Then, again with Young's inequality, we deduce

$$
\begin{aligned}
(\boldsymbol{II}) \;\leq\; & c_5 \Big\{ R^{-2} \int_{B_R(x_0)} (|p|^2 + |\partial_\gamma v - \xi|^2)\, dx \\
& + \int_\Omega |p|\, |\nabla \eta|\, \sqrt{\eta}\, \eta^{3/2} |\partial_\gamma (\partial_\gamma v - \xi)|\, dx \Big\} \\
\leq\; & c_6(\delta) R^{-2} \int_{B_R(x_0)} (|p|^2 + |\partial_\gamma v - \xi|^2)\, dx + \delta \int_\Omega \eta^3 |\partial_\gamma (\partial_\gamma v - \xi)|^2\, dx.
\end{aligned}
$$

Using

$$
\begin{aligned}
\int_\Omega \eta^3 |\partial_\gamma (\partial_\gamma v - \xi)|^2\, dx \;\leq\; & c_7 \Big\{ \int_\Omega |\partial_\gamma (\eta^3 [\partial_\gamma v - \xi])|^2\, dx + \int_\Omega |\nabla \eta|^2\, |\partial_\gamma v - \xi|^2 dx \Big\} \\
\leq\; & c_8 \Big\{ \int_\Omega |\varepsilon (\eta^3 [\partial_\gamma v - \xi])|^2\, dx + \int_\Omega |\nabla \eta|^2\, |\partial_\gamma v - \xi|^2 dx \Big\} \\
\leq\; & c_9 \Big\{ \int_\Omega \eta^3 |\varepsilon (\partial_\gamma v)|^2\, dx + \int_\Omega |\nabla \eta|^2\, |\partial_\gamma v - \xi|^2 dx \Big\}
\end{aligned}
$$

we see, combining (3.11), (3.12) and the above estimates, that (after appropriate choice of δ)

$$
\begin{aligned}
\lambda \int_{B_R(x_0)} & \eta^3 |\varepsilon (\partial_\gamma v)|^2\, dx \\
\leq\; & c_{10} \Big\{ \int_{B_R(x_0)} |g|^2 dx + R^{-2} \int_{B_R(x_0)} |\partial_\gamma v - \xi|^2 dx + R^{-2} \int_{B_R(x_0)} |p|^2 dx \Big\} - (\boldsymbol{III}).
\end{aligned}
$$

Here λ is some positive constant, and we have used the fact that all integrals have to be calculated with respect to the ball $B_R(x_0)$.

In (III) we integrate by parts to get ($A \in \mathbb{M}$, $\delta > 0$)

$$
\begin{aligned}
-(\text{III}) &= \int_\Omega \left(\frac{\partial W}{\partial E}(\varepsilon(v)) - \frac{\partial W}{\partial E}(A) \right) : \partial_\gamma (\nabla \eta^3 \odot [\partial_\gamma v - \xi]) \, dx \\
&\leq \int_\Omega \left| \frac{\partial W}{\partial E}(\varepsilon(v)) - \frac{\partial W}{\partial E}(A) \right| |\partial_\gamma \nabla \eta^3| \, |\partial_\gamma v - \xi| \, dx \\
&\quad + \int_\Omega \left| \frac{\partial W}{\partial E}(\varepsilon(v)) - \frac{\partial W}{\partial E}(A) \right| |\nabla \eta^3| \, |\partial_\gamma(\partial_\gamma v - \xi)| \, dx \\
&\leq c_{11}(\delta) \Big\{ \int_{B_R(x_0)} R^{-2} \left| \frac{\partial W}{\partial E}(\varepsilon(v)) - \frac{\partial W}{\partial E}(A) \right|^2 dx \\
&\quad + \int_{B_R(x_0)} R^{-2} |\partial_\gamma v - \xi|^2 dx \Big\} + \delta \int_{B_R(x_0)} |\varepsilon(\partial_\gamma v)|^2 dx.
\end{aligned}
$$

Putting together the various estimates, we see by letting $\eta = 1$ on $B_{R/2}(x_0)$, that

$$
(3.13) \quad
\begin{cases}
\displaystyle \int_{B_{R/2}(x_0)} |\varepsilon(\partial_\gamma v)|^2 \, dx \leq c_{12} \Big\{ \int_{B_R(x_0)} |g|^2 dx + R^{-2} \int_{B_R(x_0)} |\partial_\gamma v - \xi|^2 dx \\[2mm]
\displaystyle + R^{-2} \int_{B_R(x_0)} |p - (p)_{x_0,R}|^2 dx + R^{-2} \int_{B_R(x_0)} \left| \frac{\partial W}{\partial E}(\varepsilon(v)) - \frac{\partial W}{\partial E}(A) \right|^2 dx \Big\}.
\end{cases}
$$

Here we used that p can be replaced by $p - (p)_{x_0,R}$.

After these preparations we give a brief outline of the proof of Theorem 3.3.4.

Let now v denote the I-minimizer in $\mathbb{K}$. For notational simplicity we assume $\mu = \frac{1}{2}$, $k_* = 1/\sqrt{2}$ and define

$$
F_\delta(E) = \sqrt{\delta + |E|^2}, \ \delta > 0, \ E \in \mathbb{M}.
$$

Let $v_\delta \in \mathbb{K}$ denote the corresponding minimizer of J_δ. From Lemma 3.1.5 we have $v_\delta \to v$ as $\delta \downarrow 0$ strongly in $W_2^1(\Omega; \mathbb{R}^n)$, and by the foregoing calculations (3.13) holds for the functions v_δ with pressure p_δ and W replaced by $W_\delta(E) =$

$\frac{1}{2}|E|^2 + F_\delta(E)$. By (3.9) the pressure is estimated independent of δ, moreover, we see that

$$\left|\frac{\partial W_\delta}{\partial E}(M)\right| \leq |M| + \frac{|M|}{\sqrt{\delta + |M|^2}} \leq |M| + 1.$$

This proves that the right–hand side of (3.13) is bounded independent of δ which clearly implies $v \in W_{2,\text{loc}}^2(\Omega; \mathbb{R}^n)$. Unfortunately (3.13) does not extend to the limit $\delta \downarrow 0$ for the obvious reason that the last integral on the right–hand side of (3.13) may have not limit as $\delta \downarrow 0$.

$\square$

After these preparations we give an outline of the proof of Theorem 3.3.3 assuming from now on that all the hypothesis are satisfied. As in the proof of the previous theorem we consider the approximations $\{v_\delta\} \subset \mathbb{K}$ defined as the minimizers of the energy

$$J_\delta(u) = \int_\Omega (\mu|\varepsilon(u)|^2 + \sqrt{2}k_* \sqrt{\delta + |\varepsilon(u)|^2} - g \cdot u)\, dx.$$

Let $B_R(x_0)$ denote a disc in Ω. Then, using arguments similar to the calculations in the proof of Theorem 3.3.4, we get the following version of estimate (3.13) (compare [Se13], (2.13))

(3.14)
$$\int_{B_{R/2}(x_0)} |\nabla\varepsilon(v_\delta)|^2 dx \leq c_1 \left\{ \left(1 + \frac{1}{\sqrt{\delta + |(\varepsilon(v_\delta))_{x_0,R}|^2}}\right)^2 \right.$$

$$\left. \cdot \frac{1}{R^2} \int_{T_R(x_0)} |\varepsilon(\bar{v}_\delta)|^2 dx + R^{2\lambda} \right\},$$

$T_R(x_0) = B_R(x_0) - B_{R/2}(x_0)$, $\bar{v}_\delta = v_\delta - (\varepsilon(v_\delta))_{x_0,R}(x - x_0) - u_\times$, where $u_\times$ denotes a rigid motion which is chosen according to

$$\int_{T_R(x_0)} |\bar{v}_\delta|^2 dx \leq c_2 R^2 \int_{T_R(x_0)} |\varepsilon(\bar{v}_\delta)|^2 dx.$$

Here and in the sequel $(\cdot)_{x_0,R}$ indicates the mean value over $T_R(x_0)$. We pass to the limit $\delta \downarrow 0$ in (3.14) and obtain

(3.15)
$$\begin{cases} |(\varepsilon(v))_{x_0,R}|^2 \Phi(x_0, \tfrac{R}{2}) \leq c_3 \Big\{ \left(1 + |(\varepsilon(v))_{x_0,R}|^2\right) \\ \\ \times \frac{1}{R^2} \displaystyle\int_{T_R(x_0)} |\varepsilon(v) - (\varepsilon(v))_{x_0,R}|^2 dx + R^{2\lambda}|(\varepsilon(v))_{x_0,R}|^2 \Big\}, \end{cases}$$

$\Phi(x_0, r) = \int_{B_r(x_0)} |\nabla \varepsilon(v)|^2 dx$. It can be shown that the constant c_3 just depends on μ, k_* and the $L^{2,2\lambda}$-norm of f. By Poincaré's inequality we have

$$\int_{T_R(x_0)} |\varepsilon(v) - (\varepsilon(v))_{x_0,R}|^2 dx \le c_4 R^2 \int_{T_R(x_0)} |\nabla\varepsilon(v)|^2 dx,$$

inserting this into (3.15) we end up with

$$(3.16) \quad \begin{cases} |(\varepsilon(v))_{x_0,R}|^2 \Phi(x_0, \tfrac{R}{2}) \le c_5 \Big\{ \big(1 + |(\varepsilon(v))_{x_0,R}|^2\big) \\[2mm] \times \big(\Phi(x_0, R) - \Phi(x_0, \tfrac{R}{2})\big) + R^{2\lambda}|(\varepsilon(u))_{x_0,R}|^2 \Big\}. \end{cases}$$

In order to prove continuity of the strain velocity we will use (3.16) to obtain the following preliminary result: if $|(\varepsilon(v))_{x_0,R}|$ is rather large, then we get a growth estimate for $\Phi(x, r)$ when x is in some neighborhood of x_0. The case $\limsup_{R\downarrow 0} |(\varepsilon(v))_{x_o,R}| = 0$ has to be discussed separately.

To be precise, let $K_0(t) = \frac{c_5(t^2+1)}{t^2+c_5(t^2+1)}$, $t \ge 0$.

Using Poincaré's inequality once more, we have

$$\big| |(\varepsilon(v))_{x_o,R/2}| - |(\varepsilon(v))_{x_0,R}| \big| \le c_6 \Phi^{1/2}(x_0, R),$$

hence

$$\big| |(\varepsilon(v))_{x_0,2^{-i}R}| - |(\varepsilon(v))_{x_0,R}| \big| \le c_6 \sum_{s=0}^{i-1} \Phi^{1/2}(x_0, 2^{-s}R)$$

for any $i \in \mathbb{N}$. We fix $\ominus \in\,]1, 2^\lambda[$ and define the functions

$$K(t) = \max\big\{ K_0(t), \ominus\, 2^{-\lambda} \big\}, \quad D(t) = \frac{1}{1 - \sqrt{K(t)}}, \quad t > 0.$$

Then from estimate (3.16) we deduce (see [Se13], Lemma 3.1)

LEMMA 3.3.1 *Let $\gamma > 0$ and suppose that*

$$|(\varepsilon(v))_{x_0,R}| \geq 2\gamma$$

and

$$\Phi(x_0, R) + \frac{(2R)^{2\lambda}}{\Theta - 1} \leq \left(\frac{\gamma}{c_6 D(\gamma)}\right)^2$$

hold for some $x_0 \in \Omega$ and $R > 0$. Then, for any $i \in \mathbb{N}$, we have

$$\Phi(x_0, 2^{-i}R) \leq K^i(\gamma)\Big[\Phi(x_0, R) + \frac{(2R)^{2\lambda}}{\Theta}\sum_{s=0}^{i-1}\Theta^{-s}\Big] \leq K^i(\gamma)\Big[\frac{\gamma}{c_6 D(\gamma)}\Big]^2$$

and

$$\gamma \leq \big|(\varepsilon(v))_{x_0,2^{-i}R}\big| \leq \big|(\varepsilon(v))_{x_0,R}\big| + \gamma.$$

The proof of the lemma is obtained via induction, i.e. iteration of inequality (3.16).   □

Consider next $x_0 \in \Omega$ with the property

(3.17) $\limsup\limits_{R \to 0} |(\varepsilon(v))_{x_0,R}| > 0.$

Of course (3.17) implies the existence of $\gamma > 0$ and some small radius $R > 0$ such that

(3.18)
$$\begin{cases} |(\varepsilon(v))_{x_0,R}| > 2\gamma, \\[2mm] \Phi(x_0, R) + \frac{(2R)^{2\lambda}}{\Theta - 1} < \left(\frac{\gamma}{c_6 D(\gamma)}\right)^2. \end{cases}$$

By continuity we find a disc $B_{\rho_1}(x_0)$ such that inequalities (3.18) hold with x_0 replaced by x for any $x \in B_{\rho_1}(x_0)$; from Lemma 3.3.1 we see

$$\Phi(x, 2^{-i}R) \leq K^i(\gamma)\left(\frac{\gamma}{c_6 D(\gamma)}\right)^2, \quad x \in B_{\rho_1}(x_0), \ i \in \mathbb{N},$$

so that (by the Dirichlet–growth theorem) $\varepsilon(v)$ is Hölder continuous in a neighborhood of x_0.

So, if condition (3.17) holds, then $\varepsilon(v)$ is continuous in some open neighborhood of x_0, $\lim\limits_{R \to 0}(\varepsilon(v))_{x_0,R} = \varepsilon(v)(x_0)$ exists and $|\varepsilon(v)(x_0)| \geq \gamma$. Suppose now

$$(3.19) \qquad \limsup_{R \to 0} |(\varepsilon(v))_{x_0,R}| = 0.$$

First we note that, for any $x \in \Omega$, we have the pointwise definition

$$\varepsilon(v)(x) = \lim_{R \downarrow 0}(\varepsilon(v))_{x_0,R}.$$

Namely, if (3.17) holds at $x \in \Omega$, then the existence of the limit was proved above, in case that (3.19) holds for x we clearly have $\lim\limits_{R \to 0}(\varepsilon(v))_{x,R} = 0$. Therefore (3.19) can be written as $\varepsilon(v)(x_0) = 0$.

Let γ denote some number > 0. We fix some subdomain $\Omega' \Subset \Omega$ of Ω and calculate a small radius $R(\gamma)$ such that

$$(3.20) \qquad \Phi(x,R) + \frac{(2R)^{2\lambda}}{\Theta - 1} < \left(\frac{\gamma}{c_6 D(\gamma)}\right)^2$$

holds for any $x \in \Omega'$ and $R \leq R(\gamma)$. From (3.19) we deduce the existence of $R_1 \in]0, R(\gamma)[$ with the property

$$|(\varepsilon(v))_{x_0,R_1}| < 2\gamma,$$

hence we find $\rho_2 > 0$ such that

$$(3.21) \qquad |(\varepsilon(v))_{x,R_1}| < 2\gamma$$

for any $x \in B_{\rho_2}(x_0)$. We claim

$$(3.22) \qquad |\varepsilon(v)(x) - \varepsilon(v)(x_0)| = |\varepsilon(v)(x)| < 4\gamma$$

on $B_{\rho_2}(x_0)$. So consider $x \in B_{\rho_2}(x_0)$. If $|(\varepsilon(v))_{x,\rho}| \leq 2\gamma$ for any $\rho \in]0, R_1[$, then $|\varepsilon(v)(x)| \leq 2\gamma$ and (3.22) holds. Otherwise we find some $R_2 \in]0, R_1[$ such that $|(\varepsilon(v))_{x,R_2}| > 2\gamma$. From (3.21) we then deduce the existence of $R_3 \in]R_2, R_1[$ such that

$$(3.23) \qquad |(\varepsilon(v))_{x,R_3}| = 2\gamma.$$

By Lemma 3.3.1 - with $x_0 = x$ and $R = R_3$ - we get

$$|(\varepsilon(v))_{x,2^{-i}R_3}| \leq |(\varepsilon(v))_{x,R_3}| + \gamma.$$

Passing to the limit $i \to \infty$ and using (3.23), we get $|\varepsilon(v)(x)| \leq 3\gamma$, hence (3.22) is established. Since γ was arbitrary, continuity of $\varepsilon(v)$ at x_0 follows and the proof of Theorem 3.3.3 is complete.

$$\square$$

REMARK 3.3.5 Having proved continuity of $\varepsilon(v)$ we now can state that $[|\varepsilon(v)| > 0] = \{x \in \Omega : |\varepsilon(v)(x)| > 0\}$ is open and we may therefore define the "separating line" $\partial[|\varepsilon(v)| > 0] \cap \Omega = L$ and ask if L is a smooth curve separating the rigid zone $[\varepsilon(v) = 0]$ from the region $[\varepsilon(v) \neq 0]$.

3.4 The Bingham variational inequality in dimensions two and three

Up to now our discussion of the quasi–static flow for generalized Newtonian fluids was limited to the variational setting starting from the equation of motion (0.2)" in which the nonlinearity $(\nabla v)v = (\partial_i v^j v_i)$ is neglected. In this section we are going to investigate the "realistic case" given by equation (0.2)'. As already mentioned we are then confronted with the same difficulties as in case of the nonlinear stationary Navier–Stokes system which forces us to work in dimension 2 or 3. Most of the material presented here is taken from the paper [FS2]. For the readers convenience we start with some known existence results concerning homogeneous boundary data.

THEOREM 3.4.1 *Suppose that* $n = 2$ *or* 3*, let* $f \in L^2(\Omega; \mathbb{R}^n)$ *be given and define*

$$W(E) = \mu_\infty |E|^2 + \mu_0 \left\{ \begin{array}{c} (\delta + |E|^2)^{p/2} \\ \\ or \\ \\ \int_0^{|E|} ar \sinh t \, dt \end{array} \right\}, \quad E \in \mathbb{M},$$

μ_∞, μ_0, $\delta > 0$, $1 \le p < 2$*. Then there exists a function* $v \in V_0 = \{u \in \overset{\circ}{W}_2^1(\Omega; \mathbb{R}^n) : div\, u = 0\}$ *such that*

$$(4.1) \qquad \int_\Omega \left\{ \frac{\partial W}{\partial E}(\varepsilon(v)) : \varepsilon(\varphi) + (\nabla v)v \cdot \varphi \right\} dx = \int_\Omega f \cdot \varphi \, dx$$

for every φ *in* V_0*.*

REMARK 3.4.1 As in [Ga2] or [L1] we obtain uniqueness if the L^2–norm of the volume force term f is sufficiently small.

Proof of Theorem 3.4.1: Consider the operator $T : V_0 \ni w \mapsto v \in V_0$, v denoting the unique solution of

$$\int_\Omega \frac{\partial W}{\partial E}(\varepsilon(v)) : \varepsilon(\varphi) \, dx = \int_\Omega f \cdot \varphi \, dx - \int_\Omega (\nabla w)w \cdot \varphi \, dx,$$

$\varphi \in V_0$, which for example is obtained by minimizing

$$V_0 \ni u \mapsto \int_\Omega \{W(\varepsilon(u)) + (\nabla w)w \cdot u - f \cdot u\}\, dx.$$

It is an easy exercise to show conpactness of T, moreover, the hypothesis of the Leray–Schauder fixed point theorem hold: let $w = \sigma Tw$ for some $w \in V_0$ and $0 \le \sigma \le 1$. Then we get

$$\int_\Omega \frac{\partial W}{\partial E}(\frac{1}{\sigma}\varepsilon(w)) : \varepsilon(\varphi)\, dx = \int_\Omega f \cdot \varphi\, dx - \int_\Omega (\nabla w)w \cdot \varphi\, dx,$$

hence

$$\int_\Omega \frac{\partial W}{\partial E}(\frac{1}{\sigma}\varepsilon(w)) : \varepsilon(w)\, dx = \int_\Omega f \cdot w\, dx.$$

Using ellipticity we find that

$$\mu_\infty \frac{1}{\sigma} \int_\Omega |\varepsilon(w)|^2 dx \le \|f\|_{L^2}\|w\|_{L^2},$$

and in conclusion

$$\int_\Omega |\varepsilon(w)|^2 dx \le c \int_\Omega |f|^2 dx$$

for a positive constant independent of σ.
Clearly, any function $v \in V_0$ with $Tv = v$, is a solution of (4.1). $\qquad \square$

THEOREM 3.4.2 (the homogeneous Bingham variational inequality)

For $n = 2$ or 3 and $f \in L^2(\Omega, \mathbb{R}^n)$ there exists at least one function $v \in V_0$ satisfying the homogeneous (BVI)

(4.2)
$$\begin{cases} \int_\Omega 2\mu\varepsilon(v) : (\varepsilon(w) - \varepsilon(v)) + \sqrt{2}k_*(|\varepsilon(w)| - |\varepsilon(v)|) \\ + (\nabla v)v \cdot (w - v)\, dx \\ \ge \int_\Omega f \cdot (w - v)\, dx \quad \text{for any } w \in V_0. \end{cases}$$

Here μ, k_* denote arbitrary positive constants.

REMARK 3.4.2 Uniqueness for (4.2) holds if $\|f\|_{L^2}$ is small enough.

Proof of Theorem 3.4.2: (see [DL]) For notational simplicity we let $\mu = 1/2$ and $k_* = 1/\sqrt{2}$. With $\delta > 0$ define $W_\delta(E) = \frac{1}{2}|E|^2 + \sqrt{\delta + |E|^2}$, $E \in \mathbb{M}$, and let $v_\delta \in V_0$ denote a solution of (4.1) with $W = W_\delta$. We show that, after passing to a subsequence, we have $v_\delta \to v$ as $\delta \downarrow 0$ for some function $v \in V_0$ which is a solution of (BVI). Consider $w \in V_0$. From (4.1) we get

$$
\text{(4.3)} \quad
\begin{aligned}
&\int_\Omega \frac{\partial W_\delta}{\partial E}(\varepsilon(v_\delta)) : (\varepsilon(w) - \varepsilon(v_\delta))\, dx \\
&= \int_\Omega f \cdot (w - v_\delta)\, dx - \int_\Omega (\nabla v_\delta)v_\delta \cdot (w - v_\delta)\, dx.
\end{aligned}
$$

It is easy to see that (4.1) (with $W = W_\delta$ and $v = v_\delta$) also implies the bound

$$
\sup_{0<\delta<1} \|\varepsilon(v_\delta)\|_{L^2} < \infty,
$$

hence we may assume that weak convergence holds as $\delta \downarrow 0$, i.e. there exists $v \in V_0$ such that $v_\delta \rightharpoonup v$. This implies $(1 \leq i, j \leq n)$

$$
\partial_i v_\delta^j \rightharpoonup \partial_i v^j \text{ weakly in } L^2(\Omega),
$$

$$
v_\delta^i(w^j - v_\delta^j) \to v^i(w^j - v^j) \text{ stronly in } L^2(\Omega).
$$

Note, that for the second statement we make use of the restriction $n \leq 3$. We therefore can discuss the right–hand side of (4.3) in the limit $\delta \downarrow 0$ with the result

$$
\int_\Omega f \cdot (w - v)\, dx - \int_\Omega (\nabla v)v \cdot (w - v)\, dx.
$$

The left–hand side of (4.3) is equal to

$$\int_\Omega \varepsilon(v_\delta) : \varepsilon(w)\, dx - \int_\Omega |\varepsilon(v_\delta)|^2\, dx$$

$$+ \int_\Omega (\delta + |\varepsilon(v_\delta)|^2)^{-1/2} \varepsilon(v_\delta) : (\varepsilon(w) - \varepsilon(v_\delta))\, dx$$

$$\leq \int_\Omega \varepsilon(v_\delta) : \varepsilon(w)\, dx - \int_\Omega |\varepsilon(v_\delta)|^2\, dx$$

$$+ \int_\Omega \left\{ (\delta + |\varepsilon(w)|^2)^{1/2} - (\delta + |\varepsilon(v_\delta)|^2)^{1/2} \right\} dx$$

$$\leq \int_\Omega \varepsilon(v_\delta) : \varepsilon(w)\, dx - \int_\Omega |\varepsilon(v_\delta)|^2\, dx + \int_\Omega \left\{ (\delta + |\varepsilon(w)|^2)^{1/2} - |\varepsilon(v_\delta)| \right\} dx$$

$$= \int_\Omega \varepsilon(v_\delta) : \varepsilon(w)\, dx + \int_\Omega (\delta + |\varepsilon(w)|^2)^{1/2}\, dx - \int_\Omega \left(|\varepsilon(v_\delta)|^2 + |\varepsilon(v_\delta)| \right) dx,$$

hence

$$\limsup_{\delta \downarrow 0} \int_\Omega \frac{\partial W_\delta}{\partial E}(\varepsilon(v_\delta)) : (\varepsilon(w) - \varepsilon(v_\delta))\, dx$$

$$\leq \lim_{\delta \downarrow 0} \left\{ \int_\Omega \varepsilon(v_\delta) : \varepsilon(w)\, dx + \int_\Omega (\delta + |\varepsilon(w)|^2)^{1/2}\, dx \right\}$$

$$- \liminf_{\delta \downarrow 0} \int_\Omega \left(|\varepsilon(v_\delta)|^2 + |\varepsilon(v_\delta)| \right) dx$$

$$\leq \int_\Omega \varepsilon(v) : \varepsilon(w)\, dx + \int_\Omega |\varepsilon(w)|\, dx - \int_\Omega \left(|\varepsilon(v)|^2 + |\varepsilon(v)| \right) dx.$$

This shows that the weak limit v is a solution of (BVI). Strong convergence (at least for this subsequence) is shown as follows: (BVI) together with the equation for v_δ implies

$$\int_\Omega (\varepsilon(v) - \varepsilon(v_\delta)) : (\varepsilon(v_\delta) - \varepsilon(v))\, dx$$

$$+ \int_\Omega \left(|\varepsilon(v_\delta)| - |\varepsilon(v)| - (\delta + |\varepsilon(v_\delta)|^2)^{-1/2} \varepsilon(v_\delta) : (\varepsilon(v_\delta) - \varepsilon(v)) \right) dx$$

$$+ \int_\Omega \left\{ (\nabla v)v - (\nabla v_\delta)v_\delta \right\} \cdot (v_\delta - v)\, dx \geq 0,$$

hence

$$\int_\Omega |\varepsilon(v_\delta) - \varepsilon(v)|^2 dx$$

$$\leq \int_\Omega \{|\varepsilon(v_\delta)| - |\varepsilon(v)| - (\delta + |\varepsilon(v_\delta)|^2)^{-1/2} \varepsilon(v_\delta) : (\varepsilon(v_\delta) - \varepsilon(v))\} dx$$

$$+ \int_\Omega \{(\nabla v)v - (\nabla v_\delta)v_\delta\} \cdot (v_\delta - v) \, dx = \alpha + \beta.$$

Clearly $\beta \to 0$ as $\delta \downarrow 0$, for α we have

$$\alpha \leq \int_\Omega \{|\varepsilon(v_\delta)| - |\varepsilon(v)| + (\delta + |\varepsilon(v)|^2)^{1/2} - (\delta + |\varepsilon(v_\delta)|^2)^{1/2}\} \, dx$$

$$\leq \int_\Omega \left((\delta + |\varepsilon(v)|^2)^{1/2} - |\varepsilon(v)| \right) dx \longrightarrow 0 \text{ as } \delta \downarrow 0,$$

and therefore $v_\delta \to v$ as $\delta \downarrow 0$.

$\square$

DEFINITION 3.4.1 *Suppose that $n = 2$ or 3 and let $f \in L^2(\Omega; \mathbb{R}^n)$.*

a) *A function $v \in W_2^1(\Omega; \mathbb{R}^n)$ satisfying div $v = 0$ is said to be a weak solution of Bingham variational inequality (4.2) if and only if (4.2) holds for any w from the space V_0.*

b) *Let W denote the dissipative potential from Theorem 3.4.1. We say that a function $v \in W_2^1(\Omega; \mathbb{R}^n)$, div $v = 0$, is a weak solution of the Navier–Stokes system associated to W if and only if (4.1) holds for any $\varphi \in V_0$.*

REMARK 3.4.3 We did not try to prove existence theorems for arbitrary Dirichlet boundary data v_0 but we expect that the results are similar to the corresponding results for the Navier–Stokes system.

REMARK 3.4.4 Of course we can also define local solution $v \in W_{2,\text{loc}}^1(\Omega; \mathbb{R}^n)$ for volume forces $f \in L_{\text{loc}}^2(\Omega; \mathbb{R}^n)$ by requiring that (4.1) and (4.2) are valid for functions in V_0 with compact support.

Next we state our regularity results concerning arbitrary weak solutions of (4.1) and (4.2).

THEOREM 3.4.3 *Suppose that $n = 2$ or 3 and let $f \in L^2(\Omega; \mathbb{R}^n)$.*
Assume further that v is a weak solution of (BVI). Then we have the following statements:

a) $v \in W^2_{2,\text{loc}}(\Omega; \mathbb{R}^n)$.

b) *If $f \in L^{2,n-2}(\Omega; \mathbb{R}^n)$, then a) holds and $v \in C^{0,\alpha}(\Omega; \mathbb{R}^n)$ for any $\alpha < 1$.*

c) *In addition to the above hypothesis we assume that $f \in L^{2,n-2+2\nu}(\Omega; \mathbb{R}^n)$ for some $0 < \nu < 1$. Then:*

    i) $\varepsilon(v) \in L^\infty_{\text{loc}}(\Omega; \mathbb{M})$

    ii) *For $n = 2$ $\varepsilon(v)$ is a continuous function.*

    iii) *There exists an open subset Ω_0^+ of Ω such that $\varepsilon(v) = 0$ almost everywhere on $\Omega - \Omega_0^+$ and $\varepsilon(v) \neq 0$ on Ω_0^+. Moreover, $v \in C^{1,\alpha}(\Omega_0^+; \mathbb{R}^n)$ for any $\alpha < \nu$.*

REMARK 3.4.5 All results extend to local solutions of (BVI) with volume forces in appropriate local Morrey spaces. This is also true in the situation of the next Theorem.

THEOREM 3.4.4 *Let the assumptions of Theorem 3.4.3 hold and let v now denote a weak solution of (4.1) with W defined in Theorem 3.4.1. Then the statements a), b) and c) i) are true. In place of c) ii), iii) we obtain: there exists an open subset Ω_0 of Ω such that $\mathcal{H}^{n-2}(\Omega - \Omega_0) = 0$ and $v \in C^{1,\lambda}(\Omega_0; \mathbb{R}^n)$ for any $\lambda < \nu$. In particular, if $n = 2$, we have $\Omega_0 = \Omega$.*

The proof of Theorem 3.4.3 is based on

LEMMA 3.4.1 *Let $n = 2$ or 3 and let v denote a weak solution of (BVI) with $f \in L^2(\Omega; \mathbb{R}^n)$. Then we have $(\nabla v)v \in L^{2,n-2+2\alpha}_{\text{loc}}(\Omega; \mathbb{R}^n)$ for any $\alpha \in]0, 2 - \frac{n}{2}[$. If we assume $f \in L^{2,n-2}(\Omega; \mathbb{R}^n)$, then we have $(\nabla v)v \in L^{2,n-2+2\alpha}_{\text{loc}}(\Omega; \mathbb{R}^n)$ for any $\alpha \in]0, 1[$.*

Proof of Lemma 3.4.1: Since $n = 2$ or 3 we have by the imbedding theorem

$$(4.4) \qquad A_1 = \|v\|_{L^{2n}(\Omega)} < \infty.$$

Consider a ball $B_R(x_0) \subset \Omega$ and let v_0 denote the solution of the linear Stokes problem

$$(4.5) \quad \begin{cases} \displaystyle\int_{B_R(x_0)} \varepsilon(v_0) : \varepsilon(\varphi)\, dx = 0 \text{ for all } \varphi \in \overset{\circ}{W}^1_2(B_R(x_0); \mathbb{R}^n), \text{ div } \varphi = 0, \\[2em] v_0 \in v + \overset{\circ}{W}^1_2(B_R(x_0); \mathbb{R}^n), \text{ div } v_0 = 0 \end{cases}$$

which satisfies the estimate

$$\int_{B_\rho(x_0)} |\nabla v_0|^2 dx \le c_1(n)\left(\frac{\rho}{R}\right)^n \int_{B_R(x_0)} |\nabla v_0|^2 dx, \ 0 \le \rho \le R,$$

hence

(4.6)
$$\int_{B_\rho(x_0)} |\nabla v|^2 dx \le c_2(n)\Big\{\left(\frac{\rho}{R}\right)^n \int_{B_R(x_0)} |\nabla v|^2 dx$$
$$+ \int_{B_R(x_0)} |\nabla v - \nabla v_0|^2 dx\Big\}, \ 0 \le \rho \le R.$$

It therefore remains to estimate the last integral on the right–hand side of (4.6). Using (4.2) (w.l.o.g. we let $\mu = 1/2$, $\sqrt{2}k_* = 1$) and (4.5) we obtain

(4.7)
$$\begin{cases}
\dfrac{1}{2}\displaystyle\int_{B_R(x_0)} |\nabla v - \nabla v_0|^2 dx = \int_{B_R(x_0)} |\varepsilon(v) - \varepsilon(v_0)|^2 dx \\[2ex]
= \displaystyle\int_{B_R(x_0)} \varepsilon(v) : (\varepsilon(v) - \varepsilon(v_0))\, dx \\[2ex]
\le \displaystyle\int_{B_R(x_0)} \{|\varepsilon(v_0)| - |\varepsilon(v)| - f \cdot (v_0 - v) + (\nabla v)v \cdot (v_0 - v)\}\, dx \\[2ex]
\le \displaystyle\int_{B_R(x_0)} \{|\varepsilon(v_0 - v)| + |f|\,|v_0 - v| + |v|\,|\nabla v|\,|v_0 - v|\}\, dx.
\end{cases}$$

Clearly

(4.8)
$$\int_{B_R(x_0)} |\varepsilon(v_0) - \varepsilon(v)|\, dx \le |B_R(x_0)|^{1/2}\left(\int_{B_R(x_0)} |\varepsilon(v_0) - \varepsilon(v)|^2 dx\right)^{1/2}$$

and

(4.9)
$$\int_{B_R(x_0)} |f|\,|v_0 - v|\, dx \le c_3(n)R\left(\int_{B_R(x_0)} |f|^2 dx\right)^{1/2}$$
$$\left(\int_{B_R(x_0)} |\nabla v - \nabla v_0|^2 dx\right)^{1/2}.$$

For the remaining integral we observe

$$\int_{B_R(x_0)} |v|\,|\nabla v|\,|v_0 - v|\,dx \le \left(\int_{B_R(x_0)} |v_0 - v|^{2n}dx \right)^{1/2n}$$

$$\times \left(\int_{B_R(x_0)} [|v|\,|\nabla v|]^{\frac{2n}{2n-1}}dx \right)^{1-1/2n}$$

$$\le c_4(n)R^{-\frac{n-3}{2}} \left(\int_{B_R(x_0)} |\nabla v_0 - \nabla v|^2\,dx \right)^{1/2}$$

$$\times \left(\int_{B_R(x_0)} |\nabla v|^2 dx \right)^{1/2} \left(\int_{B_R(x_0)} |v|^{\frac{2n}{n-1}}dx \right)^{\frac{n-1}{2n}}.$$

In case $n = 2$ we have by (4.4)

$$\left(\int_{B_R(x_0)} |v|^{\frac{2n}{n-1}}dx \right)^{\frac{n-1}{2n}} \le A_1,$$

for $n = 3$ we observe

$$\left(\int_{B_R(x_0)} |v|^{\frac{2n}{n-1}}dx \right)^{\frac{n-1}{2n}} \le c_5(n)A_1\,R^{\frac{n-2}{2}},$$

hence

$$(4.10) \quad \begin{cases} \displaystyle\int_{B_R(x_0)} |v|\,|\nabla v|\,|v_0 - v|\,dx \le c_6(n)R^{1/2}A_1 \\[2mm] \displaystyle\times \left(\int_{B_R(x_0)} |\nabla v_0 - \nabla v|^2 dx \right)^{1/2} \left(\int_{B_R(x_0)} |\nabla v|^2 dx \right)^{1/2}. \end{cases}$$

Next we combine the estimates (4.6) – (4.10) to get

$$(4.11) \quad \begin{cases} \displaystyle\int_{B_\rho(x_0)} |\nabla v|^2 dx \le c_7(n)\Big\{ \big[(\tfrac{\rho}{R})^n + RA_1^2\big] \int_{B_R(x_0)} |\nabla v|^2 dx \\[2mm] \displaystyle +R^n + R^2 \int_{B_R(x_0)} |f|^2 dx \Big\}, \quad 0 < \rho \le R < \operatorname{dist}(x_0, \partial\Omega). \end{cases}$$

If f is just a function of class L^2, then (4.11) implies the growth estimate

$$\int_{B_\rho(x_0)} |\nabla v|^2 dx \le c_8(\alpha, n, A_1, A_2, \text{diam } \Omega, \|\nabla v\|_2) \rho^{n-2+2\alpha}$$

for any $\alpha < 2 - n/2$, $A_2 = \|f\|_{L^2}$, hence $v \in C^{0,\alpha}(\Omega; \mathbb{R}^n)$, is particular v in $L^\infty_{\text{loc}}(\Omega; \mathbb{R}^n)$, so that

$$(\nabla v)v \in L^{2,n-2+2\alpha}_{\text{loc}}(\Omega; \mathbb{R}^n)$$

for $\alpha < 2 - n/2$. If $f \in L^{2,n-2}(\Omega; \mathbb{R}^n)$, then (4.1) implies the same result for any $\alpha < 1$.

□

Proof of Theorem 3.4.3: Let us define the artificial volume force $\tilde{f} = f - (\nabla v)v \in L^2(\Omega; \mathbb{R}^n)$. Then the function v is the minimizer of

$$J(u) = \int_\Omega \{\mu|\varepsilon(u)|^2 + \sqrt{2}k_*|\varepsilon(u)| - \tilde{f} \cdot u\} \, dx$$

in the class $\mathbb{K} = \{w \in W^1_2(\Omega; \mathbb{R}^n) : w = v \text{ on } \partial\Omega, \text{ div } w = 0\}$, and part a) of Theorem 3.4.3 follows from Theorem 3.3.4. b) was shown in Lemma 3.4.1. Assume now that $f \in L^{2,n-2+2\nu}(\Omega; \mathbb{R}^n)$ for some $\nu < 1$. Then Lemma 3.4.1 implies $(\nabla v)v \in L^{2,n-2+2\alpha}_{\text{loc}}(\Omega; \mathbb{R}^n)$ which means $\tilde{f} \in L^{2,n-2+2\nu}_{\text{loc}}(\Omega; \mathbb{R}^n)$. Statement c)i) follows from Theorem 3.2.3 (by noting that the proof of this theorem is the same in the presence of volume forces $\tilde{f}$ as above), c)ii) is contained in Theorem 3.3.3, c)iii) is a variant of the Main Theorem of section 1 (combinded with the remark that the Main Theorem holds for volume forces in some Morrey space).

□

In a similar way we can reduce the proof of Theorem 3.4.4 to the results obtained for (local) minimizers by establishing the following version of Lemma 3.4.1.

LEMMA 3.4.2 *Let $n = 2$ or 3 and let v denote a weak solution of (4.1) with W defined in Theorem 3.4.1. If $f \in L^2(\Omega; \mathbb{R}^n)$, then we have $(\nabla v)v$ in $L^{2,n-2+2\alpha}_{\text{loc}}(\Omega; \mathbb{R}^n)$ for any $\alpha < 2 - n/2$. If $f \in L^{2,n-2}(\Omega; \mathbb{R}^n)$, then the same result holds for any $\alpha < 1$.*

Proof of Lemma 3.4.2: Let $W(E) = \frac{1}{2}|E|^2 + (\delta + |E|^2)^{p/2}$, the Powell–Eyring model is discussed in the same way. We define v_0 as in (4.5) and get estimate

(4.6); (4.7) has to be replaced by

(4.12)
$$
\begin{cases}
\dfrac{1}{2} \displaystyle\int_{B_R(x_0)} |\nabla v - \nabla v_0|^2 dx = \int_{B_R(x_0)} \varepsilon(v) : (\varepsilon(v) - \varepsilon(v_0))\, dx \\[4mm]
= \displaystyle\int_{B_R(x_0)} \dfrac{\partial W_0}{\partial E}(\varepsilon(v)) : (\varepsilon(v_0) - \varepsilon(v))\, dx + \int_{B_R(x_0)} (\nabla v)v \cdot (v_0 - v)\, dx \\[4mm]
+ \displaystyle\int_{B_R(x_0)} f \cdot (v_0 - v)\, dx \\[4mm]
\leq \displaystyle\int_{B_R(x_0)} \left| \dfrac{\partial W_0}{\partial E}(\varepsilon(v)) \right| |\varepsilon(v_0) - \varepsilon(v)|\, dx \\[4mm]
+ \displaystyle\int_{B_R(x_0)} \{|f| + |\nabla v|\, |v|\}\, |v - v_0|\, dx,
\end{cases}
$$

where $W_0(E) = (\delta + |E|^2)^{p/2}$. The last integral on the right–hand side of (4.12) is estimated as in Lemma 3.4.1. For the first term we observe

(4.13)
$$
\begin{cases}
\displaystyle\int_{B_R(x_0)} |\varepsilon(v) - \varepsilon(v_0)|\, \left| \dfrac{\partial W_0}{\partial E}(\varepsilon(v)) \right|\, dx \\[4mm]
\leq \left(\displaystyle\int_{B_R(x_0)} |\varepsilon(v) - \varepsilon(v_0)|^2 dx \right)^{1/2} \left(\int_{B_R(x_0)} \left| \dfrac{\partial W_0}{\partial E}(\varepsilon(v)) \right|^2 dx \right)^{1/2}.
\end{cases}
$$

The definition of W_0 implies (w.l.o.g. $\delta \leq 1$)

$$
\int_{B_R(x_0)} \left| \dfrac{\partial W_0}{\partial E}(\varepsilon(v)) \right|^2 dx \leq p \int_{B_R(x_0)} (\delta + |\varepsilon(v)|^2)^{p-1} dx
$$

$$
\leq
\begin{cases}
|B_R(x_0)| \qquad \text{if } p = 1 \\[6mm]
p|B_R(x_0)|^{2-p} \left(\displaystyle\int_{B_R(x_0)} (\delta + |\varepsilon(v)|^2)\, dx \right)^{p-1} \text{if } 1 < p < 2
\end{cases}
$$

$$
\leq 2 \left\{ |B_R(x_0)| + |B_R(x_0)|^{2-p} \left(\int_{B_R(x_0)} |\nabla v|^2 dx \right)^{p-1} \right\},
$$

and from (4.12), (4.13) we get

$$\int_{B_R(x_0)} |\nabla v_0 - \nabla v|^2 dx \le c_1 \left(\int_{B_R(x_0)} |\nabla v_0 - \nabla v|^2 dx \right)^{1/2} \left\{ R^{\frac{n}{2}} + R^{\frac{n}{2}(2-p)} \right.$$

$$\times \left(\int_{B_R(x_0)} |\nabla v|^2 dx \right)^{\frac{p-1}{2}} + R \left(\int_{B_R(x_0)} |f|^2 dx \right)^{1/2} + R^{\frac{1}{2}} \left. \left(\int_{B_R(x_0)} |\nabla v|^2 dx \right)^{1/2} \right\}$$

for a suitable constant c_1 independent of $B_R(x_0)$. Next we apply Young's inequality to obtain for any $t > 0$

(4.14)
$$\begin{cases} \displaystyle\int_{B_R(x_0)} |\nabla v - \nabla v_0|^2 dx \le c_2 \Big\{ (R + t) \int_{B_R(x_0)} |\nabla v|^2 dx \\[2ex] \displaystyle +R^2 \int_{B_R(x_0)} |f|^2 dx + \Big(1 + t^{\frac{p-1}{p-2}} \Big) R^n \Big\} \end{cases}$$

with c_2 independent of t and $B_R(x_0)$. By combining (4.6) and (4.14) we get the final estimate

(4.15)
$$\begin{cases} \displaystyle\int_{B_\rho(x_0)} |\nabla v|^2 dx \le c_3 \Big\{ \Big[\Big(\frac{\rho}{R}\Big)^n + R + t \Big] \int_{B_R(x_0)} |\nabla v|^2 dx \\[2ex] \displaystyle + \Big(1 + t^{\frac{p-1}{p-2}} \Big) R^n + R^2 \int_{B_R(x_0)} |f|^2 dx \Big\} \end{cases}$$

being valid for any $o < \rho < R <$ dist $(x_0, \partial\Omega)$ and $t > 0$. From (4.15) the claim follows along standard lines.

$\square$

3.5   Some open problems and comments concerning extensions

We briefly adress some questions which might be of interest for further research.

a) nonlinear systems of Stokes type

Most of the regularity results described in the foregoing sections were based on Campanato-type estimates for solutions $u \in W_2^1(\Omega; \mathbb{R}^n)$ of the linear problem

$$\begin{cases} \operatorname{div} u = 0 \text{ on } \Omega, \\ \\ \int_\Omega \varepsilon(u) : \varepsilon(\varphi) \, dx = 0, \ \varphi \in \overset{\circ}{W}_2^1(\Omega; \mathbb{R}^n), \ \operatorname{div} \varphi = 0. \end{cases}$$

Of course we would like to have a corresponding nonlinear variant: for $1 < p < \infty$ consider a function $u \in W_p^1(\Omega; \mathbb{R}^n)$ such that

$$\begin{cases} \operatorname{div} u = 0 \text{ on } \Omega, \\ \\ \int_\Omega |\varepsilon(u)|^{p-2} \, \varepsilon(u) : \varepsilon(\varphi) \, dx = 0, \ \varphi \in \overset{\circ}{W}_p^1(\Omega; \mathbb{R}^n), \ \operatorname{div} \varphi = 0, \end{cases}$$

and prove regularity estimates in the spirit of Uhlenbeck [U]. This would imply optimal regularity results for power law fluid models. Moreover, we could improve our results on minimizers of $J(u) = \int_\Omega |\varepsilon(u)|^p + \kappa |\varepsilon(u)| \, dx \ (\kappa > 0)$: minimizers are of class $C^{0,\alpha}(\Omega; \mathbb{R}^n)$ for any $0 < \alpha < 1$. We conjecture that also Theorem 3.2.3 (local boundedness of $\varepsilon(u)$) extends to the p–case.

b) the p–Bingham variational inequality

Starting from the dissipative potential $W(E) = \frac{1}{p}|E|^p + |E|, \ 1 < p < \infty$, the classical Bingham variational inequality (4.2) has to be replaced by the following more general problem: to find a function u with prescribed boundary data such that $\operatorname{div} u = 0$ and

$$\begin{cases} \int_\Omega |\varepsilon(u)|^{p-2}\varepsilon(u) : (\varepsilon(w) - \varepsilon(u)) + |\varepsilon(w)| - |\varepsilon(u)| \\ \\ +(\nabla u)u \cdot (w - u) \, dx \geq \int_\Omega f \cdot (w - u) \, dx \end{cases}$$

for every w satisfying div $w = 0$ and also $w|_{\partial\Omega} = u|_{\partial\Omega}$. The time–dependent version of p–BVI is investigated in the papers of [Ka] and [Ki]. In order to extend Theorem 3.4.3 to weak solutions of p–BVI, we first have to solve problem a). These difficulties can be avoided by making the assumption that $p > \dim\Omega$. Let us further assume that the volume forces f are in the Morrey space $L^{q,n-q+\mu q}_{loc}(\Omega;\mathbb{R}^n)$ for some $0 < \mu < 1$ where $q = p/p{-}1$, and consider a local solution $v \in W^1_{p,loc}(\Omega;\mathbb{R}^n)$ of the p–Bingham variational inequality. In this case $v \in L^\infty_{loc}(\Omega;\mathbb{R}^n)$ by Sobolev's embedding theorem and we may apply Hölder's inequality to see $(\nabla v)v \in L^{q,n-q+q(1-n/p)}_{loc}(\Omega;\mathbb{R}^n)$. Hence $\tilde{f} = f - (\nabla v)v$ is in $L^{q,n-q+\lambda q}_{loc}(\Omega,\mathbb{R}^n)$, $\lambda = \min\{\mu, 1 - n/p\}$, and v clearly is a local minimizer of $w \mapsto \int_\Omega \left(\frac{1}{p}|\varepsilon(w)|^p + |\varepsilon(w)| - \tilde{f}\cdot w\right) dx$ subject to the constraint div $w = 0$. From the Main Theorem in section 1 we get:

$$\begin{cases} \text{There is an open set } \Omega_0^+ \text{ such that } v \in C^{1,\alpha}(\Omega_0^+;\mathbb{R}^n) \text{ for any } \alpha < \lambda \text{ and} \\ \varepsilon(v)(x) \neq 0 \text{ for all } x \in \Omega_0^+; \text{ on } \Omega - \Omega_0^+ \text{ we have } \varepsilon(v) = 0 \text{ a.e.} \end{cases}$$

Hence $p > n$ together with $f \in L^{q,n-q+\mu q}_{loc}(\Omega;\mathbb{R}^n)$ is a sufficient (but unsatisfying) condition for partial C^1–regularity of solutions of the p–Bingham variational inequality. We do not know if $\varepsilon(v)$ is still in the space $L^\infty_{loc}(\Omega;\mathbb{M})$.

c) analytic properties of the separating curve

As mentioned in section 3 it is of interest to describe the separating curve $\Omega \cap \partial[\varepsilon(u) \neq o]$ where u is either a minimizer of the energy associated to the dissipative potential $W(E) = \mu|E|^2 + \sqrt{2}k_*|E|$ or just a solution of the Bingham variational inequality, provided Ω is a domain in $\mathbb{R}^2$.

d) evolution problems

Duvaut and Lions [DL] present an evolution model for the classical Bingham variational inequality and prove existence theorems, the evolution problem for p–BVI is solved in the papers of Kato [Ka] and Kim [Ki]. What are the regularity properties of these weak solutions?

e) boundary regularity

For sufficiently regular boundary data v_0 (e.g. $v_0 \in C^{1,\alpha}(\overline{\Omega};\mathbb{R}^n)$) we expect that Theorems 3.2.3, 3.4.3 c)i) and 3.4.4 can be improved to global boundedness of the strain velocity field. Some ideas can be found in the paper [FL], the details are carried out in [R] and [Sh1,2].

f) pseudoplastic fluids

are introduced in [AM]. The dissipative potential is just $W(E) = K|E|^{1+\alpha}$ with

$K > 0$ and $\alpha \in]0,1[$. As an approximation we choose $W_\delta(E) = (\delta + |E|^2)^{(1+\alpha)/2}$ with $\delta > 0$ fixed but very small. Then there holds partial C^1–regularity for local minimizers v, div $v = 0$, of $w \mapsto \int_\Omega \left(W_\delta(\varepsilon(w)) - f \cdot w \right) dx$, provided f is sufficiently regular, we refer to [R] for a further discussion.

Chapter 4

Fluids of Prandtl–Eyring type and plastic materials with logarithmic hardening law

4.0 Preliminaries

According to the discussion at the beginning of the previous chapter the slow, steady state motion of a fluid of Prandtl-Eyring type in a bounded domain $\Omega \subset \mathbb{R}^n$, $n = 2$ or $n = 3$, is governed by the following set of equations

$$(0.1) \qquad \operatorname{div} v = 0$$

$$(0.2) \qquad \operatorname{div} \tau = 0, \quad \tau = \sigma - p\mathbf{1}$$

$$(0.3) \qquad \sigma = \eta_0 \, \frac{\operatorname{ar\,sinh}(\lambda|\varepsilon(v)|)}{\lambda|\varepsilon(v)|} \, \varepsilon(v)$$

where $v : \Omega \to \mathbb{R}^n$ denotes the velocity field and η_0, λ are some fixed physical parameters. We further use the symbols τ and p for the Cauchy stress tensor and a suitable pressure function, respectively. Let us remark that, just for technical simplicity, we consider the case of zero volume forces $f : \Omega \to \mathbb{R}^n$, otherwise (0.2) has to be replaced by

$$(0.4) \qquad \operatorname{div} \tau = f, \quad \tau = \sigma - p\mathbf{1}.$$

As outlined in section 0 of chapter 3 the velocity field v is a minimizer of the variational integral

$$(0.5) \quad \begin{cases} J(v, \Omega) = \int\limits_{\Omega} W\big(\varepsilon(v)\big) \, dx, \\[2ex] W(E) = \eta_0 \int\limits_{0}^{|E|} \frac{1}{\lambda} \operatorname{ar} \sinh(\lambda t) \, dt, \end{cases}$$

in classes of functions $u : \Omega \to \mathbb{R}^n$ with div $u = 0$. Note, that under the assumption (0.4), we have to add the potential $\int\limits_{\Omega} f \cdot v \, dx$ to $J(v, \Omega)$.

From this point of view it is reasonable to discuss local minimizers of the functional J from (0.5) and to analyze their regularity properties. In contrast to our previous investigations it is not immediately obvious to which natural function space local minimizers should belong which means that we have to introduce some spaces providing also existence theorems. The correct class is

$$\overset{\circ}{V}(\Omega) = \left\{ v \in L^1(\Omega, \mathbb{R}^n) : \operatorname{div} v = 0, \int\limits_{\Omega} |\varepsilon(v)| \ln\big(1 + |\varepsilon(v)|\big) \, dx < \infty \right\}$$

and we will show (compare [FS4])

THEOREM 4.0.1 *Suppose that $v \in \overset{\circ}{V}(\Omega)$ locally minimizes $J(\cdot, \Omega)$ in this class. Then we have Hölder continuity of $\varepsilon(v)$ on an open set $\Omega_0 \subset \Omega$ whose complement is of Lebesgue measure zero. For two–dimensional domains Ω we get $\Omega_0 = \Omega$.*

REMARK 4.0.1 Our investigations of Prandtl–Eyring fluids are restricted to small values of v, i.e. we impose the restriction $(\nabla v)v = 0$. It is an open problem if the results from chapter 3, section 4, can be extended to the Prandtl–Eyring fluid model.

In order to keep notation simple we will replace J from (0.5) by the functional

$$w \mapsto \int\limits_{\Omega} G\big(\varepsilon(w)\big) \, dx, \quad G(E) = |E| \ln(1 + |E|),$$

which is no serious drawback: actually our arguments work for any integrand $g(|\varepsilon(w)|)$ with $g(t)$ being C^2–close to $t \ln(1 + t)$, $t \geq 0$, (see [FreSe], section 2, for a precise definition) and $W(E)$ from (0.5) is of this type.

Let us now turn to the plasticity case: here Ω denotes the undeformed state of an elasto–plastic body and in the case of logarithmic hardening an admissible displacement field $u : \Omega \to \mathbb{R}^n$ is sought which minimizes

$$(0.6) \qquad I(u, \Omega) = \int_\Omega \left[G\left(\varepsilon^D(u)\right) + (\text{div } u)^2 \right] dx$$

under appropriate boundary conditions (again we neglect boundary and volume force terms). The logarithmic case $G(E) = |E| \ln(1 + |E|)$ lies between perfect plasticity and plasticity with power hardening which have been discussed in chapter 1 and 2. As shown in [FreSe] the natural function space for I from (0.6) is

$$X(\Omega) = \left\{ u \in L^1(\Omega, \mathbb{R}^n) : \text{div } u \in L^2(\Omega), \int_\Omega |\varepsilon^D(u)| \ln \left(1 + |\varepsilon^D(u)|\right) dx < \infty \right\}$$

and in the two–dimensional case local $I(\cdot, \Omega)$–minimizers from $X(\Omega)$ are smooth in the interior of Ω. This result is completed by

THEOREM 4.0.2 *Consider a domain $\Omega \subset \mathbb{R}^3$ and let $u \in X(\Omega)$ denote a local $I(\cdot, \Omega)$–minimizer. Then $\varepsilon(u)$ is of class $C^{0,\alpha}$ for any $\alpha < 1$ on some open subset Ω_0 of Ω such that the measure of the complement of Ω_0 is zero.*

REMARK 4.0.2 Of course it is possible to replace $G\left(\varepsilon^D(u)\right)$ in definition (0.6) by $g_0\left(|\varepsilon^D(u)|\right)$ with g_0 being in a C^2–sense close to $t \ln(1 + t)$, again we refer to [FreSe], section 2, for a precise statement. But since in this case the calculations become more technical without significant changes we prefer to consider the simplified model.

REMARK 4.0.3 From the point of view of applications it might be interesting to consider also the case when $G\left(\varepsilon^D(u)\right)$ is replaced by a function $g_0\left(|\varepsilon^D(u)|\right)$ with g_0'' discontinuous at some point t_0 and an asymptotic behaviour like $|\varepsilon^D(u)| \ln(1 + |\varepsilon^D(u)|)$. For the necessary changes we refer to chapter 2, section 8.

Chapter 4 is organized as follows: in section 1 we introduce some natural function spaces for Prandtl–Eyring fluids and give a precise formulation of Theorem 4.0.1 (see Theorem 4.1.1). Section 2 is devoted to the proof of the fact that for a local minimizer v of (0.5) the functions $\sqrt{1 + |\varepsilon(v)|}$ and $\sigma = \frac{\partial G}{\partial E}(\varepsilon(v))$ have weak derivatives in $L^2_{\text{loc}}(\Omega)$. Here we make use of some Powell–Eyring type approximation. As a byproduct we obtain that the strain velocity is locally square–integrable on Ω, moreover, we prove a Caccioppoli–type inequality. Using a blow–up argument it is shown in section 3 that partial regularity holds for $\varepsilon(v)$ in case $n = 3$. The two–dimensional part of Theorem 4.0.1 is discussed in section 4 essentially following [FreSe]. In section 5 we give the necessary modifications for plastic materials with logarithmic hardening. A final section presents proofs of some density results for the spaces introduced in section 1.

4.1 Some function spaces related to the Prandtl–Eyring fluid model

We define the space $L \ln L(\Omega)$ in the following way: suppose that Ω is a bounded domain in $\mathbb{R}^n$. Consider an open cube Q_0 parallel to the axes of $\mathbb{R}^n$ and containing Ω. For any function $f \in L^1(\Omega)$ we let

$$\tilde{f} = \begin{cases} f & \text{in} \quad \Omega \\ 0 & \text{in} \quad Q_0 - \Omega \end{cases}$$

and introduce the norm (compare[FreSe], [Se15])

$$\|f\|_{L \ln L(\Omega)} = \int\limits_{Q_0} M_{Q_0} \tilde{f} \, dx$$

where $M_{Q_0} \tilde{f}$ is the maximal function (see [BI])

$$M_{Q_0} \tilde{f}(x) = \sup \left\{ \fint_Q |\tilde{f}| \, dx : x \in Q \subset Q_0 \right\},$$

the supremum being taken over all parallel open subcubes of Q_0 containing the point x. Let

$$L \ln L(\Omega) = \left\{ f \in L^1(\Omega) : \|f\|_{L \ln L(\Omega)} < \infty \right\}.$$

The proofs of the next results concerning the space $L \ln L(\Omega)$ can be found for example in [I], [St1,2].

LEMMA 4.1.1 *There exist positive constants $c_1(n)$, $c_2(n)$ such that for any $f \in L \ln L(\Omega)$ the following inequality is true*

(1.1) $$c_1 \|f\|_{L \ln L(\Omega)} \leq \int\limits_{\Omega} |f| \ln(2 + \frac{1}{a}|f|) \, dx \leq c_2 \|f\|_{L \ln L(\Omega)},$$

a denoting the mean value $\fint_{Q_0} |\tilde{f}| \, dx = \frac{1}{|Q_0|} \int_{Q_0} |\tilde{f}| \, dx$, $|Q_0|$ being Lebesgue's measure.

As an application of (1.1) it is easy to show that $L \ln L(\Omega)$ is a Banach space.

LEMMA 4.1.2 (De La Valée Poussin)
Suppose that the sequence $\{f_m\}$ is bounded in $L\ln L(\Omega)$. Then there exists a subsequence $\{f_{m_k}\}$ and a function $f \in L\ln L(\Omega)$ such that $f_{m_k} \rightharpoonup f$ weakly in $L^1(\Omega)$.

LEMMA 4.1.3 (see [FreSe], Lemma 2.2)
Let $Exp\,(\Omega)$ denote the normed dual of $L\ln L(\Omega)$. For any $\ell \in Exp\,(\Omega)$ there exists a unique measurable function g such that

$$\ell(f) = \int_\Omega gf\,dx, \quad f \in L\ln L(\Omega).$$

g satisfies

$$\int_\Omega e^{\lambda|g|}dx < \infty$$

with a suitable constant $\lambda > 0$. Conversely, any function g with this property generates an element of $Exp\,(\Omega)$ via integration.

According to the above lemma it is convenient to identify

$$\mathrm{Exp}(\Omega) = \{g : \Omega \to \mathbb{R} |\ g \text{ is measurable and } \int_\Omega e^{\lambda|g|}dx < \infty \text{ for some } \lambda > 0\}.$$

The next result is an easy calculation.

LEMMA 4.1.4 *For any $f \in L\ln L(\Omega)$ we have the estimates*

$$(1.2) \qquad \ln 2\Big(\int_\Omega |f|\,dx - |\Omega|\Big) \leq \int_\Omega |f|\ln(1+|f|)\,dx$$

$$(1.3) \qquad c_1\|f\|_{L\ln L(\Omega)} \leq \frac{1}{e}|Q_0| + 2\int_\Omega |f|\ln(1+|f|)\,dx + \ln 2|\Omega|.$$

By combining (1.1) and (1.3) we have

LEMMA 4.1.5 *Suppose that $f \in L^1(\Omega)$. Then f is in the space $L\ln L(\Omega)$ if and only if $\int_\Omega |f|\ln(1+|f|)\,dx < \infty$. Moreover, $\|f\|_{L\ln L(\Omega)}$ is controlled by the logarithmic integral and vice versa.*

Now we introduce some function spaces related to $L \ln L(\Omega)$. Let

$$V(\Omega) = \{u \in L^1(\Omega; \mathbb{R}^n) : |\varepsilon(u)| \in L \ln L(\Omega)\},$$

$$\overset{\circ}{V}(\Omega) = \{u \in V(\Omega) : \operatorname{div} u = \varepsilon_{ii}(u) = 0 \text{ on } \Omega\}$$

equipped with the norm

$$\|u\|_{V(\Omega)} = \|u\|_{L^1(\Omega)} + \|\varepsilon(u)\|_{L \ln L(\Omega)}.$$

The definition of the spaces $V_{\text{loc}}(\Omega)$, $\overset{\circ}{V}_{\text{loc}}(\Omega)$ should be obvious. Let us further define

$$\left\{ \begin{array}{l} V_0(\Omega) = \text{closure of } C_0^\infty(\Omega; \mathbb{R}^n) \text{ in } V(\Omega) \text{ w.r.t.} \| \cdot \|_{V(\Omega)}, \\[2em] \overset{\circ}{C}_0^\infty(\Omega; \mathbb{R}^n) = \{u \in C_0^\infty(\Omega; \mathbb{R}^n) : \operatorname{div} u = 0\}, \\[2em] \overset{\circ}{V}_0(\Omega) = \text{closure of } \overset{\circ}{C}_0^\infty(\Omega; \mathbb{R}^n) \text{ in } V(\Omega) \text{ w.r.t.} \| \cdot \|_{V(\Omega)}. \end{array} \right.$$

The following result is demonstrated in the Appendix.

LEMMA 4.1.6 *Suppose that Ω is a Lipschitz domain. Then*

a) $V(\Omega) = $ closure of $C^\infty(\bar{\Omega}; \mathbb{R}^n)$ in $V(\Omega)$ w.r.t. $\| \cdot \|_{V(\Omega)}$

b) $\overset{\circ}{V}(\Omega) = $ closure of $\{u \in C^\infty(\bar{\Omega}; \mathbb{R}^n) : \operatorname{div} u = 0\}$ in $V(\Omega)$ w.r.t. $\| \cdot \|_{V(\Omega)}$

c) $V_0(\Omega) = \{u \in V(\Omega) : u = 0 \text{ on } \partial\Omega\}$

d) $\overset{\circ}{V}_0(\Omega) = \{u \in \overset{\circ}{V}(\Omega) : u = 0 \text{ on } \partial\Omega\}$.

In c), d) of Lemma 4.1.6 $u = 0$ on $\partial\Omega$ has to be understood in the trace sense. Using the obvious inclusion

$$(1.4) \qquad V(\Omega) \subset BD(\Omega; \mathbb{R}^n)$$

of $V(\Omega)$ into the space of functions having bounded deformation (see Appendix to chapter 2) we can associate to $u \in V(\Omega)$ a trace in $L^1(\partial\Omega; \mathbb{R}^n)$ (compare [ST2], Theorem 1.1). From [ST2] or [AG1] (see also [MM1]) we get by quoting (1.4) again

LEMMA 4.1.7 *Let Ω denote a Lipschitz domain*

a) *For $p \leq \frac{n}{n-1}$ we have $V(\Omega) \subset L^p(\Omega, \mathbb{R}^n)$ and the inclusion is compact for $p < \frac{n}{n-1}$.*

b) *Let Γ be a subset of $\partial\Omega$ with positive $(n-1)$–dimensional measure, then there is a constant $c = c(n, \Omega, \Gamma)$ such that*

$$\|u\|_{L^{n/n-1}(\Omega)} \leq c \int\limits_{\Omega} |\varepsilon(u)| \, dx$$

for all $u \in V(\Omega)$ with $u|_\Gamma = 0$. In particular, this result holds for u in $V_0(\Omega)$.

We introduce the energy density

$$G(E) = |E| \ln(1 + |E|)$$

for matrices $E \in \mathbb{M}$ and observe

(1.5) $\quad \dfrac{\partial G}{\partial E}(F) = \left(\ln(1 + |F|) + \dfrac{|F|}{1+|F|} \right) \dfrac{F}{|F|}, \quad \left| \dfrac{\partial G}{\partial E}(F) \right| \leq 2\ln(1 + |F|)$

(1.6)

$$\begin{cases} D^2G(F)(E, H) = & \dfrac{g'(|F|)}{|F|} \left[E : H - \dfrac{(E:F)(H:F)}{|F|^2} \right] \\[2mm] & + g''(|F|) \dfrac{(E:F)(H:F)}{|F|^2} \\[2mm] E, F, H \in \mathbb{M}, & g(t) = t\ln(1 + t), t \geq 0. \end{cases}$$

Using (1.6) it is easy to see that

(1.7) $\quad \dfrac{|E|^2}{1 + |F|} \leq D^2G(F)(E, E) \leq 2\dfrac{\ln(1 + |F|)}{|F|}|E|^2, \quad |D^2G(F)| \leq 2$

holds for $E, F \in \mathbb{M}$.

According to Lemma 4.1.5 the variational integral

$$J(u, \Omega) = \int\limits_{\Omega} G\big(\varepsilon(u)\big) \, dx$$

is well defined on the space $V(\Omega)$ and exactly as in [FreSe], Theorem 2.1, one can show: if $\partial\Omega$ is Lipschitz and if u_0 denotes a given function in $\overset{\circ}{V}(\Omega)$, then the variational problem

$$J(\cdot, \Omega) \to \min \text{ in } u_0 + \overset{\circ}{V}_0(\Omega)$$

admits a unique solution. From this point of view it is reasonable to introduce the following concept.

DEFINITION 4.1.1 *A function* $u \in \overset{\circ}{V}_{\text{loc}}(\Omega)$ *is said to be a local minimizer of the energy J if and only if*

$$J(u, \text{spt}\varphi) \leq J(u + \varphi, \text{spt}\varphi)$$

for all $\varphi \in \overset{\circ}{C}{}_0^\infty(\Omega, \mathbb{R}^n)$.

REMARK 4.1.1 Let $u \in \overset{\circ}{V}_{\text{loc}}(\Omega)$ denote a local minimizer. Then we have

$$J(u, \omega) \leq J(u + v, \omega)$$

for any open subregion ω with $\bar{\omega} \subset \Omega$ and any $v \in \overset{\circ}{V}_0(\omega)$. In case that $\partial\omega$ is Lipschitz the latter inequality is equivalent to $J(u, \omega) \leq J(\hat{u}, \omega)$ for all $\hat{u} \in \overset{\circ}{V}(\omega)$ such that $u = \hat{u}$ on $\partial\omega$.

For local minimizers u we have the variational identity

$$\int_\Omega \sigma : \varepsilon(\varphi)\, dx = 0$$

being valid for all $\varphi \in \overset{\circ}{C}_0^\infty(\Omega; \mathbb{R}^n)$ where $\sigma := \frac{\partial G}{\partial E}(\varepsilon(u))$ is in the space $\text{Exp}(\omega)$ for any open set ω such that $\bar{\omega} \subset \Omega$. This is a consequence of (1.5). Let us now state the main result of this section already announced in Theorem 4.0.1.

THEOREM 4.1.1 *Suppose that* $n \leq 3$ *and let* $v \in \overset{\circ}{V}_{\text{loc}}(\Omega)$ *denote a local J–minimizer.*

  a) *If* $n = 3$, *then there exists an open set* $\Omega_0 \subset \Omega$ *such that* $|\Omega - \Omega_0| = 0$ *and* $\varepsilon(v) \in C^{0,\alpha}(\Omega_0; \mathbb{M})$ *for any* $\alpha < 1$.

  b) *In case* $n = 2$ *the result of a) holds with* $\Omega_0 = \Omega$.

REMARK 4.1.2 For $n = 3$ the following problems seem to be of some interest. 1) Do singularities actually occur? If yes, optimal estimates for $\Omega - \Omega_0$ should replace statement a). 2) Are local minimizers of class $C^{0,\beta}(\Omega; \mathbb{R}^3)$ for some $\beta < 1$? 3) Can one prove local boundedness of $\varepsilon(v)$? Of course, 3) would imply 2).

4.2   Existence of higher order weak derivatives and a Caccioppoli–type inequality

In this section we show for fluids of Prandtl–Eyring type that the deviator of the Cauchy stress tensor and also certain scalar functions involving the strain velocity have weak derivatives in the space $L^2_{\mathrm{loc}}(\Omega)$.

THEOREM 4.2.1 *Let* $u \in \overset{\circ}{V}_{\mathrm{loc}}(\Omega)$ *denote a local minimizer of* $J(v, \Omega) = \int_\Omega G(\varepsilon(v))\, dx$. *Then the following statements hold:*

    a) The deviator $\sigma = \frac{\partial G}{\partial E}(\varepsilon(u))$ *of the Cauchy stress tensor* τ *is in the space* $W^1_{2,\mathrm{loc}}(\Omega; \mathbb{M})$.

    b) The function $\sqrt{1 + |\varepsilon(u)|}$ *is of class* $W^1_{2,\mathrm{loc}}(\Omega)$.

From b) and the imbedding theorem it immediately follows

COROLLARY 4.2.1 *Under the assumptions of Theorem 2.1 we have* $\varepsilon(u) \in L^3_{\mathrm{loc}}(\Omega; \mathbb{M})$, *if* $n = 3$, *and* $\varepsilon(u) \in L^p_{\mathrm{loc}}(\Omega; \mathbb{M})$ *for any finite* p, *if* $n = 2$.

Proof of Theorem 4.2.1: We fix two subregions $\omega_1 \subset\subset \omega_2 \subset\subset \Omega$ having smooth boundaries. In order to prove

$$\sqrt{1 + |\varepsilon(u)|} \in W^1_2(\omega_1)$$

we may assume that $\varepsilon(u)$ does not vanish identically on ω_1. According to the density lemma 4.1.6 b) we choose a sequence $\{\bar{u}_m\}$ in $C^\infty(\bar{\omega}_2; \mathbb{R}^n)$ with the property div $\bar{u}_m = 0$ and

$$(2.1) \qquad \bar{u}_m \longrightarrow u \text{ in } V(\omega_2)$$

satisfying in addition

$$\|\bar{u}_m - u\|_{V(\omega_2)} > 0, \quad \|\varepsilon(\bar{u}_m)\|_{L^2(\omega_2)} > 0.$$

Let

$$\rho_m = \|\varepsilon(\bar{u}_m)\|^{-2}_{L^2(\omega_2)} \|\bar{u}_m - u\|_{V(\omega_2)}$$

and

$$J_m(v, \omega_2) = \frac{1}{2}\rho_m \int_{\omega_2} |\varepsilon(v)|^2 \, dx + J(v, \omega_2).$$

Moreover, we let u_m denote the unique minimizer of $J_m(\cdot, \omega_2)$ in the class $\{w \in W_2^1(\omega_2; \mathbb{R}^n) : \operatorname{div} w = 0 \text{ on } \omega_2,\ w = \bar{u}_m \text{ on } \partial\omega_2\}$. From (2.1) we deduce

$$(2.2) \quad \begin{cases} J_m(u_m, \omega_2) \leq J_m(\bar{u}_m, \omega_2) = \tfrac{1}{2}\|\bar{u}_m - u\|_{V(\omega_2)} + J(\bar{u}_m, \omega_2) \\[2mm] \longrightarrow \int_{\omega_2} G(\varepsilon(u))\, dx \qquad \text{as } m \to \infty, \end{cases}$$

hence

$$(2.3) \quad \sup_m J_m(u_m, \omega_2) < \infty, \quad \sup_m J(\bar{u}_m, \omega_2) < \infty.$$

According to (2.2) we have

$$\frac{1}{2}\rho_m \int_{\omega_2} |\varepsilon(\bar{u}_m)|^2 dx + J(\bar{u}_m, \omega_2) \longrightarrow J(u, \omega_2).$$

The left–hand side is bounded from below by

$$\frac{1}{2}\rho_m J(\bar{u}_m, \omega_2) + J(\bar{u}_m, \omega_2),$$

hence $J(\bar{u}_m, \omega_2) \to J(u, \omega_2)$ together with $J(u, \omega_2) > 0$ implies

$$(2.4) \quad \lim_{m \to \infty} \rho_m = 0.$$

Next we combine (2.3) and estimate (1.3) of Lemma 4.1.4 to see

$$(2.5) \quad \sup_m \|\varepsilon(u_m)\|_{L \ln L(\omega_2)} < \infty.$$

Since $\partial\omega_2$ is smooth, we have

$$u_m - \bar{u}_m \in W_2^1(\omega_2; \mathbb{R}^n) \cap \overset{\circ}{V}_0(\omega_2) \subset V_0(\omega_2);$$

on the other hand

$$\sup_m \|\bar{u}_m\|_{V(\omega_2)} < \infty$$

so that on account of (2.5) and Lemma 4.1.7 b) the sequence $\{u_m\}$ is uniformly bounded in $V(\omega_2)$. Using Lemma 4.1.7 a) and Lemma 4.1.2 we find a function $\hat{u} \in \overset{\circ}{V}(\omega_2)$ and a subsequence of $\{u_m\}$ such that

$$(2.6) \quad u_m \to \hat{u} \text{ in } L^1(\omega_2; \mathbb{R}^n)$$

(2.7) $\varepsilon(u_m) \rightharpoonup \varepsilon(\hat{u})$ weakly in $L^1(\omega_2; \mathbb{M})$.

Moreover, since $u_m - \bar{u}_m$ has zero trace on $\partial\omega_2$, we can state that

(2.8) $\hat{u} - u \in \overset{\circ}{V}_0(\omega_2)$.

Recalling that u is a local J–minimizer the assumption $u \neq \hat{u}$ together with (2.8) would imply (by the strict convexity of J)

(2.9) $J(u, \omega_2) < J(\hat{u}, \omega_2)$.

But from (2.7) it follows that

$$J(\hat{u}, \omega_2) \leq \liminf_{m \to \infty} J(u_m, \omega_2),$$

whereas (2.2) gives the estimate

$$\limsup_{m \to \infty} J(u_m, \omega_2) \leq J(u, \omega_2),$$

which is in contradiction to (2.9). So we have $u = \hat{u}$ and the convergence properties stated in (2.6), (2.7) hold not only for a subsequence. Moreover, we deduce from the above calculations (using (2.2) again)

(2.10) $\displaystyle\lim_{m \to \infty} J_m(u_m, \omega_2) = J(u, \omega_2) = \lim_{m \to \infty} J(u_m, \omega_2)$

(2.11) $\displaystyle\lim_{m \to \infty} \rho_m \int_{\omega_2} |\varepsilon(u_m)|^2 dx = 0.$

After these preparations we are going to prove differentiability of $\sqrt{1 + |\varepsilon(u)|}$ following ideas of the papers [Se 3–7, 9, 10], [FreSe] and [FS3]. First of all, we recall that u_m is in the space $W^2_{2,\text{loc}}(\omega_2; \mathbb{R}^n)$ (a proof is given in [FS2], Theorem 3.1, compare also chapter 3, Theorem 3.3.1), hence

(2.12) $\sigma_m := \dfrac{\partial G}{\partial E}\big(\varepsilon(u_m)\big) + \rho_m \varepsilon(u_m) \in W^1_{2,\text{loc}}(\omega_2; \mathbb{M}).$

With this notation the Euler–Lagrange equation takes the form

$$\int_{\omega_2} \sigma_m : \varepsilon(v)\, dx = 0, \quad v \in \overset{\circ}{C}{}^\infty_0(\omega_2; \mathbb{R}^n),$$

hence there exists a pressure function (observe $\sigma_m \in L^2(\omega_2; \mathbb{M})$)

$$p_m \in L^2(\omega_2)$$

satisfying the condition

$$\int_{\omega_2} p_m \, dx = 0$$

such that for the tensor

$$\tau_m := \sigma_m - p_m \, 1$$

we have

$$(2.13) \qquad \int_{\omega_2} \tau_m : \varepsilon(v) \, dx = 0, \quad v \in C_0^\infty(\omega_2; \mathbb{R}^n).$$

From (2.12) and (2.13) it follows that

$$(2.14) \qquad \operatorname{div} \tau_m = 0 \qquad \text{a.e. on } \omega_2$$

and

$$(2.15) \qquad \nabla p_m = \operatorname{div} \sigma_m \in L^2_{\text{loc}}(\omega_2; \mathbb{R}^n).$$

We also deduce from (2.13) the identity

$$(2.16) \qquad \int_{\omega_2} \partial_\alpha \tau_m : \varepsilon(v) \, dx = 0, \quad v \in C_0^\infty(\omega_2; \mathbb{R}^n), \; 1 \le \alpha \le n.$$

Let $\varphi \in C_0^\infty(\omega_2)$, $0 \le \varphi \le 1$, and insert $v = \varphi^6 \partial_\alpha u_m$ into (2.16) where α is some number $\le n$. From now on it will be convenient to drop the index m, i.e. u, τ, σ and p denote the quantities u_m, τ_m, σ_m and p_m, respectively. (2.16) then implies (we will use summation convention, i.e. the sum is taken w.r.t. Greek or Latin indices occuring twice)

$$0 = \int_{\omega_2} \partial_\alpha \tau : \varepsilon(\varphi^6 \partial_\alpha u) \, dx$$

$$= \int_{\omega_2} \varphi^6 \, \partial_\alpha \sigma : \varepsilon(\partial_\alpha u) \, dx + \int_{\omega_2} \partial_\alpha \tau_{ij} \, \partial_\alpha u^i \, \partial_j \varphi^6 \, dx$$

where we have used the fact that div $u = 0$. This equation can be rewritten in the form

(2.17)
$$
\begin{cases}
\displaystyle\int_{\omega_2} \varphi^6\, \partial_\alpha \sigma : \varepsilon(\partial_\alpha u)\, dx = -\int_{\omega_2} \partial_\alpha \tau_{ij}\, \partial_\alpha u^i\, \partial_j \varphi^6 dx = \\[3mm]
\displaystyle -2\int_{\omega_2} \partial_\alpha \tau_{ij} \varepsilon_{i\alpha}(u)\partial_j \varphi^6 dx + \int_{\omega_2} \partial_\alpha \tau_{ij}\, \partial_i u^\alpha\, \partial_j \varphi^6 dx = \\[3mm]
-2I_1 + I_2,
\end{cases}
$$

and we next give estimates for I_1 and I_2. We begin by discussing I_1.

$$
I_1 = \int_{\omega_2} (\partial_\alpha \sigma_{ij} - \delta_{ij}\partial_\alpha p)\varepsilon_{i\alpha}(u)\partial_j \varphi^6 dx = (2.15)
$$

$$
= \int_{\omega_2} \left[\partial_\alpha \sigma_{ij}\, \varepsilon_{i\alpha}(u)\partial_j \varphi^6 - \partial_j \sigma_{j\alpha}\, \varepsilon_{i\alpha}(u)\partial_i \varphi^6 \right] dx
$$

$$
= \int_{\omega_2} \left[\partial_\alpha \sigma_{ij}\, \varepsilon_{i\alpha}(u)\partial_j \varphi^6 - \partial_\alpha \sigma_{j\alpha}\, \varepsilon_{\ell j}(u)\partial_\ell \varphi^6 \right] dx
$$

where the last identity follows by index permutation. Hence we may write

$$
I_1 = \int_{\omega_2} \partial_\alpha \sigma_{ij} \left(\varepsilon_{i\alpha}(u)\partial_j \varphi^6 - \delta_{i\alpha}\varepsilon_{\ell j}(u)\partial_\ell \varphi^6 \right) dx.
$$

We introduce the tensors

$$
S^{(\alpha)} = \tfrac{1}{2}\big(Q^{(\alpha)} + (Q^{(\alpha)})^T\big),
$$

$$
Q_{ij}^{(\alpha)} = \varepsilon_{i\alpha}(u)\partial_j \varphi - \delta_{i\alpha}\varepsilon_{\ell j}(u)\partial_\ell \varphi
$$

and get

$$
I_1 = 6\int_{\omega_2} \varphi^5\, \partial_\alpha \sigma : S^{(\alpha)} dx.
$$

Using

$$
|S^{(\alpha)}| \le |Q^{(\alpha)}| \le c_1 |\varepsilon(u)|\,|\nabla \varphi|
$$

for some suitable constant $c_1 = c_1(n)$, we find (recall the definition of σ and write $\rho = \rho_m$)

$$
(2.18) \quad
\begin{cases}
|I_1| \;\leq\; c_2 \Big[\rho \displaystyle\int_{\omega_2} |\nabla\varepsilon(u)|\, |\varepsilon(u)|\varphi^5 |\nabla\varphi|\, dx \Big] \\[4mm]
\qquad +\; 6 \displaystyle\int_{\omega_2} \varphi^5 D^2 G\big(\varepsilon(u)\big) \big(\varepsilon(\partial_\alpha u), S^{(\alpha)}\big)\, dx \\[4mm]
\quad\leq\; c_2 \Big(\displaystyle\int_{\omega_2} \varphi^6 \rho |\nabla\varepsilon(u)|^2 dx \Big)^{1/2} \Big(\displaystyle\int_{\omega_2} \varphi^4 \rho |\nabla\varphi|^2\, |\varepsilon(u)|^2 dx \Big)^{1/2} \\[4mm]
\qquad +\; 6 \Big(\displaystyle\int_{\omega_2} \varphi^6 D^2 G\big(\varepsilon(u)\big) \big(\varepsilon(\partial_\alpha u), \varepsilon(\partial_\alpha u)\big)\, dx \Big)^{1/2} \\[4mm]
\qquad\times\; \Big(\displaystyle\int_{\omega_2} \varphi^4 D^2 G\big(\varepsilon(u)\big) \big(S^{(\alpha)}, S^{(\alpha)}\big)\, dx \Big)^{1/2} \;\leq\; (1.7) \\[4mm]
\quad\leq\; c_3 \Big(\displaystyle\int_{\omega_2} \varphi^6 \partial_\alpha \sigma : \varepsilon(\partial_\alpha u)\, dx \Big)^{1/2} \\[4mm]
\qquad\times\; \Big(\|\nabla\varphi\|_\infty^2 \displaystyle\int_{\omega_2} \big[\rho |\varepsilon(u)|^2 + |\varepsilon(u)| \ln\big(1 + |\varepsilon(u)|\big) \big]\, dx \Big)^{1/2}.
\end{cases}
$$

We further have

$$
\begin{aligned}
I_2 \;&=\; \int_{\omega_2} \partial_\alpha \tau_{ij}\, \partial_i u^\alpha \partial_j \varphi^6\, dx \;=\; -\int_{\omega_2} \tau_{ij}\partial_\alpha(\partial_i u^\alpha \partial_j \varphi^6)\, dx \\[3mm]
&=\; -\int_{\omega_2} \tau_{ij}\partial_i\partial_\alpha u^\alpha \partial_j \varphi^6\, dx \\[3mm]
&\quad\; -\int_{\omega_2} \tau_{ij}\partial_i u^\alpha \partial_\alpha \partial_j \varphi^6\, dx \\[3mm]
&=\; -\int_{\omega_2} \tau_{ij}\partial_i u^\alpha \partial_\alpha \partial_j \varphi^6\, dx \;=\; (2.14) \\[3mm]
&=\; \int_{\omega_2} \tau_{ij} u^\alpha \partial_\alpha \partial_i \partial_j \varphi^6\, dx
\end{aligned}
$$

and therefore

$$|I_2| \leq c_4(n,\varphi) \Big(\int_{\omega_2} |\varphi^3 \tau|^n dx \Big)^{1/n} \Big(\int_{\omega_2} |u|^{\frac{n}{n-1}} dx \Big)^{1-1/n}.$$

Let us now introduce the index m again. Then (2.17), (2.18) and the latter estimate imply

$$\int_{\omega_2} \varphi^6 \, \partial_\alpha \sigma_m : \varepsilon(\partial_\alpha u_m) \, dx \leq$$

$$c_5(n,\varphi) \Big[\Big(\int_{\omega_2} \varphi^6 \partial_\alpha \sigma_m : \varepsilon(\partial_\alpha u_m) \, dx \Big)^{1/2} J_m(u_m,\omega_2)^{1/2}$$

$$+ \Big(\int_{\omega_2} |\varphi^3 \tau_m|^n dx \Big)^{1/n} \Big(\int_{\omega_2} |u_m|^{\frac{n}{n-1}} dx \Big)^{1-1/n} \Big].$$

Since $\{u_m\}$ is bounded in the space $V(\omega_2)$, the imbedding lemma shows

$$\sup_m \int_{\omega_2} |u_m|^{\frac{n}{n-1}} dx < \infty,$$

(2.10) guarantees boundedness of $J_m(u_m,\omega_2)$, therefore we get

$$(2.19) \qquad \int_{\omega_2} \varphi^6 \, \partial_\alpha \sigma_m : \varepsilon(\partial_\alpha u_m) \, dx \leq c_5 \Big[1 + \Big(\int_{\omega_2} |\varphi^3 \tau_m|^n dx \Big)^{1/n} \Big]$$

with c_5 independent of m.

We recall that τ_m is in the space $W^1_{2,\mathrm{loc}}(\omega_2; \mathbb{M}) \cap L^2(\omega_2; \mathbb{M})$ and observe $n \leq 3$ to get the estimate

$$\Big(\int_{\omega_2} |\varphi^3 \tau_m|^n dx \Big)^{1/n} \leq c_6 \Big(\int_{\omega_2} |\nabla(\varphi^3 \tau_m)|^2 dx \Big)^{1/2}$$

$$\leq c_7 \Big[\Big(\int_{\omega_2} |\tau_m|^2 dx \Big)^{1/2} + \Big(\int_{\omega_2} \varphi^6 |\nabla \tau_m|^2 dx \Big)^{1/2} \Big].$$

Standard results of [L1] or [LS] concerning the presssure function combined with (2.13) show

$$\int_{\omega_2} |\tau_m|^2 \, dx \le c_8 \int_{\omega_2} \left(|\sigma_m|^2 + |p_m|^2 \right) dx \le c_9 \int_{\omega_2} |\sigma_m|^2 dx.$$

The definition (2.12) of σ_m together with the bounds for $\frac{\partial G}{\partial E}$ gives boundedness of $\int_{\omega_2} |\sigma_m|^2 dx$ independent of m, moreover, by (2.15)

$$|\nabla \tau_m|^2 \le c_{10} |\nabla \sigma_m|^2 \le c_{11} \, \partial_\alpha \sigma_m : \varepsilon(\partial_\alpha u_m).$$

Inserting these estimates into (2.19) and using Young's inequality we finally obtain

$$(2.20) \qquad \sup_m \int_{\omega_2} \varphi^6 \, \partial_\alpha \sigma_m : \varepsilon(\partial_\alpha u_m) \, dx < \infty.$$

We now specify $\varphi = 1$ on ω_1. Then (2.20) shows

$$\int_{\omega_1} \left[\rho_m |\nabla \varepsilon(u_m)|^2 + \frac{1}{1 + |\varepsilon(u_m)|} |\nabla \varepsilon(u_m)|^2 \right] dx \le c_{12}$$

or

$$\int_{\omega_1} \left| \nabla \sqrt{1 + |\varepsilon(u_m)|} \right|^2 dx \le c_{13}.$$

After passing to a subsequence we find a function $q \in W_2^1(\omega_1)$ such that

$$(2.21) \qquad \sqrt{1 + |\varepsilon(u_m)|} \rightharpoonup q \quad \text{in } W_2^1(\omega_1),$$

$$(2.22) \qquad \sqrt{1 + |\varepsilon(u_m)|} \to q \quad \text{in } L^2(\omega_1),$$

$$(2.23) \qquad \sqrt{1 + |\varepsilon(u_m)|} \to q \quad \text{a.e. on } \omega_1.$$

From this the desired claim $\sqrt{1 + |\varepsilon(u)|} \in W_2^1(\omega_1)$ will follow as soon as we can show $q = \sqrt{1 + |\varepsilon(u)|}$. To this purpose we write

$$J_m(u_m, \omega_2) - J(u, \omega_2)$$

$$= \tfrac{1}{2}\rho_m \|\varepsilon(u_m)\|_{L^2(\omega_2)} + J(u_m, \omega_2) - J(u, \omega_2)$$

$$= \tfrac{1}{2}\rho_m \|\varepsilon(u_m)\|_{L^2(\omega_2)} + \int_{\omega_2} \frac{\partial G}{\partial E}(\varepsilon(u)) : (\varepsilon(u_m) - \varepsilon(u))\, dx$$

$$+ \int_{\omega_2} \left(\int_0^1 D^2 G((1-t)\varepsilon(u) + t\varepsilon(u_m)) \right.$$

$$\left. (\varepsilon(u_m) - \varepsilon(u), \varepsilon(u_m) - \varepsilon(u))(1-t)\, dt \right) dx$$

and

$$\int_{\omega_2} \frac{\partial G}{\partial E}(\varepsilon(u)) : (\varepsilon(u_m) - \varepsilon(u))\, dx$$

$$= \int_{\omega_2} \frac{\partial G}{\partial E}(\varepsilon(u)) : (\varepsilon(u_m) - \varepsilon(\bar{u}_m))\, dx$$

$$+ \int_{\omega_2} \frac{\partial G}{\partial E}(\varepsilon(u)) : (\varepsilon(\bar{u}_m) - \varepsilon(u))\, dx = I_m' + I_m''.$$

From (1.5) we get $\frac{\partial G}{\partial E}(\varepsilon(u)) \in \mathrm{Exp}\,(\omega_2; \mathbb{M})$, (2.1) implies

$$\lim_{m \to \infty} I_m'' = 0.$$

Clearly $I_m' = 0$ which just follows from the Euler equation for u together with $u_m = \bar{u}_m$ on $\partial\omega_2$. Using (2.10) and (1.7) we deduce

$$\int_{\omega_2} \frac{1}{1 + |\varepsilon(u)| + |\varepsilon(u_m)|} |\varepsilon(u_m) - \varepsilon(u)|^2 dx \to 0, \quad m \to \infty,$$

in particular (at least for a subsequence)

(2.24) $$\frac{1}{1 + |\varepsilon(u)| + |\varepsilon(u_m)|} |\varepsilon(u_m) - \varepsilon(u)|^2 \to 0 \quad \text{a.e. on } \omega_1.$$

According to $q \in W_2^1(\omega_1)$ the function q is finite a.e. and (see (2.23))

$$|\varepsilon(u_m)| \to q^2 - 1 \quad \text{a.e. on } \omega_1.$$

In view of (2.24) this implies

$$\varepsilon(u_m) \to \varepsilon(u) \quad \text{a.e. on } \omega_1,$$

hence $q = \sqrt{1 + |\varepsilon(u)|}$, and the proof of Theorem 4.2.1 b) is complete.

For part a) of Theorem 4.2.1 we just recall that according to (2.20)

$$\sup_m \int_{\omega_1} |\nabla \sigma_m|^2 dx < \infty$$

so that there exists a tensor $\tilde{\sigma} \in W_2^1(\omega_1)$ with the property

$$\begin{cases} \sigma_m \rightharpoonup \tilde{\sigma} & \text{in } W_2^1(\omega_1), \\\\ \sigma_m \to \tilde{\sigma} & \text{a.e. on } \omega_1 \end{cases}$$

at least for a subsequence. But the definition (2.12) of σ_m together with (2.4) and the pointwise convergence $\varepsilon(u_m) \to \varepsilon(u)$ clearly gives $\tilde{\sigma} = \sigma$.

$\square$

During the proof of partial regularity given in section 3 we will make essential use of the following result.

THEOREM 4.2.2 *Let* $u \in \overset{\circ}{V}_{\mathrm{loc}}(\Omega)$ *denote a local minimizer of the energy* $J(\cdot, \Omega)$. *Then, for any ball* $B_R(x_0) \subset \Omega$, *any* $t \in]0, 1[$ *and any symmetric matrix* A, *we have*

$$(2.25) \qquad \int_{B_{tR}(x_0)} \left|\nabla \sqrt{1 + |\varepsilon(u)|}\right|^2 dx \le c(1-t)^{-2} R^2 \int_{B_R(x_0)} |\varepsilon(u) - A|^2 dx$$

with $c = c(n)$ *denoting a finite constant.*

Proof of Theorem 4.2.2: With notation introduced before we let

$$\omega_1 = B_{\frac{t+1}{2}R}(x_0), \quad \omega_2 = B_R(x_0)$$

and recall the following facts

$$(2.26) \qquad \left\{\sqrt{1 + |\varepsilon(u_m)|}\right\} \text{ is bounded in } W_2^1(\omega_1)$$

$$(2.27) \qquad \sqrt{1 + |\varepsilon(u_m)|} \rightharpoonup \sqrt{1 + |\varepsilon(u)|} \text{ in } W_2^1(\omega_1)$$

(2.28) $\varepsilon(u_m) \to \varepsilon(u)$ a.e. in ω_1.

Since $n = 2$ or $n = 3$, (2.27) clearly implies

(2.29) $\varepsilon(u_m) \to \varepsilon(u)$ in $L^2(\omega_1; \mathbb{M})$.

Let $\varphi \in C_0^1(\omega_1)$, $\varphi = 1$ on $B_{tR}(x_0)$, and

$$|\nabla\varphi| \le \frac{c_1}{(1-t)R} \quad \text{in } \omega_1.$$

Then (2.16) gives

(2.30) $$\int_{\omega_1} \partial_\alpha \tau_m : \varepsilon(\varphi^2 \partial_\alpha \hat{u}_m)\, dx = 0$$

where $\hat{u}_m(x) = u_m(x) - A(x - x_0) - \gamma_m$, γ_m denoting a rigid motion which is chosen to satisfy

(2.31) $$\int_{\omega_1} |\nabla\hat{u}_m|^2 dx \le c_2(n) \int_{\omega_1} |\varepsilon(\hat{u}_m)|^2 dx = c_2(n) \int_{\omega_1} |\varepsilon(u_m) - A|^2 dx.$$

The existence of γ_m with (2.31) can be deduced from chapter 3, Lemma 3.0.3. From (2.30) we infer

$$\int_{\omega_1} \varphi^2 \partial_\alpha \sigma_m : \varepsilon(\partial_\alpha u_m)\, dx = \int_{\omega_1} \varphi^2 \partial_\alpha \sigma_m : \varepsilon(\partial_\alpha \hat{u}_m)\, dx$$

$$= \int_{\omega_1} \varphi^2 \partial_\alpha \tau_m : \varepsilon(\partial_\alpha \hat{u}_m)\, dx + \int_{\omega_1} \varphi^2 \partial_\alpha p_m \mathbf{1} : \varepsilon(\partial_\alpha \hat{u}_m)\, dx$$

$$= -2 \int_{\omega_1} \varphi \partial_\alpha \tau_m : (\nabla\varphi \odot \partial_\alpha \hat{u}_m)\, dx + \int_{\omega_1} \varphi^2 \partial_\alpha p_m \partial_\beta(\partial_\alpha \hat{u}_m^\beta)\, dx$$

$$= -2 \int_{\omega_1} \varphi \partial_\alpha \tau_m : (\nabla\varphi \odot \partial_\alpha \hat{u}_m)\, dx$$

$$\le c_3 \left(\int_{\omega_1} \varphi^2 |\nabla\tau_m|^2 dx \right)^{1/2} \left(\int_{\omega_1} |\nabla\varphi|^2 |\nabla\hat{u}_m|^2 dx \right)^{1/2}$$

$$\le c_4 \frac{1}{1-t} \frac{1}{R} \left(\int_{\omega_1} \varphi^2 |\nabla\tau_m|^2 dx \right)^{1/2} \left(\int_{\omega_1} |\nabla\hat{u}_m|^2 dx \right)^{1/2}$$

$$\underset{(2.31)}{\le} c_5 \frac{1}{1-t} \frac{1}{R} \left(\int_{\omega_1} \varphi^2 |\nabla\tau_m|^2 dx \right)^{1/2} \left(\int_{B_R(x_0)} |\varepsilon(u_m) - A|^2 dx \right)^{1/2}.$$

Using the definition of τ_m and equation (2.15) again we have as before

$$|\nabla \tau_m|^2 \leq c_6 |\nabla \sigma_m|^2 \leq c_7 \partial_\alpha \sigma_m : \varepsilon(\partial_\alpha u_m).$$

This together with the above estimates implies

$$\int_{B_{tR}(x_0)} \partial_\alpha \sigma_m : \varepsilon(\partial_\alpha u_m) \, dx \leq c_8 (1-t)^{-2} R^{-2} \int_{B_R(x_0)} |\varepsilon(u_m) - A|^2 dx.$$

This shows

$$\int_{B_{tR}(x_0)} \left|\nabla \sqrt{1 + |\varepsilon(u_m)|}\right|^2 dx \leq c_9 (1-t)^{-2} R^{-2} \int_{B_R(x_0)} |\varepsilon(u_m) - A|^2 dx.$$

On the left–hand side we may apply (2.27), on the right–hand side we use (2.29) to get inequality (2.25).

$\square$

REMARK 4.2.1 In the variational setting discussed in [FS3] we established a slightly stronger version of (2.25) which was needed to handle the case $n = 4$. So in the framework of Prandtl–Eyring fluids we do not have to prove this refined version of (2.25).

4.3 Blow–up:
the proof of Theorem 4.1.1 for $n = 3$

We are now going to prove the first part of Theorem 4.1.1. So let us assume that $n = 3$ and consider a local minimizer $u \in \overset{\circ}{V}_{loc}(\Omega)$ of $J(\cdot, \Omega)$. Recall that by Corollary 4.2.1 we already know that $\varepsilon(u)$ is in the space $L^2_{loc}(\Omega; \mathbb{M})$. Partial regularity is a consequence of

LEMMA 4.3.1 *Fix some $L > 0$ and calculate $C_0 = C_0(L)$ as indicated in the proof. Then, for all $t \in]0, 1[$, we find a number $\varepsilon = \varepsilon(t, L) > 0$ such that*

$$\left| (\varepsilon(u))_{x_0, R} \right| < L \text{ and } \fint_{B_R(x_0)} \left| \varepsilon(u) - (\varepsilon(u))_{x_0, R} \right|^2 dx < \varepsilon^2$$

imply

$$\fint_{B_{tR}(x_0)} \left| \varepsilon(u) - (\varepsilon(u))_{x_0, tR} \right|^2 dx \leq C_0 t^2 \fint_{B_R(x_0)} \left| \varepsilon(u) - (\varepsilon(u))_{x_0, R} \right|^2 dx$$

for any ball $B_R(x_0) \subset \Omega$.

Proof of Lemma 4.3.1: Suppose that the claim is false. Then there exists a sequence of balls $B_{R_k}(x_k) \subset \Omega$ such that

$$|A_k| \leq L, \quad A_k = (\varepsilon(u))_{x_k, R_k}, \quad \fint_{B_{R_k}(x_k)} \left| \varepsilon(u) - (\varepsilon(u))_{x_k, R_k} \right|^2 dx = \varepsilon_k^2 \to 0$$

and

$$\fint_{B_{tR}(x_k)} \left| \varepsilon(u) - (\varepsilon(u))_{x_k, tR} \right|^2 dx > C_0 t^2 \varepsilon_k^2.$$

Let

$$v_k(z) = \frac{1}{\varepsilon_k R_k} \left(u(x_k + R_k z) - R_k A_k z - \gamma_k(z) \right), \quad z \in B_1,$$

where γ_k is a rigid motion with the property

$$\int_{B_1} |v_k|^2 dz \leq c_1 \int_{B_1} |\varepsilon(v_k)|^2 dz.$$

After passing to subsequences we find a symmetric matrix A and a function $v \in W_2^1(B_1; \mathbb{R}^3)$, div $v = 0$, such that

$$A_k \to A, \qquad |A| \leq L,$$

$$v_k \to v \qquad \text{in } L^2(B_1; \mathbb{R}^3),$$

$$\varepsilon(v_k) \rightharpoonup \varepsilon(v) \quad \text{in } L^2(B_1; \mathbb{M}),$$

$$\varepsilon_k \varepsilon(v_k) \to 0 \quad \text{in } L^2(B_1; \mathbb{M}) \text{ and a.e.}$$

Using these facts it is easy to show that

$$\int_{B_1} D^2 G(A) \big(\varepsilon(v), \varepsilon(\varphi) \big) \, dx = 0$$

holds for any $\varphi \in \overset{\circ}{C}{}^\infty_0(B_1; \mathbb{R}^3)$, hence v is of class $C^\infty(B_1; \mathbb{R}^3)$ and satisfies the Campanato–type estimate (compare chapter 3, Lemma 3.0.5)

$$\fint_{B_t} |\varepsilon(v) - (\varepsilon(v))_t|^2 \, dz \leq C_1(L) t^2 \fint_{B_1} |\varepsilon(v)|^2 \, dz.$$

We set $C_0 = 2C_1$ and obtain the desired contradiction as soon as we can show

(3.1) $\qquad \varepsilon(v_k) \to \varepsilon(v)$ in $L^2_{\text{loc}}(B_1; \mathbb{M})$.

Let $\varphi \in C^1_0(B_1)$, $\varphi \geq 0$, and consider functions $\bar{v}_k \in \overset{\circ}{W}{}^1_2(B_1; \mathbb{R}^3)$ with the properties (see chapter 3, Lemma 3.0.4)

(3.2) $\qquad \operatorname{div} \bar{v}_k = \operatorname{div} \big(v_k + \varphi(v - v_k) \big) = \nabla\varphi \cdot (v - v_k)$ in B_1

(3.3) $\qquad \displaystyle\int_{B_1} |\nabla\bar{v}_k|^2 \, dx \leq c_1 \int_{B_1} |\nabla\varphi|^2 \, |v - v_k|^2 \, dz, \quad c_1 = c_1(n).$

The minimizing property of u in $B_{R_k}(x_k)$ gives after scaling

$$\int_{B_1} G(A_k + \varepsilon_k \varepsilon(v_k))\, dz \leq \int_{B_1} G\Big(A_k + \varepsilon_k \varepsilon\big(v_k + \varphi(v - v_k) - \bar{v}_k\big)\Big)\, dz$$

$$= \int_{B_1} G\Big((1 - \varphi)(A_k + \varepsilon_k \varepsilon(v_k)) + \varphi(A_k + \varepsilon_k \varepsilon(v))\Big)\, dz$$

$$+ \int_{B_1} \varepsilon_k \frac{\partial G}{\partial E}\Big(A_k + \varepsilon_k(\varepsilon(v_k) + \varphi \varepsilon(v - v_k))\Big) : \big((v - v_k) \odot \nabla \varphi - \varepsilon(\bar{v}_k)\big)\, dz$$

$$+ \int_{B_1} \int_0^1 \varepsilon_k^2 (1 - s) D^2 G(\xi_k)\big((v - v_k) \odot \nabla \varphi - \varepsilon(\bar{v}_k),$$

$$(v - v_k) \odot \nabla \varphi - \varepsilon(\bar{v}_k)\big)\, ds\, dz$$

$$\leq \int_{B_1} (1 - \varphi) G(A_k + \varepsilon_k \varepsilon(v_k))\, dz + \int_{B_1} \varphi G(A_k + \varepsilon_k \varepsilon(v))\, dz$$

$$+ \int_{B_1} \varepsilon_k \frac{\partial G}{\partial E} \ldots dz + \int_{B_1} \int_0^1 \varepsilon_k^2 (1 - s) D^2 G \ldots ds\, dz$$

where we have used the fact that G is a convex function. ξ_k just denotes the tensor

$$A_k + \varepsilon_k\big(\varepsilon(v_k) + \varphi \varepsilon(v - v_k)\big) + s \varepsilon_k\big((v - v_k) \odot \nabla \varphi - \varepsilon(\bar{v}_k)\big).$$

Recalling (3.3) and the growth of $D^2 G$ we infer

$$(3.4) \quad \left\{ \begin{array}{l} \displaystyle\int_{B_1} \varphi \Big\{ G(A_k + \varepsilon_k \varepsilon(v_k)) - G(A_k + \varepsilon_k \varepsilon(v)) \Big\}\, dz \\[2em] \displaystyle\leq \int_{B_1} \varepsilon_k \frac{\partial G}{\partial E}\Big(A_k + \varepsilon_k\big(\varepsilon(v_k) + \varphi \varepsilon(v - v_k)\big)\Big) : \\[2em] \big((v - v_k) \odot \nabla \varphi - \varepsilon(\bar{v}_k)\big)\, dz \\[2em] \displaystyle+ \int_{B_1} \varepsilon_k^2 c_2 |\nabla \varphi|^2 |v_k - v|^2\, dz. \end{array} \right.$$

The left–hand side of (3.4) equals

$$\int_{B_1} \varepsilon_k \varphi \frac{\partial G}{\partial E} \big(A_k + \varepsilon_k \varepsilon(v)\big) : \varepsilon(v_k - v)\, dz$$

$$+ \int_{B_1} \int_0^1 (1-s)\varphi \varepsilon_k^2 D^2 G\big(A_k + \varepsilon_k \varepsilon(v) + s\varepsilon_k \varepsilon(v_k - v)\big)$$

$$\big(\varepsilon(v_k - v), \varepsilon(v_k - v)\big)\, ds\, dz,$$

hence

$$\int_{B_1} \int_0^1 (1-s)\varphi \varepsilon_k^2 D^2 G\big(A_k + \varepsilon_k \varepsilon(v) + s\varepsilon_k \varepsilon(v_k - v)\big)$$

$$\big(\varepsilon(v_k - v), \varepsilon(v_k - v)\big)\, ds\, dz$$

$$\leq \int_{B_1} \varepsilon_k \frac{\partial G}{\partial E} \big(A_k + \varepsilon_k \big(\varepsilon(v_k) + \varphi \varepsilon(v - v_k)\big)\big) :$$

$$\big((v - v_k) \odot \nabla \varphi - \varepsilon(\bar{v}_k)\big)\, dz$$

$$- \int_{B_1} \varepsilon_k \varphi \frac{\partial G}{\partial E} \big(A_k + \varepsilon_k \varepsilon(v)\big) : \varepsilon(v_k - v)\, dz$$

$$+ \int_{B_1} c_2 \varepsilon_k^2 |\nabla \varphi|^2 |v - v_k|^2 dz$$

$$= \int_{B_1} \varepsilon_k \Big\{ \frac{\partial G}{\partial E} \big(A_k + \varepsilon_k \big(\varepsilon(v_k) + \varphi \varepsilon(v - v_k)\big)\big)$$

$$- \frac{\partial G}{\partial E} \big(A_k + \varepsilon_k \varepsilon(v)\big) \Big\} : \big((v - v_k) \odot \nabla \varphi\big)\, dz$$

$$+ \int_{B_1} \varepsilon_k \frac{\partial G}{\partial E} \big(A_k + \varepsilon_k \varepsilon(v)\big) : \varepsilon\big(\varphi(v - v_k)\big)\, dz -$$

$$\int\limits_{B_1} \varepsilon_k \frac{\partial G}{\partial E}\Big(A_k + \varepsilon_k\big(\varepsilon(v_k) + \varphi\varepsilon(v - v_k)\big)\Big) : \varepsilon(\bar{v}_k)\, dz$$

$$+ \int\limits_{B_1} c_2\varepsilon_k^2 |\nabla\varphi|^2 |v - v_k|^2 dz$$

$$= \varepsilon_k I_k^1 + \varepsilon_k I_k^2 - \varepsilon_k I_k^3 + c_2\varepsilon_k^2 \int\limits_{B_1} |\nabla\varphi|^2 |v - v_k|^2 dz.$$

We discuss the quantities I_k^j, $j = 1, 2, 3$:

$$|I_k^1| = \varepsilon_k \left| \int\limits_{B_1} \int\limits_0^1 D^2 G\big(A_k + \varepsilon_k\varepsilon(v) + s\varepsilon_k(1 - \varphi)\varepsilon(v_k - v)\big)\right.$$

$$\left.\big(\varepsilon(v_k - v), (v - v_k) \odot \nabla\varphi\big)\, ds\, dz \right|$$

$$\leq 2\varepsilon_k \int\limits_{B_1} |\varepsilon(v_k - v)||\nabla\varphi|\, |v - v_k|\, dz$$

and by the convergence properties of $\{v_k\}$ we get

$$\lim_{k\to\infty} \frac{1}{\varepsilon_k} I_k^1 = 0.$$

Next we write

$$\frac{1}{\varepsilon_k} I_k^2 = \frac{1}{\varepsilon_k} \int\limits_{B_1} \left\{ \frac{\partial G}{\partial E}\big(A_k + \varepsilon_k\varepsilon(v)\big) - \frac{\partial G}{\partial E}(A_k) \right\} : \varepsilon\big(\varphi(v - v_k)\big)\, dz$$

$$= \int\limits_{B_1} \int\limits_0^1 D^2 G\big(A_k + s\varepsilon_k\varepsilon(v)\big)\Big(\varepsilon(v), \varepsilon\big(\varphi(v - v_k)\big)\Big)\, dz$$

and again

$$\lim_{k\to\infty} \frac{1}{\varepsilon_k} I_k^2 = 0$$

which follows from $\int\limits_0^1 D^2G\big(A_k + s\varepsilon_k\varepsilon(v)\big)(\varepsilon v,\,\cdot\,) \to D^2G(A)\big(\varepsilon(v),\,\cdot\,)$ in $L^2_{\mathrm{loc}}(B_1;\mathbb{M})$

together with $\varepsilon\big(\varphi(v - v_k)\big) \to 0$ in $L^2(B_1;\mathbb{M})$.

Finally we observe

$$\tfrac{1}{\varepsilon_k}|I_k^3| \;=\; \left| \tfrac{1}{\varepsilon_k}\int\limits_{B_1} \left\{ \frac{\partial G}{\partial E}\Big(A_k + \varepsilon_k\big(\varepsilon(v_k) + \varphi\varepsilon(v - v_k)\big) \Big) - \frac{\partial G}{\partial E}(A_k) \right\} : \right.$$

$$\left. \varepsilon(\bar{v}_k)\, dz \right|$$

$$\leq\; c_3 \int\limits_{B_1} |\varepsilon(\bar{v}_k)|\, |\varepsilon(v_k) + \varphi\varepsilon(v - v_k)|\, dz$$

$$\leq\; c_4 \left(\int\limits_{B_1} |\varepsilon(\bar{v}_k)|^2 dz \right)^{1/2} \left(\int\limits_{B_1} \big\{ |\varepsilon(v)|^2 + |\varepsilon(v_k)|^2 \big\}\, dz \right)^{1/2},$$

the second factor staying bounded, whereas

$$\int\limits_{B_1} |\varepsilon(\bar{v}_k)|^2 dz \longrightarrow 0, \quad k \to \infty,$$

according to (3.3).

Putting together our results we arrive at

$$\lim_{k\to\infty} \int_{B_1} \varphi \int_0^1 (1 - s) D^2 G\big(A_k + \varepsilon_k\varepsilon(v) + s\varepsilon_k\varepsilon(v_k - v) \big)$$

$$\big(\varepsilon(v_k - v), \varepsilon(v_k - v) \big)\, ds\, dz = 0,$$

and the growth of D^2G implies

$$(3.5) \qquad \lim_{k\to\infty} \int_{B_1} \varphi\, \frac{|\varepsilon(v - v_k)|^2}{1 + |A_k| + \varepsilon_k\big(|\varepsilon(v_k)| + |\varepsilon(v)| \big)}\, dz = 0.$$

We are now going to establish (3.1). Fix some radius $r \in\,]0, 1[$ and some number $M > 0$. Then

$$\int\limits_{B_r} |\varepsilon(v_k - v)|^2 dz = \int\limits_{B_r \cap [\varepsilon_k|\varepsilon(v_k)| \leq M]} |\varepsilon(v_k - v)|^2 dz + \Theta_k$$

where Θ_k is the integral over $B_r \cap [\varepsilon_k |\varepsilon(v_k)| > M]$. If we choose $\varphi = 1$ on B_r, then (3.5) obviously implies (recall $\varepsilon(v) \in L_{\text{loc}}^\infty(B_1; \mathbb{M})$)

$$\lim_{k \to \infty} \int_{B_r \cap [\varepsilon_k |\varepsilon(v_k)| \leq M]} |\varepsilon(v_k - v)|^2 dz = 0.$$

For discussing the behaviour of Θ_k we introduce the auxiliary function

$$\eta_k = \varepsilon_k^{-1} \left(\sqrt{1 + |A_k + \varepsilon_k \varepsilon(v_k)|} - \sqrt{1 + |A_k|} \right).$$

We have $|\eta_k| \leq \frac{1}{2} |\varepsilon(v_k)|$ and after scaling we get from (2.25)

$$\int_{B_r} |\nabla \eta_k|^2 dz \leq c(1 - r)^{-2} \int_{B_1} |\varepsilon(v_k)|^2 dz = c(1 - r)^{-2}$$

so that $\{\eta_k\}$ is uniformly bounded in $W_2^1(B_r)$.

The definition of η_k implies: if M is sufficiently large (independent of k), then we have

$$\eta_k \geq \frac{1}{2} \frac{1}{\varepsilon_k} \sqrt{\varepsilon_k |\varepsilon(v_k)|}$$

a.e. on $B_r \cap [\varepsilon_k |\varepsilon(v_k)| > M]$. This shows

(3.6) $$\int_{B_r \cap [\varepsilon_k |\varepsilon(v_k)| > M]} |\varepsilon(v_k)|^2 dz \leq 2^4 \varepsilon_k^2 \int_{B_r} \eta_k^4 \, dz \xrightarrow[k \to \infty]{} 0$$

since $\|\eta_k\|_{L^4(B_r)}$ stays bounded. Clearly

$$\int_{B_r \cap [\varepsilon_k |\varepsilon(v_k)| > M]} |\varepsilon(v)|^2 dz \longrightarrow 0, \quad k \to \infty,$$

on account of $\varepsilon_k \varepsilon(v_k) \to 0$ a.e. which together with (3.6) gives $\lim_{k \to \infty} \Theta_k = 0$. This completes the proof of (3.1).

$\square$

REMARK 4.3.1 Using Lemma 4.3.1 it is easy to show that $\varepsilon(u)$ is Hölder continuous near some point $x_0 \in \Omega$ iff

$$\sup_{r > 0} |\varepsilon(u)_{x_0,r}| < \infty \quad \text{and} \quad \fint_{B_r(x_0)} |\varepsilon(u) - (\varepsilon(u))_{x_0,r}|^2 dx \to 0, \ r \downarrow 0.$$

Moreover, Hölder continuity of $\varepsilon(u)$ on some open set ω in Ω implies that u is of class $C^{1,\alpha}(\omega; \mathbb{R}^3)$. All results remain valid if we consider the case $n = 2$.

4.4   The two–dimensional case

In this section we discuss Theorem 4.1.1 for the case $n = 2$ applying arguments from [FreSe]. So let us assume that $u \in \overset{\circ}{V}_{\text{loc}}(\Omega)$ locally minimizes $J(\cdot, \Omega)$ and fix some smooth subdomain $\omega_2 \subset\subset \Omega$. Consider a disc $B_R(x_0) \subset\subset \omega_2$ and define u_m exactly as in section 2. Using also the other notation introduced there we recall (see (2.16))

$$(4.1) \qquad \int_{\omega_2} \partial_\alpha \tau_m : \varepsilon(v)\, dx = 0, \ v \in C_0^\infty(\omega_2; \mathbb{R}^2), \ \alpha = 1, 2.$$

Let $T_R(x_0) = B_R(x_0) - \overline{B}_{R/2}(x_0)$, $A_m = \fint_{T_R(x_0)} \varepsilon(u_m)\, dx$ and

$$\tilde{u}_m(x) = u_m(x) - A_m(x - x_0) - \Gamma_m(x),$$

Γ_m being a rigid motion which is chosen according to

$$(4.2) \qquad \int_{T_R(x_0)} |\nabla \tilde{u}_m|^2 dx \le c_1 \int_{T_R(x_0)} |\varepsilon(\tilde{u}_m)|^2 dx.$$

We use (4.1) with $v = \varphi^2 \partial_\alpha \tilde{u}_m$, $\varphi \in C_0^1(B_R(x_0))$, $0 \le \varphi \le 1$, $\varphi = 1$ in $B_{R/2}(x_0)$ and $|\nabla \varphi| \le 4/R$.
Introducing the function $Q_m = \partial_\alpha \sigma_m : \varepsilon(\partial_\alpha u_m)$ (from now on the sum is taken w.r.t. to Greek indices) we get just as in the proof of Theorem 4.2.2

$$\int_{B_R(x_0)} \varphi^2 Q_m dx = -2 \int_{B_R(x_0)} \varphi \partial_\alpha \tau_m : (\nabla \varphi \odot \partial_\alpha \tilde{u}_m)\, dx \le$$

$$2 \left(\int_{T_R(x_0)} \varphi^2 |\nabla \tau_m|^2 dx \right)^{1/2} \left(\int_{T_R(x_0)} |\nabla \varphi|^2 |\nabla \tilde{u}_m|^2 dx \right)^{1/2}.$$

We also have $|\nabla \tau_m|^2 \le c_2 Q_m$ so that by (4.2) and the above inequality

$$(4.3) \qquad \int_{B_R(x_0)} \varphi^2 Q_m dx \le \frac{1}{R} c_3 \left(\int_{T_R(x_0)} Q_m dx \right)^{1/2} \left(\int_{T_R(x_0)} |\varepsilon(\tilde{u}_m)|^2 dx \right)^{1/2}.$$

Clearly $\varepsilon(\tilde{u}_m) = \varepsilon(u_m) - A_m$ and $\nabla \varepsilon(\tilde{u}_m) = \nabla \varepsilon(u_m)$ so that application of the Sobolev–Poincaré inequality yields

$$(4.4) \qquad \left(\int_{T_R(x_0)} |\varepsilon(\tilde{u}_m)|^2 dx \right)^{1/2} \le c_4 \int_{T_R(x_0)} |\nabla \varepsilon(u_m)|\, dx.$$

Let $h_m = \sqrt{1 + |\varepsilon(u_m)|}$. Then it is easy to check that

$$|\nabla \varepsilon(u_m)| \le c_5 h_m \sqrt{Q_m}$$

and by combining (4.3) and (4.4) we arrive at

$$(4.5) \qquad \int\limits_{B_{R/2}(x_0)} Q_m \, dx \le c_6 \frac{1}{R} \Big(\int\limits_{T_R(x_0)} Q_m dx \Big)^{1/2} \int\limits_{T_R(x_0)} h_m \sqrt{Q_m} \, dx$$

being valid for any disc $B_R(x_0) \subset\subset \omega_2$. (4.5) is exactly the hypothesis of [FreSe], Lemma 4.1, and as demonstrated there we deduce from (4.5): for any $q \ge 1$ and any compact subdomain ω_1 of ω_2 there exists a constant $K = K(\omega_1, q)$ such that

$$(4.6) \qquad \int\limits_{B_R(x_0)} Q_m \, dx \le K \,|\ln R|^{-q}$$

is true for any disc $B_R(x_0) \subset\subset \omega_1$. In place of (4.6) we may write $(K' = K'(\omega_1, q))$

$$(4.7) \qquad \int\limits_{B_R(x_0)} |\nabla \tau_m|^2 dx \le K' |\ln R|^{-q}.$$

If we choose $q > 2$ in (4.7), then the modification of the Dirichlet–growth lemma given in [Fre], p. 287, shows: τ_m is continuous on ω_1 with modulus of continuity independent of m.

Next we recall the formulas from section 2

$$\tau_m = \sigma_m - p_m \mathbf{1}, \quad \sigma_m = \rho_m \varepsilon(u_m) + \frac{\partial G}{\partial E}\big(\varepsilon(u_m)\big) =$$

$$= \rho_m \varepsilon(u_m) + \frac{1}{|\varepsilon(u_m)|}\Big(\ln\big(1 + |\varepsilon(u_m)|\big) + \frac{|\varepsilon(u_m)|}{1+|\varepsilon(u_m)|} \Big) \varepsilon(u_m).$$

div $u_m = 0$ implies $tr\,\sigma_m = 0$ and therefore $-2p_m = tr\,\tau_m$ so that uniform continuity of τ_m gives the same for the pressure functions p_m. Hence σ_m is continuous on ω_1 with modulus of continuity independent of m. Using $\varepsilon(u_m) \to \varepsilon(u)$ a.e. we get continuity of $\frac{\partial G}{\partial E}(\varepsilon(u))$ by Arcela's theorem. But $\frac{\partial G}{\partial E}$ is a homeomorphism $\mathbb{M} \to \mathbb{M}$ which shows $\varepsilon(u) \in C^0(\omega_1; \mathbb{M})$. In this case the criterion for partial regularity from Remark 4.3.1 holds at any point. The proof of part b) of Theorem 4.1.1 is complete.

$\square$

4.5 Partial regularity for plastic materials with logarithmic hardening

As mentioned in the beginning this type of plastic material behaviour has been completely investigated in the twodimensional case by Frehse and Seregin [FreSe] who also gave definitions of reasonable energy spaces in which solutions should be located. Our main concern in this section is to prove partial regularity for three dimensions (compare Theorem 4.5.1 below) but for sake of completeness we first recall from [FreSe] the necessary background material.

Let Ω denote a bounded domain in $\mathbb{R}^n$ (no restriction on n) and define

$$X(\Omega) = \left\{ v \in L^1(\Omega; \mathbb{R}^n) : \operatorname{div} v \in L^2(\Omega), |\varepsilon^D(v)| \in L \ln L(\Omega) \right\}$$

with norm

$$\|v\|_{X(\Omega)} = \|v\|_{L^1(\Omega)} + \|\operatorname{div} v\|_{L^2(\Omega)} + \|\varepsilon^D(v)\|_{L \ln L(\Omega)}.$$

We further let

$$X_{\text{loc}}(\Omega) = \{v : \Omega \to \mathbb{R}^n | v \in X(\omega) \text{ for any open } \omega \text{ with } \overline{\omega} \subset \Omega\},$$

$$X_0(\Omega) = \text{closure of } C_0^\infty(\Omega; \mathbb{R}^n) \text{ in } X(\Omega) \text{ w.r.t.} \|\cdot\|_{X(\Omega)}.$$

Clearly $X(\Omega)$ is a subspace of $BD(\Omega; \mathbb{R}^n)$, hence for Lipschitz Ω there exists a trace operator $X(\Omega) \to L^1(\partial\Omega; \mathbb{R}^n)$. Quoting [FreSe] we have

LEMMA 4.5.1 *Suppose Ω is a Lipschitz domain. Then*

a) $C^\infty(\overline{\Omega}; \mathbb{R}^n)$ *is dense in* $X(\Omega)$.

b) $u \in X_0(\Omega)$ *iff* $u \in X(\Omega)$ *with* $u = 0$ *on* $\partial\Omega$.

and from Theorem 2.1 of [FreSe] we also get that for smooth boundary $\partial\Omega$ and a given function u_0 there exists a unique minimizer of the variational integral

$$I(u, \Omega) = \int_\Omega \left[\frac{1}{2}(\operatorname{div} u)^2 + G(\varepsilon^D(u)) \right] dx$$

in the class $u_0 + X_0(\Omega)$. We therefore define

DEFINITION 4.5.1 *A function $u \in X_{\text{loc}}(\Omega)$ is a local extremal of the functional I iff for all $\varphi \in C_0^\infty(\Omega; \mathbb{R}^n)$*

$$I(u, \text{spt } \varphi) \leq I(u + \varphi, \text{spt } \varphi).$$

Of course Remark 4.1.1 holds with obvious modifications. For local minimizers $u \in X_{\text{loc}}(\Omega)$ it is easy to show that

$$\int_\omega \left[\text{div } u \text{ div } \varphi + \frac{\partial G}{\partial E}(\varepsilon^D(u)) : \varepsilon^D(\varphi) \right] dx = 0$$

holds for any smooth subdomain $\omega \subset \Omega$ with compact closure in Ω and any function $\varphi \in X_0(\omega)$. Moreover, we have

$$\frac{\partial G}{\partial E}(\varepsilon^D(u)) \in \text{Exp } (\omega; \mathbb{M}).$$

THEOREM 4.5.1 *Assume that $n = 3$ and let $u \in X_{\text{loc}}(\Omega)$ denote a local I-minimizer. Then there exists an open set $\Omega_0 \subset \Omega$ such that $|\Omega - \Omega_0| = 0$ and $\varepsilon(u) \in C^{0,\alpha}(\Omega_0; \mathbb{M})$ for any $\alpha < 1$, i.e. $u \in C^{1,\alpha}(\Omega_0; \mathbb{R}^3)$.*

REMARK 4.5.1 In [FreSe] it has been shown that $\Omega_0 = \Omega$ in case that $n = 2$.

Since the proof of Theorem 4.5.1 is very similar to the Prandtl–Eyring case, we will concentrate on the necessary changes.

THEOREM 4.5.2 *Suppose that $u \in X_{\text{loc}}(\Omega)$ is a local I-minimizer. Then we have*

a) *$\text{div } u \in W_{2,\text{loc}}^1(\Omega)$*

b) *$\sqrt{1 + |\varepsilon^D(u)|} \in W_{2,\text{loc}}^1(\Omega)$.*

If $n = 3$, then $\varepsilon(u) \in L_{\text{loc}}^3(\Omega; \mathbb{M})$. For $n = 2$ we have $\varepsilon(u) \in L_{\text{loc}}^p(\Omega; \mathbb{M})$ for any finite p.

Proof of Theorem 4.5.2: We argue as in Theorem 4.2.1 by choosing smooth domains $\omega_1 \Subset \omega_2 \Subset \Omega$. The approximating sequence $\{\bar{u}_m\}$ is located in the space $C^\infty(\bar{\omega}_2; \mathbb{R}^n)$ (see Lemma 4.5.1 a)) and satisfies

$$\|\bar{u}_m - u\|_{X(\omega_2)} \to 0$$

as well as all other requirements with $V(\omega_2)$ replaced by $X(\omega_2)$.

Let

$$I_m(v, \omega_2) = \frac{1}{2}\rho_m \int_{\omega_2} |\varepsilon(v)|^2 dx + I(v, \omega_2)$$

and let u_m denote the unique $I_m(\cdot, \omega_2)$–minimizer in the class $\bar{u}_m + \overset{\circ}{W}{}_2^1(\omega_2; \mathbb{R}^n)$. Then we have as in section 2

(5.1) $\quad u_m \to u$ in $L^1(\omega_2; \mathbb{R}^n)$,

(5.2) $\quad \text{div } u_m \rightharpoonup \text{div } u$ in $L^2(\omega_2)$,

(5.3) $\quad \varepsilon(u_m) \rightharpoonup \varepsilon(u)$ in $L^1(\omega_2; \mathbb{M})$,

(5.4) $\quad \rho_m \|\varepsilon(u_m)\|_{L^2(\omega_2)}^2 \to 0$,

(5.5) $\quad J_m(u_m, \omega_2) \to J(u, \omega_2)$.

u_m is of class $W_{2,\text{loc}}^2(\omega_2; \mathbb{R}^n)$, hence

$$\sigma_m = \text{div } u_m \mathbf{1} + \frac{\partial G}{\partial E}(\varepsilon^D(u_m)) + \rho_m \varepsilon(u_m) \in W_{2,\text{loc}}^1(\omega_2; \mathbb{M}),$$

and the Euler equation for u_m implies ($1 \le \alpha \le n$)

(5.6) $\quad \displaystyle\int_{\omega_2} \partial_\alpha \sigma_m : \varepsilon(v)\, dx = 0, \; v \in C_0^\infty(\omega_2; \mathbb{R}^n)$.

The following calculations can be found in [FreSe] but for completeness we give an outline of the arguments. We insert $v = \varphi^6 \partial_\alpha u_m$ for some $\varphi \in C_0^\infty(\omega_2)$, $\varphi \ge 0$, into (5.6) and use summation convention with respect to Greek and Latin indices repeated twice. For notational simplicity we also drop the index m. Then (5.6) implies

$$0 = \int_{\omega_2} \partial_\alpha \sigma : \varepsilon(\varphi^6 \partial_\alpha u)\, dx =$$

$$\int_{\omega_2} \varphi^6 \partial_\alpha \sigma : \varepsilon(\partial_\alpha u)\, dx + \int_{\omega_2} \partial_\alpha \sigma_{ij} \partial_\alpha u^i\, \partial_j \varphi^6\, dx$$

or equivalently

$$\int_{\omega_2} \varphi^6 \, \partial_\alpha \sigma : \varepsilon(\partial_\alpha u) \, dx = -\int_{\omega_2} \partial_\alpha \sigma_{ij} \, \partial_\alpha u^i \, \partial_j \varphi^6 \, dx =$$

$$-2 \int_{\omega_2} \partial_\alpha \sigma_{ij} \, \varepsilon_{i\alpha}(u) \partial_j \varphi^6 \, dx + \int_{\omega_2} \partial_\alpha \sigma_{ij} \, \partial_i u^\alpha \, \partial_j \varphi^6 \, dx =$$

$$-2(I)_1 + (I)_2.$$

In order to discuss $(I)_1$ we observe $\partial_k \sigma_{k\beta} = 0$, $\beta = 1, \ldots, n$. Then

$$(I)_1 = \int_{\omega_2} \partial_\alpha \sigma_{ij} \left(\varepsilon_{i\alpha}^D(u) + \delta_{i\alpha} \frac{1}{n} \text{ div } u \right) \partial_j \varphi^6 dx$$

$$= \int_{\omega_2} \partial_\alpha \sigma_{ij} \varepsilon_{i\alpha}^D(u) \partial_j \varphi^6 dx$$

$$= \int_{\omega_2} \left(\partial_\alpha \sigma_{ij}^D + \frac{1}{n} \delta_{ij} \text{ tr } (\partial_\alpha \sigma) \right) \varepsilon_{i\alpha}^D(u) \partial_j \varphi^6 dx$$

$$= \int_{\omega_2} \partial_\alpha \sigma_{ij}^D \varepsilon_{i\alpha}^D(u) \partial_j \varphi^6 dx + \frac{1}{n} \int_{\omega_2} \text{ tr } (\partial_\alpha \sigma) \varepsilon_{i\alpha}^D(u) \partial_i \varphi^6 dx.$$

We have

$$0 = \partial_j \sigma_{jk} = \partial_j \sigma_{jk}^D + \frac{1}{n} \delta_{jk} \text{ tr } (\partial_j \sigma),$$

i.e.

$$\frac{1}{n} \text{ tr } (\partial_k \sigma) = -\partial_j \sigma_{jk}^D$$

and therefore

$$(I)_1 = \int_{\omega_2} \left[\partial_\alpha \sigma_{ij}^D \varepsilon_{i\alpha}^D(u) \partial_j \varphi^6 - \partial_j \sigma_{j\alpha}^D \varepsilon_{i\alpha}^D(u) \partial_i \varphi^6 \right] dx$$

$$= \int_{\omega_2} \left[\partial_\alpha \sigma_{ij}^D \varepsilon_{i\alpha}^D(u) \partial_j \varphi^6 - \partial_\alpha \sigma_{j\alpha}^D \varepsilon_{j\ell}^D(u) \partial_\ell \varphi^6 \right] dx,$$

the last identity trivially follows by index permutation. This may also be written as

$$(I)_1 = \int_{\omega_2} \partial_\alpha \sigma_{ij}^D \left[\varepsilon_{i\alpha}^D(u) \partial_j \varphi^6 - \delta_{i\alpha} \varepsilon_{\ell j}^D(u) \partial_\ell \varphi^6 \right] dx.$$

Let

$$T^{(\alpha)} = \tfrac{1}{2} \left(P^{(\alpha)} + (P^{(\alpha)})^T \right),$$

$$P_{ij}^{(\alpha)} = \varepsilon_{i\alpha}^D(u) \partial_j \varphi - \delta_{i\alpha} \varepsilon_{\ell j}^D(u) \partial_\ell \varphi, \ \alpha = 1, \dots, n,$$

hence

$$(I)_1 = 6 \int_{\omega_2} \varphi^5 \partial_\alpha \sigma^D : T^{(\alpha)} dx.$$

For the tensors $T^{(\alpha)}$ we observe

$$T^{(\alpha)} : T^{(\alpha)} \le P^{(\alpha)} : P^{(\alpha)}$$

$$= \left[\varepsilon_{i\alpha}^D(u) \partial_j \varphi - \delta_{i\alpha} \varepsilon_{\ell j}^D(u) \partial_\ell \varphi \right] \left[\varepsilon_{i\alpha}^D(u) \partial_j \varphi - \delta_{i\alpha} \varepsilon_{rj}^D(u) \partial_r \varphi \right]$$

$$\le c_1 |\varepsilon^D(u)|^2 |\nabla \varphi|^2$$

with a suitable positive constant $c_1(n)$.

Next recall the definition of σ (ρ denotes the quantity ρ_m) which implies

$$\sigma^D = \frac{\partial G}{\partial E} (\varepsilon^D(u)) + \rho \varepsilon^D(u),$$

hence

$$|(I)_1| = \ 6 \left| \int_{\omega_2} \varphi^5 D^2 G(\varepsilon^D(u)) (\partial_\alpha \varepsilon^D(u), T^{(\alpha)}) \, dx + \rho \int_{\omega_2} \varphi^5 \partial_\alpha \varepsilon^D(u) : \right.$$

$$\left. T^{(\alpha)} dx \right| \le c_2 \left[\left| \int_{\omega_2} \varphi^5 D^2 G(\varepsilon^D(u)) (\partial_\alpha \varepsilon^D(u), T^{(\alpha)}) \, dx \right| \right.$$

$$\left. + \rho \int_{\omega_2} |\nabla \varepsilon^D(u)| \, |\varepsilon^D(u)| \, |\nabla \varphi| \varphi^5 dx \right].$$

Splitting $\varepsilon^D(u) = \varepsilon(u) - \frac{1}{n}\,\mathrm{div}\,u\,\mathbf{1}$ and using Cauchy–Schwarz inequality for the bilinear form $D^2G\big(\varepsilon^D(u)\big)(\cdot,\cdot)$ we deduce

$$|(I)_1| \leq c_3\Big[\int_{\omega_2}\varphi^6\Big(\rho|\nabla\varepsilon(u)|^2 + D^2G(\varepsilon^D(u))\big(\partial_\alpha\varepsilon^D(u),\partial_\alpha\varepsilon^D(u)\big)$$

$$+\,\mathrm{div}\,\partial_\alpha u\,\mathrm{div}\,\partial_\alpha u\Big)\,dx\Big]^{1/2}$$

$$\times\Big[\int_{\omega_2}\varphi^4\Big(\rho|\nabla\varphi|^2|\varepsilon(u)|^2 + D^2G(\varepsilon^D(u))\big(T^{(\alpha)},T^{(\alpha)}\big)\Big)\,dx\Big]^{1/2}$$

and get the final estimate for $(I)_1$

$$(5.7)\qquad |(I)_1| \leq c_4(n,\varphi)\Big(\int_{\omega_2}\varphi^6\partial_\alpha\sigma_m : \varepsilon(\partial_\alpha u_m)\,dx\Big)^{1/2}J_m(u_m,\omega_2)^{1/2}$$

where the index m has been introduced again. We need to estimate $(I)_2$. After integration by parts we have (dropping the index m)

$$(I_2) = -\int_{\omega_2}\sigma_{ij}\,\mathrm{div}\,\partial_i u\partial_j\varphi^6\,dx$$

$$-\int_{\omega_2}\sigma_{ij}\partial_i u^\alpha\,\partial_\alpha\partial_j\varphi^6\,dx = A_1 + A_2,$$

$$|A_1| \leq c_5\Big(\int_{\omega_2}\varphi^6|\nabla\,\mathrm{div}\,u|^2dx\Big)^{1/2}\Big(\int_{\omega_2}\varphi^4|\nabla\varphi|^2|\sigma|^2dx\Big)^{1/2}$$

$$\leq c_5\Big(\int_{\omega_2}\varphi^6\partial_\alpha\sigma : \varepsilon(\partial_\alpha u)\,dx\Big)^{1/2}\Big(\int_{\omega_2}\varphi^4|\nabla\varphi|^2|\sigma|^2dx\Big)^{1/2},$$

$$|A_2| = \Big|\int_{\omega_2}\big[\partial_i\sigma_{ij}u^\alpha\partial_\alpha\partial_j\varphi^6 + \sigma_{ij}u^\alpha\partial_\alpha\partial_i\partial_j\varphi^6\big]\,dx\Big|$$

$$= \Big|\int_{\omega_2}\sigma_{ij}u^\alpha\partial_\alpha\partial_i\partial_j\varphi^6\,dx\Big|$$

$$\leq c_6(n,\varphi)\Big(\int_{\omega_2}|\varphi^3\sigma|^n dx\Big)^{1/n}\Big(\int_{\omega_2}|u|^{\frac{n}{n-1}}dx\Big)^{1-1/n}.$$

The remaining integrals are handled as in section 2 using the convergences (5.1) – (5.5), the boundedness of the imbedding

$$X(\omega_2) \longrightarrow L^{n/n-1}(\omega_2; \mathbb{R}^n)$$

together with estimates like

$$\left(\int_{\omega_2} |\varphi^3 \sigma_m|^n dx\right)^{1/n} \le c_7 \left(\int_{\omega_2} |\nabla(\varphi^3 \sigma_m)|^2 dx\right)^{1/2},$$

$$|\nabla \sigma_m|^2 \le c_8 \partial_\alpha \sigma_m : \varepsilon(\partial_\alpha u_m).$$

In conclusion we deduce from (5.7) and the latter inequalities the final bound

$$(5.8) \qquad \int_{\omega_2} \varphi^6 \, \partial_\alpha \sigma_m : \varepsilon(\partial_\alpha u_m) \, dx \le c_9 < \infty$$

with c_9 not depending on m.

Choosing $\varphi = 1$ on ω_1 in (5.8) we arrive at

$$\sup_m \int_{\omega_1} \left[|\nabla \operatorname{div} u_m|^2 dx + |\nabla \sqrt{1 + |\varepsilon^D(u_m)|}|^2 \right] dx < \infty$$

and the proof of Theorem 4.5.2 can be completed along the lines of Theorem 4.2.1, in particular, we see

$$(5.9) \qquad \begin{cases} \sqrt{1 + |\varepsilon^D(u_m)|} \rightharpoonup \sqrt{1 + |\varepsilon^D(u)|} & \text{in } W_2^1(\omega_1), \\[2mm] \varepsilon^D(u_m) \to \varepsilon^D(u) & \text{a.e. in } \omega_1, \\[2mm] \operatorname{div} u_m \rightharpoonup \operatorname{div} u & \text{in } W_2^1(\omega_1). \end{cases}$$

$\square$

REMARK 4.5.2 Going through the proof of Theorem 4.5.2 we see that $\sqrt{1 + |\varepsilon^D(u)|} \in W_{2,\mathrm{loc}}^1(\Omega)$ holds up to $n = 4$. Also Theorem 4.2.1 is valid for $n = 4$ but the four–dimensional situation seems to be of no physical interest.

The analogue of Theorem 4.2.2 is

THEOREM 4.5.3 *Consider a local I–minimizer u in the space $X_{\text{loc}}(\Omega)$. Then, for any symmetric matrix A, any ball $B_R(x_0)$ and any $t \in\,]0,1[$, we have*

$$\int_{B_{tR}(x_0)} \left(|\nabla \,\operatorname{div} u|^2 + |\nabla\sqrt{1 + |\varepsilon^D(u)|}\,|^2 \right) dx$$

$$\leq c(n)(1-t)^{-2}R^{-2} \int_{B_R(x_0)} |\varepsilon(u) - A|^2 dx.$$

Proof of Theorem 4.5.3: Let $\omega_1 = B_{\frac{t+1}{2}R}(x_0)$, $\omega_2 = B_R(x_0)$ and define u_m as in the proof of Theorem 4.5.2. We use (5.6) with

$$\begin{cases} v = \varphi^2 \partial_\alpha \hat{u}_m, \\[2mm] \varphi \in C_0^1(\omega_1),\ \varphi = 1 \text{ on } B_{tR}(x_0),\ |\nabla\varphi| \leq c_1/(1-t)R, \\[2mm] \hat{u}_m(x) = u_m(x) - A(x - x_0) - \gamma_m(x), \end{cases}$$

where γ_m is chosen according to (2.31). Then

$$\int_{\omega_1} \varphi^2 \partial_\alpha \sigma_m : \varepsilon(\partial_\alpha u_m)\, dx =$$

$$-2\int_{\omega_1} \varphi \partial_\alpha \sigma_m : (\nabla\varphi \odot \partial_\alpha \hat{u}_m)\, dx$$

which leads to the estimate

$$\int_{\omega_1} \varphi^2 \partial_\alpha \sigma_m : \varepsilon(\partial_\alpha u_m)\, dx$$

$$\leq c_2 \frac{1}{R} \frac{1}{1-t} \left(\int_{\omega_1} \varphi^2 |\nabla\sigma_m|^2 dx \right)^{1/2} \left(\int_{\omega_1} |\varepsilon(u_m) - A|^2 dx \right)^{1/2}.$$

We observe

$$|\nabla\sigma_m|^2 \leq c_3 \partial_\alpha \sigma_m : \varepsilon(\partial_\alpha u_m)$$

so that (using the estimates given at the end of the proof of Theorem 4.5.2)

$$\int\limits_{B_{tR}(x_0)} \left[|\nabla \operatorname{div} u_m|^2 + \left|\nabla\sqrt{1+|\varepsilon^D(u_m)|}\right|^2\right] dx$$

$$\leq c_3 \frac{1}{R^2} \frac{1}{(1-t)^2} \int\limits_{B_R(x_0)} |\varepsilon(u_m) - A|^2 dx.$$

On the left–hand side we use (5.9) together with weak lower–semicontinuity w.r.t. weak convergence in L^2, for the right–hand side we observe that (5.9) implies $\varepsilon(u_m) \to \varepsilon(u)$ in $L^2(\omega_1; \mathbb{M})$ since $n \leq 3$.

$\square$

Now we are in the position to prove Theorem 4.5.1: consider some local I–minimizer in the space $X_{\text{loc}}(\Omega)$. We want to establish the claim of Lemma 4.3.1 by contradiction and use the same notation as introduced in the proof of this lemma. The limit function v now satisfies the equation

$$\int\limits_{B_1} \left(\operatorname{div} v \operatorname{div}\varphi + D^2G(A^D)\left(\varepsilon^D(v), \varepsilon^D(\varphi)\right)\right) dx = 0$$

for all $\varphi \in \overset{\circ}{W}{}^1_2(B_1; \mathbb{R}^3)$, hence v is smooth, and an appropriate Campanato–type estimate holds. Again it remains to show

(5.10) $\varepsilon(v_k) \to \varepsilon(v)$ in $L^2_{\text{loc}}(B_1; \mathbb{M})$.

Let

$$G_0(E) = \frac{1}{2}(\operatorname{tr} E)^2 + G(E^D).$$

For any $\varphi \in C^1_0(B_1)$, $\varphi \geq 0$, we then get (compare (3.4))

$$\int\limits_{B_1} \varphi\left\{G_o\left(A_k + \varepsilon_k\varepsilon(v_k)\right) - G_0\left(A_k + \varepsilon_k\varepsilon(v)\right)\right\} dz$$

$$\leq \int\limits_{B_1} \varepsilon_k \frac{\partial G_0}{\partial E}\left(A_k + \varepsilon_k\left(\varepsilon(v_k) + \varphi\varepsilon(v - v_k)\right)\right) : \left((v - v_k) \odot \nabla\varphi\right) dz$$

$$+ \int\limits_{B_1} \varepsilon_k^2 |\nabla\varphi|^2 |v_k - v|^2 dz.$$

Using Taylor expansion for the left hand side and also bounds for D^2G_0 we find that

$$\varepsilon_k^2 \frac{1}{2} \int_{B_1} \varphi \left[\text{div } (v_k - v)^2 + \frac{|\varepsilon^D(v_k - v)|^2}{1 + |A_k^D| + \varepsilon_k (|\varepsilon^D(v_k)| + |\varepsilon^D(v)|)} \right] dz$$

$$\leq \int_{B_1} \varepsilon_k^2 |\nabla\varphi|^2 |v - v_k|^2 dz$$

$$+ \int_{B_1} \varepsilon_k \left\{ \frac{\partial G_0}{\partial E} \left(A_k + \varepsilon_k (\varepsilon(v_k) + \varphi\varepsilon(v - v_k)) \right) - \frac{\partial G_0}{\partial E} \left(A_k + \varepsilon_k \varepsilon(v) \right) \right\} :$$

$$((v - v_k) \odot \nabla\varphi) \, dz$$

$$+ \int_{B_1} \varepsilon_k \frac{\partial G_0}{\partial E} \left(A_k + \varepsilon_k \varepsilon(v) \right) : \varepsilon \big(\varphi(v - v_k) \big) \, dz.$$

Proceeding exactly as in the proof of Lemma 4.3.1 we get

$$\varepsilon_k^{-2} \cdot (\text{right–hand side}) \to 0, \quad k \to \infty,$$

in conclusion (after specifying φ)

$$(5.11) \qquad \int_{B_r} \frac{|\varepsilon^D(v - v_k)|^2}{1 + |A_k^D| + \varepsilon_k (|\varepsilon^D(v_k)| + |\varepsilon^D(v)|)} \, dz \to 0,$$

$$(5.12) \qquad \int_{B_r} |\text{div } v_k - \text{div } v|^2 dz \to 0, \quad k \to \infty,$$

for any $r < 1$. Clearly (5.10) will follow from (observe (5.12))

$$\int_{B_r} |\varepsilon^D(v_k - v)| \, dz \to 0.$$

Let

$$\eta_k(z) = \frac{1}{\varepsilon_k} \left(\sqrt{1 + |A_k^D + \varepsilon_k \varepsilon^D(v_k)|} - \sqrt{1 + |A_k^D|} \right).$$

The scaled version of Theorem 4.5.3 implies

$$\sup_k \|\eta_k\|_{W_2^1(B_r)} < \infty$$

for any $r < 1$ and exactly the same argument as in the proof of Lemma 4.3.1 gives the desired result. Hence the blow–up lemma is also established for local I–minimizers in $X_{\mathrm{loc}}(\Omega)$, partial regularity follows.

$\square$

4.6 A general class of constitutive relations

Without being complete we indicate how to treat other classes of generalized Newtonian fluids and also elastic–plastic materials with general hardening laws. For simplicity we just discuss the mathematical problem of minimizing the variational integral

$$J(u) = \int_\Omega G(\nabla u)dx$$

among functions $u : \Omega \to \mathbb{R}^M$ for some bounded domain $\Omega \subset \mathbb{R}^n$ with $n \geq 2$ and $M \geq 1$. The following assumptions are imposed on the density $G : \mathbb{R}^{nM} \to \mathbb{R}$.

(6.1) $C_1\left(A(|E|) - 1\right) \leq G(E) \leq C_2\left(A(|E|) + 1\right)$

(6.2) $G(E) \leq C_3(|E|^2 + 1)$

(6.3) $|E|^2 |D^2G(E)| \leq C_4(G(E) + 1)$

(6.4) $D^2G(Q)(E, E) \geq \lambda(1 + |Q|)^{-\mu}|E|^2$

(6.5) $A^*(|DG(E)|) \leq C_5(A(|E|) + 1).$

Here C_1, C_2, C_3, C_4, C_5 and μ denote non–negative constants, λ is some positive number, and (6.1)–(6.5) are required to be valid for all matrices $E, Q \in \mathbb{R}^{nM}$. $A : [0, \infty[\to [0, \infty[$ is a N–function with conjugate function A^* which means that A is strictly increasing and convex satisfying in addition

$$\lim_{t \downarrow 0} A(t)/t = 0, \quad \lim_{t \to \infty} A(t)/t = \infty,$$

$$A(2t) \leq k\, A(t) \quad \forall t \geq t_0$$

for suitable constants t_0 and k. Then the energy J is well–defined on the Orlicz–Sobolev space $W_A^1(\Omega; \mathbb{R}^M)$ (see [A] for a definition), and in [FO] we proved using the techniques of the foregoing sections.

THEOREM 4.6.1 *Let $u \in W_A^1(\Omega; \mathbb{R}^M)$ denote a local J–minimizer.*

a) *If $n = 2$ and $\mu \leq 1$, then $u \in C^{1,\alpha}(\Omega; \mathbb{R}^M)$ for any $\alpha < 1$.*

b) *Let $n \geq 2$ and assume that $\mu < 4/n$. Then there is an open subset Ω_0 of Ω such that $|\Omega - \Omega_0| = 0$ and $u \in C^{1,\alpha}(\Omega_0; \mathbb{R}^M)$ for all $\alpha < 1$.*

Let us consider some examples of N–functions A and corresponding integrands G.

1) $A_1(t) = t^p \ln(1 + t)$, $1 \le p < 2$

In this case we may define

$$G(E) = \begin{cases} A_1(|E|) & \text{if } |E| \ge 1 \\ \\ g(|E|) & \text{if } |E| \le 1 \end{cases}, E \in \mathbb{R}^{nM},$$

where g is a quadratic polynomial which has to be chosen such that G is of class C^2. It is easy to see that (6-1)–(6.5) hold, in particular, (6.4) is satisfied with $\mu = 2 - p$, and we obtain

COROLLARY 4.6.1 *If $n = 2$, then the singular set is empty. In case $n \ge 2$ and $p > 2 - 4/n$ we have partial regularity of local J–minimizers.*

2) $A_2(t) = t \ln(1 + \ln(1 + t))$, $G(E) = A_2(|E|)$

Now (6.4) is valid for any number $\mu > 1$, Theorem 4.6.1 implies

COROLLARY 4.6.2 *Local J–minimizers are partially regular provided $n = 2$ or $n = 3$.*

3) $A_3(t) = \int\limits_0^t s^{1-\alpha}(\text{ar sinh } s)^\alpha ds$, $0 < \alpha \le 1$, $G(E) = A_3(|E|)$

Condition (6.4) is valid for $\mu = \alpha$, and we get

COROLLARY 4.6.3 *If Ω is a domain in $\mathbb{R}^2$, then local J–minimizers u are smooth in the interior of Ω, partial regularity holds in case $n \ge 3$ together with $\alpha < 4/n$.*

The third example is related to the Sutterby fluid model (see [BAH]), and for $\alpha = 1$ it reduces to the Prandtl–Eyring fluid: in the physical setting of the Sutterby model we have to discuss the variational problem ($n = M$, $n = 2$ or 3)

$$J(v) = \int\limits_\Omega \int\limits_0^{|\varepsilon(v)|} s^{1-\alpha}(\text{ar sinh } s)^\alpha ds\, dx \to \min$$

subject to the constraint $\operatorname{div} v = 0$ where v has to be taken from the space

$$\{u \in L^1(\Omega; \mathbb{R}^n) : |\varepsilon(u)| \in L_A(\Omega)\}.$$

THEOREM 4.6.2 *Let v denote a locally minimizing velocity field for the Sutterby model. If $n = 2$, then v is smooth, in case $n = 3$ the measure of the singular set vanishes.*

The proof of Theorem 4.6.2 essentially follows the lines of [FO], the necessary adjustments are carried out in [R]. The formulation and proofs of corresponding results for plastic materials, i.e. for minimizers of $\int_{\Omega} \{G(\varepsilon^D(u)) + (\operatorname{div} u)^2\}dx$ with G satisfying (6.1)–(6.5), are left to the reader.

REMARK 4.6.1 Very recently the results of Theorem 4.6.1 have been improved in various directions. In the papers [BiF], [BiFM], [FM] and [MS], for example, the set of assumptions (6.1)–(6.5) can be replaced by the so–called (s, μ, q)–condition still implying partial regularity, whereas in the special case $G(\nabla u) = g(|\nabla u|^2)$ full regularity holds. With some additional work the result of [BiF] should extend to the setting of fluid mechanics replacing ∇u by $\varepsilon(u)$ and taking care of the constraint $\operatorname{div} u = 0$.

Appendix B

B.1 Density results

For completeness we give an outline of the proof of Lemma 4.1.6. Part a) and c) can be deduced from [FreSe], Lemma A 2.6, we concentrate on the proof of d).

For $r \geq 1$ let

$$D^r_{L \ln L}(\Omega) = \{v \in L^1(\Omega; \mathbb{R}^n) : \text{div } v \in L^r(\Omega), |\varepsilon(v)| \in L \ln L(\Omega)\},$$

$$\overset{\circ}{D}{}^r_{L \ln L}(\Omega) = \text{closure of } C^\infty_0(\Omega; \mathbb{R}^n) \text{ with respect to the norm}$$

$$\|v\| = \|v\|_{L^1(\Omega)} + \|\text{div } v\|_{L^r(\Omega)} + \|\varepsilon(v)\|_{L \ln L(\Omega)}.$$

Then it was shown in [FreSe] (recall Ω is a Lipschitz domain) assuming $r \leq \frac{n}{n-1}$:

(1.1) $\qquad D^r_{L \ln L}(\Omega) = \text{closure of } C^\infty(\overline{\Omega}; \mathbb{R}^n) \text{ w.r.t. the norm } \| \cdot \| \text{ in } D^r_{L \ln L}(\Omega)$

(1.2) $\qquad \overset{\circ}{D}{}^r_{L \ln L}(\Omega) = \{v \in D^r_{L \ln L}(\Omega) : v = 0 \text{ on } \partial\Omega\}.$

Let us abbreviate

$$W = \overset{\circ}{V}_0(\Omega), \quad \tilde{W} = \{v \in \overset{\circ}{V}(\Omega) : v = 0 \text{ on } \partial\Omega\}.$$

We assume that there is a velocity field $u_0 \in \tilde{W}$, $u_0 \notin W$. In order to get a contradiction we select a linear functional $\ell \in \tilde{W}^*$ such that

(1.3) $\qquad \ell(u_0) = 1, \quad \ell(v) = 0 \text{ for } v \in W$

and introduce the operator

$$A : \tilde{W} \to \overset{\circ}{\Sigma}(\Omega) = \{ \varkappa \in L^1(\Omega; \mathbb{M}) : \text{tr } \varkappa = 0, \ |\varkappa| \in L \ln L(\Omega) \},$$

$$A(v) = \varepsilon(v),$$

the norm on $\overset{\circ}{\Sigma}(\Omega)$ being defined through

$$\|\varkappa\|_{\overset{\circ}{\Sigma}(\Omega)} = \|\varkappa\|_{L \ln L(\Omega)}.$$

With suitable positive constants c_1, c_2 we have

$$c_1 \|v\|_{V(\Omega)} \leq \|A(v)\|_{\overset{\circ}{\Sigma}(\Omega)} \leq c_2 \|v\|_{V(\Omega)}$$

and therefore

$$g : A(\tilde{W}) \to \mathbb{R}, \quad g(\tau) = \ell(A^{-1}\tau),$$

is a continuous linear functional on the subspace $A(\tilde{W})$ of $\overset{\circ}{\Sigma}(\Omega)$ satisfying

$$|g(\tau)| \leq \|\ell\| \, \|A^{-1}\tau\|_{V(\Omega)} \leq \frac{1}{c_1} \|\ell\| \, \|\tau\|_{\overset{\circ}{\Sigma}(\Omega)}.$$

Let $G \in \overset{\circ}{\Sigma}(\Omega)^*$ denote some extension of g with the property

$$\|G\| = \|g\| \leq \frac{1}{c_1} \|\ell_1\|.$$

Recalling the definition of Exp (Ω) and also the statement of Lemma 4.1.3 we have the representation

$$G(\tau) = \int_\Omega \tau : \sigma \, dx, \ \tau \in \overset{\circ}{\Sigma}(\Omega),$$

with a unique tensor $\sigma \in L^1(\Omega; \mathbb{M})$, tr $\sigma = 0$, $\sigma \in$ Exp $(\Omega; \mathbb{M})$.

From the construction we then deduce

$$\ell(v) = g(A(v)) = G(A(v)) = \int_\Omega \sigma : \varepsilon(v) \, dx$$

for any $v \in \tilde{W}$. The functional ℓ vanishes on the space W, quoting [L1] or [LS] we find a pressure function $p \in L^2(\Omega)$, $\int_\Omega p\, dx = 0$, such that

$$(1.3) \qquad \int_\Omega p \operatorname{div} v\, dx = \int_\Omega \sigma : \varepsilon(v)\, dx$$

for any $v \in \overset{\circ}{W}{}^1_2(\Omega; \mathbb{R}^n)$. $\sigma \in \operatorname{Exp}(\Omega; \mathbb{M})$ implies $\sigma \in L^q(\Omega; \mathbb{M})$ for any finite q so that

$$(1.4) \qquad p \in L^q(\Omega), \quad q < \infty.$$

Taking $q > n$ we get from (1.3)

$$(1.5) \qquad \int_\Omega p \operatorname{div} v\, dx = \int_\Omega \sigma : \varepsilon(v)\, dx, \quad v \in \overset{\circ}{D}{}^r_{L\ln L}(\Omega),$$

where $r = \frac{q}{q-1} < \frac{n}{n-1}$. Clearly $u_0 \in D^r_{L\ln L}(\Omega)$ and $u_0 = 0$ on $\partial\Omega$ (recall $\operatorname{div} u_0 = 0$), by (1.2) a sequence $u_m \in C^\infty_0(\Omega; \mathbb{R}^n)$ exists such that

$$u_m \to u_0 \text{ in } D^r_{L\ln L}(\Omega),$$

(1.5) implies

$$\int_\Omega p \operatorname{div} u_m\, dx = \int_\Omega \sigma : \varepsilon(u_m)\, dx.$$

Due to $\operatorname{div} u_0 = 0$ we must have $\|\operatorname{div} u_m\|_{L^r(\Omega)} \to 0$, in conclusion

$$\int_\Omega \sigma : \varepsilon(u_0)\, dx = 0,$$

hence $\ell(u_0) = 0$ which is the desired contradiction. $\qquad\qquad\qquad\square$

Notation and tools from functional analysis

The following definitions and results are standard and can be found in any textbook on functional analysis.

Consider a real vectorspace X with a given norm $\| \cdot \|$. X is a Banach space (w.r.t. the norm $\| \cdot \|$) iff each Cauchy–sequence has a limit in X, and a Banach space is called a Hilbert space iff the norm is induced by a scalar product "$\cdot$" via $\|x\| = \sqrt{x \cdot x}$. Let X^* denote the normed dual of the normed space X, i.e.

$$X^* := \{\varphi : X \to \mathbb{R} : \varphi \text{ is linear and continuous}\}.$$

We use the symbol $< \varphi, x > := \varphi(x)$, $\varphi \in X^*$, $x \in X$, for the dual pairing.

Let $\{x_k\}$ denote a sequence in X. We say that $\{x_n\}$ converges weakly to $x \in X$ iff

$$\lim_{n \to \infty} \varphi(x_n) = \varphi(x)$$

holds for all $\varphi \in X^*$. In this case we write $x_n \stackrel{n \to \infty}{\rightharpoonup} x$. Note that $x_n \to x$ (i.e. $\|x_n - x\| \to 0$) implies $x_n \rightharpoonup x$ but not vice versa. Moreover, the weak limit is unique, and we have (weak) lower semicontinuity of the norm w.r.t. weak convergence:

$$\|x\| \le \liminf_{n \to \infty} \|x_n\|.$$

More generally, weak lower semicontinuity holds for any convex function Φ: $X \to \mathbb{R}$ which is continuous w.r.t. the given norm $\| \cdot \|$.

There is another concept of weak convergence: a sequence $\{u_n\}$ in X^* converges in the weak-$*$ sense towards $u \in X^*$ iff

$$\lim_{n \to \infty} < u_n, x > = < u, x >$$

holds for all $x \in X$. In this case we write $u_n \overset{*}{\rightharpoonup} u$ as $n \to \infty$.

Let us consider the isometric embedding

$$J: \ X \to X^{**}, \ < J(x), \varphi >:=< \varphi, x >.$$

X is called a reflexive space iff $J(X) = X^{**}$. The most important property of these spaces is the following compactness result: any bounded sequence in a reflexive space X contains a weakly convergent subsequence. This fact gives rise to the following basic existence theorem in variational calculus: let X denote a reflexive space and consider a convex, (norm–) closed subset K of X. Further let $\Phi: \ K \to \mathbb{R}$ denote a convex, continuous function which is coercive, i.e. we have $\Phi(x) \geq \alpha \|x\| + \beta$ for all $x \in K$ with constants $\alpha > 0$, $\beta \in \mathbb{R}$. Then the variational problem

$$\begin{cases} \text{to find } u_0 \in K \text{ such that} \\ \Phi(u_0) = \inf_K \Phi \end{cases}$$

has a solution which is also unique provided Φ is strictly convex. We sketch the proof: consider a minimizing sequence $\{u_n\}$, i.e. $\Phi(u_n) \overset{n \to \infty}{\to} \inf_K \Phi$. The coercivity condition implies $\sup_n \|u_n\| < +\infty$, hence $u_n \rightharpoonup u_0$ in X for a subsequence. Since K is also weakly closed, we have $u_0 \in K$, the minimality of u_0 follows from the weak lower semicontinuity of Φ.

An important class of Banach spaces are the Lebesgue spaces $L^p(\Omega)$ defined on a bounded open set $\Omega \subset \mathbb{R}^n$. Let $1 \leq p < \infty$. Then $L^p(\Omega)$ consists of all (equivalence classes of) Lebesgue measurable functions $u: \ \Omega \to \mathbb{R}$ such that

$$\|u\|_p := \|u\|_{L^p(\Omega)} := \left(\int_\Omega |u|^p \, dx \right)^{\frac{1}{p}} < +\infty.$$

A function $u: \ \Omega \to \mathbb{R}$ is in the Lebesgue space $L^\infty(\Omega)$ iff

$$\|u\|_\infty := \|u\|_{L^\infty(\Omega)} := (ess) \sup_\Omega |u| < \infty.$$

By $L^p_{loc}(\Omega)$ we denote the space of all Lebesque measurable functions with the property $u_{|\Omega'} \in L^p(\Omega')$ for each open set Ω' with compact closure in Ω. For the basic notions from measure theory we refer to the monograph [Fe], here the reader will find further information concerning the spaces $L^p(M, \lambda)$ for arbitrary sets M and measures λ on M. Let us remark that the spaces $L^p_{(loc)}(\Omega; \mathbb{R}^k)$ of Lebesgue measurable vectorfunctions $\Omega \to \mathbb{R}^k$ are defined componentwise.

LEMMA 1   a) $L^p(\Omega)$ is a Banach space for $1 \leq p \leq \infty$.
$L^2(\Omega)$ is a Hilbert space with scalar product

$$f \cdot g = \int_\Omega fg \, dx.$$

b) For $1 < p < \infty$ $L^p(\Omega)$ is a reflexive space, $L^1(\Omega)$ and $L^\infty(\Omega)$ are not of this type.

c) We have $(L^p(\Omega))^* \cong L^{p'}(\Omega)$, $p' = \dfrac{p}{p-1}$, for any $1 \leq p < \infty$ $(1' := \infty)$.
More precisely:

$$L^{p'}(\Omega) \ni u \quad \to \quad J(u) \in (L^p(\Omega))^*,$$

$$< J(u), v > \quad := \quad \int_\Omega uv \, dx, \; v \in L^p(\Omega),$$

is an isometric isomorphism.

d) Consider a bounded sequence $\{u_m\}$ in $L^\infty(\Omega)$. Then, for some subsequence $\{u'_m\}$ and some $u \in L^\infty(\Omega)$, we have $u'_m \overset{*}{\rightharpoonup} u$ in $L^\infty(\Omega)$, i.e.

$$\int_\Omega u'_m v \, dx \to \int_\Omega uv \, dx \text{ for all } v \in L^1(\Omega).$$

e) If $\{u_n\}$ converges in $L^p(\Omega)$ to a function u, then $u'_n(x) \to u(x)$ a.e. on Ω for a subsequence.

f) $C^\infty(\Omega)$ is dense in $L^p(\Omega)$ for $1 \leq p < +\infty$.

g) Hölder's inequality:

$$\int_\Omega uv \, dx \leq \|u\|_p \|v\|_{p'} \text{ for } u \in L^p(\Omega), \; v \in L^{p'}(\Omega), \; p' = \frac{p}{p-1}.$$

Note that g) is a consequence of Young's inequality

$$ab \leq \varepsilon \frac{a^p}{p} + \varepsilon^{-\frac{p'}{p}} \frac{b^{p'}}{p'}, \; a, b \geq 0, \; p, p' > 1, \; p' = \frac{p}{p-1}, \; \varepsilon > 0.$$

It is well-known that for a function $u \in L^1(\Omega)$ almost every point x in Ω is a Lebesgue point of u, i.e.

$$u^*(x) = \lim_{r\downarrow 0} \fint_{B_r(x)} u \, dy$$

exists for almost all $x \in \Omega$ and defines a representative of u. Moreover, we have

$$\lim_{r \downarrow 0} \fint_{B_r(x)} |u - u^*(x)|^p \, dy = 0 \text{ a.e. on } \Omega,$$

provided u is in the space $L^p_{loc}(\Omega)$, and this implies

$$\lim_{r \downarrow 0} \fint_{B_r(x)} |u - (u)_{x,r}|^p \, dy = 0$$

for a.a. $x \in \Omega$. Here we abbreviated $(u)_{x,r} = \fint_{B_r(x)} u \, dy$. The quantity $\fint_{B_r(x)} |u - (u)_{x,r}|^p \, dy$ is termed the mean oscillation of the function u on the ball $B_r(x)$, and there is an important characterization of Hölder continuity in terms of the mean oscillation (see, e.g. [Gi])

LEMMA 2 *Suppose that $u \in L^p_{loc}(\Omega)$ satisfies*

$$\fint_{B_\rho(x)} |u - (u)_{x,\rho}|^p \, dy \le C\rho^{\lambda-n}$$

for some number λ such that $n < \lambda \le n+p$ and for all balls $B_\rho(x)$ with center in a compact subdomain Ω' of Ω and radius $\rho < \frac{1}{2} \text{dist}(\Omega', \partial\Omega)$. Then u is α–Hölder continuous on Ω' with exponent $\alpha = \frac{\lambda-n}{p}$.

Next we briefly discuss the basic properties of the Sobolev spaces $W^k_p(\Omega)$, $k \in \mathbb{N}$, $1 \le p \le +\infty$, for details we refer to [A] or [LU]. Let

$$W^k_p(\Omega) = \{u \in L^p(\Omega) : \partial^\alpha u \in L^p(\Omega) \text{ for all } |\alpha| \le k\},$$

where $\partial^\alpha u$ denotes the distributional derivative of order α. We use the norm

$$\|u\|_{W^k_p(\Omega)} = \begin{cases} \max_{|\alpha| \le k} \|\partial^\alpha u\|_{L^\infty(\Omega)}, & p = \infty, \\[2mm] \left(\sum_{|\alpha| \le k} \|\partial^\alpha u\|_{L^p(\Omega)} \right)^{\frac{1}{p}}, & p < \infty. \end{cases}$$

Then $W^k_p(\Omega)$ is a Banach space and from the definition it is immediate that

$$u_m \rightharpoonup u \text{ in } W^k_p(\Omega)$$

if and only if

$$\partial^\alpha u_m \rightharpoonup \partial^\alpha u \text{ in } L^p(\Omega) \text{ for all } |\alpha| \le k.$$

Local variants $W_{p,loc}^k(\Omega)$ or Sobolev spaces $W_{p,loc}^k(\Omega; \mathbb{R}^L)$ of vectorfunctions are defined in the same manner as we did for Lebesgue spaces. Further observe that $W_p^k(\Omega)$ is reflexive iff $1 < p < +\infty$. In this case $\sup_m \|u_m\|_{W_p^k(\Omega)} < +\infty$ implies the existence of a weakly convergent subsequence. In order to formulate boundary value problems we let

$$\overset{\circ}{W}_p^k(\Omega) := \text{closure of the test functions with compact support in } \Omega \text{ in } W_p^k(\Omega).$$

By definition, $\overset{\circ}{W}_p^k(\Omega)$ consists of all Sobolev functions with zero trace on $\partial\Omega$. There is a more explicit description of the boundary behaviour of Sobolev functions.

LEMMA 3 *Suppose that $\partial\Omega$ is Lipschitz. Then there exits a unique continuous linear operator $B: W_p^1(\Omega) \to L^p(\partial\Omega)$ such that $Bu = u_{|\partial\Omega}$ for all $u \in W_p^1(\Omega) \cap C^0(\overline{\Omega})$. Moreover, we have*

$$\overset{\circ}{W}_p^1(\Omega) = \{u \in W_p^1(\Omega) : Bu = 0\}.$$

Here $L^p(\partial\Omega)$ is the Lebesgue space of functions $u: \partial\Omega \to \mathbb{R}$ w.r.t. the $(n-1)-$ dimensional Hausdorff measure on $\partial\Omega$.

Sobolev functions can always be approximated by smooth functions, precisely

LEMMA 4 *Let $1 \le p < +\infty$. Then*

a) *$\{u \in C^\infty(\Omega) : \|u\|_{W_p^k(\Omega)} < +\infty\}$ is dense in $W_p^k(\Omega)$.*

b) *If Ω is Lipschitz, then $C^\infty(\Omega)$ can be replaced by $C^\infty(\overline{\Omega})$.*

The next result is a simplified version of Sobolev's imbedding theorem

LEMMA 5   a) *We have continuous imbeddings*

$$\overset{\circ}{W}_p^1(\Omega) \hookrightarrow \begin{cases} L^{\frac{np}{n-p}}(\Omega) & \text{for } p < n, \\ C^0(\overline{\Omega}) & \text{for } p > n. \end{cases}$$

b) *Furthermore, there exists a constant $c = c(n,p)$ such that for any $u \in \overset{\circ}{W}_p^1(\Omega)$*

$$\|u\|_{L^{\frac{np}{n-p}}(\Omega)} \le c\|\nabla u\|_{L^p(\Omega)} \text{ for } p < n,$$

$$\sup_\Omega |u| \le c|\Omega|^{\frac{1}{n} - \frac{1}{p}} \|\nabla u\|_{L^p(\Omega)} \text{ for } p > n.$$

c) *For $p < n$ and $s < \frac{np}{n-p}$ the imbedding $\overset{\circ}{W}^1_p(\Omega) \hookrightarrow L^p(\Omega)$ is compact.*

d) *If $\partial\Omega$ is Lipschitz, then in a) and c) $\overset{\circ}{W}^1_p(\Omega)$ can be replaced by $W^1_p(\Omega)$.*

Of great importance in regularity theory are the following inequalities for Sobolev functions.

LEMMA 6   a) *Poincaré's inequality. Let $1 \le p < +\infty$. Then there is a constant $c = c(n)$ such that*

$$\|u\|_{L^p(\Omega)} \le c|\Omega|^{\frac{1}{n}}\|\nabla u\|_{L^p(\Omega)}$$

holds for any function $u \in \overset{\circ}{W}^1_p(\Omega)$. For balls Ω the same is true for any $u \in W^1_p(\Omega)$ such that $\fint_\Omega u\, dx = 0$.

b) *Sobolev–Poincaré inequality. Let $u \in W^1_p(B_R)$ s.t. $\fint_{B_R} u\, dx = 0$. Then, if $p < n$, we have for a constant $c = c(n,p)$*

$$\|u\|_{L^{\frac{np}{n-p}}(B_R)} \le c\|\nabla u\|_{L^p(B_R)}.$$

Another tool from regularity theory is the description of Sobolev functions in terms of the behaviour of their difference quotients:

LEMMA 7 *Let $u \in L^p_{loc}(\Omega)$ and define for $i = 1, \ldots, n$, $h \neq 0$*

$$\Delta^i_h u(x) = \frac{1}{h}\{u(x + he_i) - u(x)\},$$

where e_i is the ith unit vector in $\mathbb{R}^n$. Suppose further that $1 < p < +\infty$ and that

$$\|\Delta^i_h u\|_{L^p(K)} \le c(K)$$

holds for $i = 1, \ldots, n$, any compact subset K of Ω and all $|h| < \mathrm{dist}(K, \partial\Omega)$. Then u is of class $W^1_{p,loc}(\Omega)$.

Bibliography

[A]   Adams, R.A., Sobolev spaces, Academic Press, New York 1975.

[An]   Anzellotti, G., On the existence of the rates of stress and displacement
         for Prandtl–Reuss plasticity, Quarterly J. of Appl.Math. 41 (1983), 181–
         208.

[AG1]   Anzellotti, G., Giaquinta, M., Existence of the displacement field for
         an elastic–plastic body subject to Hencky's law and von Mises yield
         condition, Manus.Math. 32 (1980), 101–136.

[AG2]   Anzellotti, G., Giaquinta, M., On the existence of the fields of stresses
         and displacements for an elastic–perfectly plastic body in static equilib-
         rium, J.Math. Pure Appl. 61 (1982), 219–244.

[AG3]   Anzellotti, G., Giaquinta, M., Convex functionals and partial regularity,
         Arch.Rational Mech.Anal. 102 (3) (1988), 243–272.

[AM]   Astarita, G., Marrucci, G., Principles of non–Newtonian fluid mechan-
         ics, McGraw–Hill, London, 1974.

[BAH]   Bird, R., Armstrong, R., Hassager, O., Dynamics of polymeric liquids,
         Volume 1 Fluid mechanics, John Wiley, Second Edition 1987.

[BF]   Bensoussan, A., Frehse, J., Asymptotic Behaviour of Norton–Hoff's
         Law in Plasticity Theory and H^1–regularity, Res.Notes Appl.Math. 29
         (1993), 3–25.

[BiF]   Bildhauer, M., Fuchs, M., Partial regularity for variational integrals
         with (s, μ, q)–growth, Preprint No. 635, SFB 256, Universität Bonn
         (2000).

[BiFM]   Bildhauer, M., Fuchs, M., Mingione, G., Apriori gradient bounds and
         local $C^{1,\alpha}$–estimates for (double) obstacle problems under nonstandard
         growth conditions, Preprint No. 647, SFB 256, Universität Bonn (2000).

[BI]   Bojarski, B., Iwaniec, T., Analytical foundations of the theory of quasi-
         conformal mappings in $\mathbb{R}^n$, Ann.Acad.Sci.Fenn.Ser.A.I. 8 (1983), 257–
         324.

[BS]   Brézis, H., Stampacchia, G., Sur la régularité de la solution d'inequations elliptiques, Bull.Soc.Math.France 96 (1968), 153–180.

[CR]   Caffarelli, L.A., Riviere, N.M., On the Lipschitz character of the stress tensor when twisting an elastic–plastic bar, Arch.Rational Mech.Anal. 69 (1979), 31–36.

[C]   Campanato, S., Proprieta di Hölderianita di alcune classi di funzioni, Ann. Scuola Norm.Sup. Pisa 17 (1963), 175–188.

[CLT]   Carriero, M., Leaci, A., Tomarelli, F., Strong solutions for an elastic plastic plate, Calc. of Var. 2 (1994), 219–240.

[D]   Demengel, F., Compactness theorems for spaces of functions with bounded derivatives and applications to limit analysis problems in plasticity, Arch.Rational Mech.Anal. 105 (1989), 123–161.

[DT]   Demengel, F., Temam, R., Convex functions of a measure and applications, Indiana Univ.Math.J. 33 (1984), 673–709.

[DL]   Duvaut, G., Lions, J.L., Inequalities in Mechanics and Physics, Springer Grundlehren 219, Springer Verlag, Berlin, 1976.

[DS]   Dunford, N., Schwartz, J.T., Linear Operators, Vol.1, General Theory, Wiley, New York, 1957.

[ET]   Ekeland, I., Temam, R., Convex Analysis and Variational Problems, North–Holland, Amsterdam, 1976.

[EG]   Evans, L.C., Gariepy, R.F., Blow–up, compactness and partial regularity in the calculus of variations, Ind. Univ. Math. J. 36 No.2, (1987), 361–371.

[EK]   Evans, L.C., Knerr, B., Elastic–plastic plane stress problems, Appl.Math.Optim. 5 (1979), 331–348.

[E]   Eyring, H.J., Viscosity, plasticity, and diffusion as examples of absolute reaction rates, J. Chemical Physics 4 (1936), 283–291.

[Fe]   Federer, H., Geometric measure theory, Grundlehren der math. Wiss. in Einzeldarstellungen 153, Springer–Verlag, New York, 1969.

[Fi]   Fichera, G., Existence theorems in elasticity, and unilateral constraints in elasticity, Handbuch der Physik VI a, 347–424, Springer–Verlag, Berlin, 1972.

[Fre]   Frehse, J., Twodimensional variational problems with thin obstacles, Math.Z. 143 (1975), 279–288.

[FreSe] Frehse, J., Seregin, G., Regularity for solutions of variational problems
         in the deformation theory of plasticity with logarithmic hardening, Proc.
         St.Petersburg Math.Soc. 5 (1998), 184–222 (in Russian). English trans-
         lation: Transl.Amer.Math.Soc, II, 193 (1999), 127–152.

[F]   Friedman, A., Variational principles and free boundary problems, New
         York, Wiley–Interscience, 1972.

[Fri]   Friedrichs, K.O., On the boundary value problems of the theory of elas-
         ticity and Korn's inequality, Ann. Math. 48 (1947), 441–471.

[Fu1]   Fuchs, M., Regularity for a class of variational integrals motivated by
         nonlinear elasticity, Asymp. Analysis 9 (1994), 23–38.

[Fu2]   Fuchs, M., On stationary incompressible Norton fluids and some exten-
         sions of Korn's inequality, J. Analysis Appl. Z.A.A. 13 (1994), 191–197.

[Fu3]   Fuchs, M., On quasi–static non–Newtonian fluids with power law, Math.
         Meth. Appl. Sciences 19 (1996), 1225–1231.

[Fu4]   Fuchs, M., Quasi–static non–Newtonian fluids, Vorlesungsreihe No. 38,
         SFB 256, Universität Bonn.

[Fu5]   Fuchs, M., On a class of variational problems related to plasticity with
         polynomial hardening, Applicable Analysis 60 (1996), 269–279.

[Fu6]   Fuchs, M., Variational methods for quasi–static non–Newtonian fluids,
         Zap. Nauchn. Sem. St.Petersburg Odtel Mat. Inst. Steklov (POMI) 233
         (1995), 55–62.

[FGR]   Fuchs, M., Grotowski, J.F., Reuling, J., On variational models for qua-
         sistatic Bingham fluids, Math. Meth. Appl. Sciences 19 (1996), 991–
         1015.

[FL]   Fuchs, M., Li, G., Global gradient bounds for relaxed variational prob-
         lems, Manus. Math. 92 (1997), 287–302.

[FM]   Fuchs, M., Mingione, G., Full $C^{1,\alpha}$–regularity for free and constrained
         local minimizers of elliptic variational integrals with nearly linear
         growth, Manus.Math. 102, 2 (2000), 227–250.

[FO]   Fuchs, M., Osmolovski, V.; Variational integrals on Orlicz–Sobolev
         spaces, J. Anal. Appl. Z.A.A. 17,2 (1998), 393–415.

[FR]   Fuchs, M., Reuling, J., Partial regularity for certain classes of polycon-
         vex functionals related to non–linear elasticity, Manus.Math. 87 (1995),
         13–26.

[FS1]   Fuchs, M., Seregin, G., Some remarks on non–Newtonian fluids includ-
         ing nonconvex perturbations of the Bingham and Powell–Eyring model
         for viscoplastic fluids, M^3AS 7 (1997), 405–433.

[FS2]   Fuchs, M., Seregin, G., Regularity results for the quasi–static Bingham variational inequality in dimensions two and three, Math.Z. 227 (1998), 525–541.

[FS3]   Fuchs, M., Seregin, G., A regularity theory for variational integrals with $L \ln L$–growth, Calc. of Var. 6 (1998) 2, 171–187.

[FS4]   Fuchs, M., Seregin, G., Variational methods for fluids of Prandtl–Eyring type and plastic materials with logarithmic hardening, Math.Meth.Appl.Sciences 22 (1999), 317–351.

[Ga1]   Galdi, G., An introduction to the mathematical theory of the Navier–Stokes equations, Vol. I, Springer Tracts in Natural Philosophy Vol. 38, Springer Verlag, New York, 1994.

[Ga2]   Galdi, G., An introduction to the mathematical theory of the Navier–Stokes equations, Vol. II, Springer Tracts in Natural Philosophy Vol.. 39, Springer Verlag, New York, 1994.

[Gi]   Giaquinta, M., Multiple integrals in the calculus of variations and non-linear elliptic systems, Ann.Math.Studies, No.105, Princeton University Press, Princeton – N.Y., 1983.

[GM]   Giaquinta, M., Modica, G., Nonlinear systems of the type of the stationary Navier–Stokes system, J.Reine Angew.Math.330 (1982), 173–214.

[HK]   Hardt, R., Kinderlehrer, D., Elastic–Plastic deformation, Appl.Math. Optim. 10 (1983), 203–246.

[IS]   Ionescu, I.R., Sofonea, M., Functional and numerical methods in viscoplasticity, Oxford University Press, Oxford, 1993.

[I]   Iwaniec, T., L^p–theory of quasiregular mappings, Quasiconformal space mappings. A collection of surveys 1960–1990, Springer Lecture Notes in Math. Vol. 1508, Springer Verlag 1992.

[K]   Kachanov, L.M., Foundations of the theory of plasticity, North–Holland Publishing Company, Amsterdam–London, 1971.

[Ka]   Kato, Y., Variational inequalities of the Bingham type in three dimensions, Nagoya Math.J. 129 (1993), 53–95.

[Ki]   Kim, J.U., On the initial–boundary value problem for a Bingham fluid in a three–dimensional domain, Trans. A.M.S. 304 (1987), 751–770.

[Kl]   Klyushnikov, V.D., The mathematical theory of plasticity, Izd.Moskov. Gos.Univ., Moscov, 1979.

[KT]   Kohn, R., Temam, R., Dual spaces of stresses and strains with applications to Hencky plasticity, Appl.Math.Optim. 10 (1983), 1–35.

[Ko1] Korn, A., Die Eigenschwingungen eines elastischen Körpers mit ruhender Oberflche, Akad.Wiss. München, Math.–Phys. Kl., Ber. 36 (1906), 351–401.

[Ko2] Korn, A., Über einige Ungleichungen, welche in der Theorie der elastischen und elektrischen Schwingungen eine Rolle spielen, Bull. ist. Cracovie Akad. umiejet, Classe sci.math.nat. (1909), 705–724.

[L1] Ladyzhenskaya, O.A., The mathematical theory of viscous incompressible flow, Gordon and Breach, 1969.

[L2] Ladyzhenskaya, O.A., On nonlinear problems of continuum mechanics, Proc.Internat.Congr.Math. (Moscow 1966), "Nauka", Moscow l968, 560–573; English transl. in Amer.Math.Soc.Transl. (2) 70 (1968).

[L3] Ladyzhenskaya, O.A., New equations for the description of motion of viscous incompressible fluids and global solvability of boundary value problems for them, Trudy Mat.Inst.Steklov 102 (1967), 85–104; English transl.Proc.Steklov Inst.Math. 102 (1967).

[L4] Ladyzhenskaya, O.A., On some modifications of the Navier–Stokes equations for large gradients of velocity, Zap.Nauchn.Sem. Leningrad Odtel.Mat.Inst.Steklov (LOMI) 7 (1968), 126–154; English transl. in Sem.Math.Inst.Leningrad 7 (1968).

[LS] Ladyzhenskaya, O.A., Solonnikov, V.A., Some problems of vector analysis, and generalized formulations of boundary value problems for the Navier–Stokes equations, Zap.Nauchn.Sem. Leningrad Odtel. Mat.Inst.Steklov (LOMI) 59, 81–116 (1976); Engl.transl. in J.Soviet Math. 10 No.2 (1978).

[LU] Ladyzhenskaya, O.A., Ural'tseva, N.N. Linear and Quasilinear Elliptic Equations. Nauka, Moskow, 1964. English translation: Academic Press, New York 1968. Second Russian edition: Nauka, Moscow, 1973.

[MNRR] Málek, J., Necǎs, J., Rokyta, M., Růžička, M., Weak and Measure-valued Solutions to Evolution Partial Differential Equations. Applied Mathematic and Mathematical Computation vol. 13, Chapman and Hall, 1996.

[MSC] Mathies, H., Strang, G., Christiansen, E., The saddle point of a differential program, Energy Methods in Finite Element Analysis, volume dedicated to Professor Fraejs de Veubeke / Eds. R. Glowinski, E. Rodin, O.C. Zienkiewicz. New York: Wiley, 1978.

[MS] Mingione, G., Siepe, F., Full $C^{1,\alpha}$ regularity for minimizers of integral functionals with $L \log L$ growth, Z.Anal.Anw. 18 (1999), 1083–1100.

[MM1]   Mosolov, P.P., Mjasnikov, V.P., On well–posedness of boundary value problems in the mechanics of continuous media, Mat. Sbornik 88 (130) (1972), 256–284, Engl.translation in Math. USSR Sbornik 17, no. 2 (1972), 257–268.

[MM2]   Mosolov, P.P., Mjasnikov,V.P., Mechanics of rigid plastic media, „Nauka", Moscow, 1981 (Russian).

[NH]   Necăs, J., Hlaváček, I., Mathematical theory of elastic and elasto–plastic bodies: an introduction. Elsevier Publishing Company, Amsterdam–Oxford–New York, 1981.

[N]   Norton, M., The creep of steel at high temperature, McGraw Hill, New York, 1929.

[P]   Pileskas, K.I., On spaces of solenoidal vectors, Zap.Nauch.Sem. Leningrad Otdel.Mat.Inst.Steklov (LOMI) 96 (1980), 237–239. English trans. in J.Soviet Math. 21 (1983), no.5.

[Pr]   Prager, W., Introduction to the mechanics of continua, Dover Publications, New York 1973.

[PE]   Powell, R.E., Eyring, H., Mechanism for relaxation theory of viscosity, Nature 154 (1944), 427–428.

[RS]   Repin, S.I., Seregin, G.A., Existence of a weak solution of the minimax problem arising in Coulomb–Mohr plasticity, Amer.Math. Soc.Transl.(2) Vol.164 (1995), 189–220.

[R]   Reuling, J., thesis Saarbrücken 1997.

[Se1]   Seregin, G.A., Variational–difference scheme for problems in the mechanics of ideally elastoplastic media, Zh.Vychisl.Mat. i Mat.Fiz. 25 (1985), 237–352 (in Russian). English translation: U.S.S.R. Comput.Math. and Math.Phys. 25 (1985), 153–165.

[Se2]   Seregin, G.A., Differential properties of weak solutions of nonlinear elliptic systems arising in plasticity theory, Mat.Sb. (N.S.) 130 (3), 172 (7) (1986), 291–309 (in Russian). English translation: Math. USSR-Sb. 58 (1987), 289–309.

[Se3]   Seregin, G.A., Differentiability of local extremals of variational problems in the mechanics of perfect elastoplastic media, Differentsial'nye Uravneniya 23 (11) (1987), 1981–1991; English transl. Differential Equations 23 (1987), 1349–1358.

[Se4]   Seregin, G.A., On differential properties of extremals of variational problems arising in plasticity theory, Differentsial'nye Uravneniya 26 (1990), 1033–1043 (in Russian). English translation: Differential Equations 26 (1990).

[Se5]   Seregin, G.A., On regularity of weak solutions of variational problems
         in plasticity theory, Dokl.Acad.Sci. 314 (1990), 1344–1349 (in Russian).
         English translation: Soviet Math.Dokl. 42 (1991).

[Se6]   Seregin, G.A., On the regularity of weak solutions of variational prob-
         lems in plasticity theory, Algebra i Analiz 2 (1990), 121–140 (in Rus-
         sian). English translation: Leningrad Math. J. 2 (1991).

[Se7]   Seregin, G.A., On regularity of minimizers of certain variational prob-
         lems in plasticity theory, Algebra i Analiz 4 (1992), 181–218 (in Rus-
         sian). English translation: St.Petersburg Math.J. 4 (1993), 989–1020.

[Se8]   Seregin, G.A., On differentiability properties of the stress tensor in
         Coulomb–Mohr plasticity, Algebra i Analiz 4 (1992), 234–252 (in Rus-
         sian). English translation: St. Petersburg Math. J. 4 (1993), 1257–1272.

[Se9]   Seregin, G.A., Differentiability properties of weak solutions of cer-
         tain variational problems in the theory of perfect elasticplastic plates,
         Appl.Math.Optim. 28 (1993), 307–335.

[Se10]   Seregin, G.A., Twodimensional variational problems in plasticity the-
         ory, Izv.Russian Academy of Sciences 60 (1996), 175–210 (in Russian).
         English translation in Izvestiya: Mathematics 60, no. 1 (1996), 179–216.

[Se11]   Seregin, G.A., On the differentiability of local extremals of vari-
         ational problems in the mechanics of rigidly viscoplastic media,
         Izv.Vyssh.Uchebn.Zaved Mat.No. 10 (305) (1987), 23–30 (in Russian).
         Engl.translation: Sov.Math. (Iz.VUZ) 31 (1987).

[Se12]   Seregin, G.A., On differential properties of extremals of variational
         problems of the mechanics of viscoplastic media, Proc. Steklov
         Inst. Math. 3 (1991), 147–157.

[Se13]   Seregin, G.A., Continuity for the strain velocity tensor in two–
         dimensional variational problems from the theory of the Bingham fluid,
         Preprint No.402, SFB 256, Universitt Bonn.

[Se14]   Seregin, G.A., A local estimate of maximum of the module of
         the deviator of strain tensor in elastic body with linear hardening,
         Zap.Nauchn.Sem. St.Petersburg Otdel. Mat.Inst. Steklov (POMI) 200
         (1992), 167–176.

[Se15]   Seregin, G.A., Some remarks on variational problems for functionals
         with $L \ln L$–growth, Zapiski Nauchn.Sem. POMI, Petersburg Odtel.
         Steklov Math.Inst. 213 (1994), 164–174.

[Ser]   Serrin, J., Mathematic principles of classical fluid mechanics. In the
         Encyclopedia of Physics, Vol. VIII/1, edited by S. Flügge, Springer
         Verlag, Berlin 1959.

[Sh1]   Shilkin, T.N., Regularity up to the boundary for solutions to some boundary–value problems of the generalized Newtonian fluid theory. Problemy Mat.Analiza 16 (1997), 239–265 (in Russian).

[Sh2]   Shilkin, T.N., On problems of the generalized Newtonian fluids theory with dissipative potential of subquadratic growth. Problemy Mat.Analiza 17 (1998) (in Russian).

[So1]   Sobolev, S.L., Introduction to the Theory of Cubic Formulas, Nauka, Moscow, 1974 (in Russian).

[So2]   Sobolev, S.L., Some Applications of Functional Analysis to Mathematical Physics, Nauka, Moscow, 1988 (in Russian).

[St1]   Stein, E.M., Note on the class $L \log L$, Studia Math. 32 (1969), 305–310.

[St2]   Stein, E., Singular integrals and differentiability properties of functions, Princeton U.P., Princeton, 1970.

[Str]   Strauss, M.J., Variations of Korn's and Sobolev's inequality, Berkeley symp. on P.D.E., AMS Symposia 23 (1971), 207–214.

[ST1]   Strang, G., Temam, R., Duality and relaxations in the theory of plasticity, J.Méchanique 19 (1980), 1–35.

[ST2]   Strang, G., Temam, R., Functions of bounded deformation, Arch. Rational Mech.Anal. 75 (1981), 7–21.

[Su]   Suquet, P., Existence et régularité des solutions des equations de la plasticité parfaite. Thèse de 3e Cycle, Université de Paris–VI, 1978. Also C.R.Acad.Sci.Paris, Ser.D. 286 (1978), 1201–1204.

[T]   Temam, R., Problèmes mathématiques en plasticité. Paris: Gauthier–Villars, 1985.

[U]   Uhlenbeck, K., Regularity for a class of nonlinear elliptic systems, Acta Math. 138 (1977), 219–240.

[Z]   Zeidler, E., Nonlinear functional analysis and its applications, Vol. IV, Springer Verlag, Berlin, 1987.

Index

admissible stress tensor, 8
Arcela's theorem, 236

Banach space, 101, 254
$BD(\Omega; \mathbb{R}^n)$, 116
Bingham fluid, 132, 136
Bingham variational inequality, 193
blow–up equation, 147
blow–up lemma, 144
bounded measure, 33

Caccioppoli inequality, 170
Caccioppoli–type estimate, 57, 65, 216
Campanato–type estimate, 138, 163, 204, 229, 245
Cauchy stress tensor, 216
coercivity, 18, 44
conjugate function, 9, 16, 42, 248
constitutive equations, 7
constitutive relations, 248
convexity, 34, 161

$D^r_{L \ln L}(\Omega), \overset{\circ}{D}^r_{L \ln L}(\Omega)$, 251
$D^{p,q}(\Omega)$, 28, 111
$D^{p,q}_0(\Omega), \overline{D}^{p,q}_0(\Omega)$, 112
deformation relations, 7
deformation theory, 5, 40, 42, 100
difference quotient, 259
Dirichlet boundary condition, 134
Dirichlet–growth theorem, 190
dissipative potential, 132, 135
dom, 18
duality, 15

Egorov's theorem, 149
elastic domain, 104, 105
elastic zone, 45, 51

elasto–plastic boundary, 105
equilibrium equations for the stresses, 7, 43
Euler–Lagrange equation, 147
$Exp(\Omega)$, 212

Fatou's lemma, 83
finite difference method, 57
function of bounded deformation, 33

generalized Newtonian fluid, 131–133

Hölder continuity, 133, 208, 257
Hölder continuous, 97
Hölder's inequality, 140, 256
Haar–Karman principle, 9
Hahn–Banach theorem, 24
Hausdorff φ–measure, 98
Hausdorff dimension, 98
Hausdorff measure, 180
Hencky–Il'yushin plasticity, 7, 14, 27, 37, 51
Hilbert space, 254
hole–filling trick, 98

imbedding theorem, 111
imbedding, compact, 124
imbedding, continuous, 124
int dom, 18

Jensen's inequality, 55, 91, 126

kinematically admissible displacement, 8
Korn type inequality, 133
Korn's inequality, 75, 137, 139, 141, 163
Korn's inequality, L^p, 111

Lln $L(\Omega)$, 211
Lagrange's formula, 89
Lagrangian, 8, 9, 15, 27
Lebesgue measure, 56, 98
Lebesgue point, 143, 161, 256
Lebesgue space, 8, 255
Lebesgue's theorem, 83
Legendre transformation, 37, 49, 104
linear elasticity, 51, 104
linear hardening, 101
logarithmic hardening, 207, 209, 237
lower semicontinuity, 161

maximal function, 211
mean oscillation, 257
minimax problem, 8, 10, 23, 27, 28, 33
minimizing sequence, 24
Morrey space, 143, 198

N–function, 248, 249
Navier–Stokes system, 135
Newton fluid, 136
non reflexive space, 15
nonlinear system of Stokes type, 204
normed dual, 254
Norton fluid, 136

Orlicz–Sobolev space, 248

p–Bingham variational inequality, 204
partial regularity, 50
perfect elastoplasticity, 5
perturbed functional, 16
plastic deformation, 105
plastic domain, 104, 105
plastic zone, 45
plasticity with hardening, 14
plasticity with power hardening, 100
Poincaré's inequality, 259
Powell–Eyring model, 136
Prandtl–Eyring fluid, 136, 208, 211
Prandtl–Eyring model, 136
precompactness, 124

quasi–static case, 132
quasi–static fluid, 131

Radon measure, 55, 116
reflexive space, 18, 102, 255
relaxation, 15, 33
relaxed Lagrangian, 33
relaxed minimax problem, 44
rigid body, 133
rigid zone, 192

saddle point, 10, 24
safe load condition, 10, 30, 35, 44
smoothing kernel, 108
Sobolev space, 8, 257
Sobolev's imbedding theorem, 258
Sobolev's inequality, 74
Sobolev–Poincaré inequality, 259
solenoidal vector field, 115
solenoidal vector–valued function, 110
star–shaped domain, 117
stationary case, 132
steady state motion, 132
Stokes system, 134, 138
stress deviator, 132
subdifferential, 132
superlinear growth, 101
Sutterby fluid model, 249

Taylor's formula, 81
trace, 258

upper semicontinuity, 54

$V(\Omega)$, $\overset{\circ}{V}(\Omega)$, 213
$V_0(\Omega)$, $\overset{\circ}{V}_0(\Omega)$, 213
$V_0^{p,q}(\Omega)$, 112
viscoplastic fluid, 132
von Mises yield condition, 7

weak convergence, 254
weak limit, 254
weak lower semicontinuity, 254
weak solution, 5, 27
weak–$*$ topology, 54

$X(\Omega)$, $X_{\mathrm{loc}}(\Omega)$, $X_0(\Omega)$, 237

yield surface, 7
Young's inequality, 63, 140, 256

Printing: Weihert-Druck GmbH, Darmstadt
Binding: Buchbinderei Schäffer, Grünstadt

Lecture Notes in Mathematics

For information about Vols. 1–1560
please contact your bookseller or Springer-Verlag

Vol. 1561: I. S. Molchanov, Limit Theorems for Unions of Random Closed Sets. X, 157 pages. 1993.

Vol. 1562: G. Harder, Eisensteinkohomologie und die Konstruktion gemischter Motive. XX, 184 pages. 1993.

Vol. 1563: E. Fabes, M. Fukushima, L. Gross, C. Kenig, M. Röckner, D. W. Stroock, Dirichlet Forms. Varenna, 1992. Editors: G. Dell'Antonio, U. Mosco. VII, 245 pages. 1993.

Vol. 1564: J. Jorgenson, S. Lang, Basic Analysis of Regularized Series and Products. IX, 122 pages. 1993.

Vol. 1565: L. Boutet de Monvel, C. De Concini, C. Procesi, P. Schapira, M. Vergne. D-modules, Representation Theory, and Quantum Groups. Venezia, 1992. Editors: G. Zampieri, A. D'Agnolo. VII, 217 pages. 1993.

Vol. 1566: B. Edixhoven, J.-H. Evertse (Eds.), Diophantine Approximation and Abelian Varieties. XIII, 127 pages. 1993.

Vol. 1567: R. L. Dobrushin, S. Kusuoka, Statistical Mechanics and Fractals. VII, 98 pages. 1993.

Vol. 1568: F. Weisz, Martingale Hardy Spaces and their Application in Fourier Analysis. VIII, 217 pages. 1994.

Vol. 1569: V. Totik, Weighted Approximation with Varying Weight. VI, 117 pages. 1994.

Vol. 1570: R. deLaubenfels, Existence Families, Functional Calculi and Evolution Equations. XV, 234 pages. 1994.

Vol. 1571: S. Yu. Pilyugin, The Space of Dynamical Systems with the C^0-Topology. X, 188 pages. 1994.

Vol. 1572: L. Göttsche, Hilbert Schemes of Zero-Dimensional Subschemes of Smooth Varieties. IX, 196 pages. 1994.

Vol. 1573: V. P. Havin, N. K. Nikolski (Eds.), Linear and Complex Analysis – Problem Book 3 – Part I. XXII, 489 pages. 1994.

Vol. 1574: V. P. Havin, N. K. Nikolski (Eds.), Linear and Complex Analysis – Problem Book 3 – Part II. XXII, 507 pages. 1994.

Vol. 1575: M. Mitrea, Clifford Wavelets, Singular Integrals, and Hardy Spaces. XI, 116 pages. 1994.

Vol. 1576: K. Kitahara, Spaces of Approximating Functions with Haar-Like Conditions. X, 110 pages. 1994.

Vol. 1577: N. Obata, White Noise Calculus and Fock Space. X, 183 pages. 1994.

Vol. 1578: J. Bernstein, V. Lunts, Equivariant Sheaves and Functors. V, 139 pages. 1994.

Vol. 1579: N. Kazamaki, Continuous Exponential Martingales and *BMO*. VII, 91 pages. 1994.

Vol. 1580: M. Milman, Extrapolation and Optimal Decompositions with Applications to Analysis. XI, 161 pages. 1994.

Vol. 1581: D. Bakry, R. D. Gill, S. A. Molchanov, Lectures on Probability Theory. Editor: P. Bernard. VIII, 420 pages. 1994.

Vol. 1582: W. Balser, From Divergent Power Series to Analytic Functions. X, 108 pages. 1994.

Vol. 1583: J. Azéma, P. A. Meyer, M. Yor (Eds.), Séminaire de Probabilités XXVIII. VI, 334 pages. 1994.

Vol. 1584: M. Brokate, N. Kenmochi, I. Müller, J. F. Rodriguez, C. Verdi, Phase Transitions and Hysteresis. Montecatini Terme, 1993. Editor: A. Visintin. VII. 291 pages. 1994.

Vol. 1585: G. Frey (Ed.), On Artin's Conjecture for Odd 2-dimensional Representations. VIII, 148 pages. 1994.

Vol. 1586: R. Nillsen, Difference Spaces and Invariant Linear Forms. XII, 186 pages. 1994.

Vol. 1587: N. Xi, Representations of Affine Hecke Algebras. VIII, 137 pages. 1994.

Vol. 1588: C. Scheiderer, Real and Étale Cohomology. XXIV, 273 pages. 1994.

Vol. 1589: J. Bellissard, M. Degli Esposti, G. Forni, S. Graffi, S. Isola, J. N. Mather, Transition to Chaos in Classical and Quantum Mechanics. Montecatini Terme, 1991. Editor: 2S. Graffi. VII, 192 pages. 1994.

Vol. 1590: P. M. Soardi, Potential Theory on Infinite Networks. VIII, 187 pages. 1994.

Vol. 1591: M. Abate, G. Patrizio, Finsler Metrics – A Global Approach. IX, 180 pages. 1994.

Vol. 1592: K. W. Breitung, Asymptotic Approximations for Probability Integrals. IX, 146 pages. 1994.

Vol. 1593: J. Jorgenson & S. Lang, D. Goldfeld, Explicit Formulas for Regularized Products and Series. VIII, 154 pages. 1994.

Vol. 1594: M. Green, J. Murre, C. Voisin, Algebraic Cycles and Hodge Theory. Torino, 1993. Editors: A. Albano, F. Bardelli. VII, 275 pages. 1994.

Vol. 1595: R.D.M. Accola, Topics in the Theory of Riemann Surfaces. IX, 105 pages. 1994.

Vol. 1596: L. Heindorf, L. B. Shapiro, Nearly Projective Boolean Algebras. X, 202 pages. 1994.

Vol. 1597: B. Herzog, Kodaira-Spencer Maps in Local Algebra. XVII, 176 pages. 1994.

Vol. 1598: J. Berndt, F. Tricerri, L. Vanhecke, Generalized Heisenberg Groups and Damek-Ricci Harmonic Spaces. VIII, 125 pages. 1995.

Vol. 1599: K. Johannson, Topology and Combinatorics of 3-Manifolds. XVIII, 446 pages. 1995.

Vol. 1600: W. Narkiewicz, Polynomial Mappings. VII, 130 pages. 1995.

Vol. 1601: A. Pott, Finite Geometry and Character Theory. VII, 181 pages. 1995.

Vol. 1602: J. Winkelmann, The Classification of Three-dimensional Homogeneous Complex Manifolds. XI, 230 pages. 1995.

Vol. 1603: V. Ene, Real Functions – Current Topics. XIII, 310 pages. 1995.

Vol. 1604: A. Huber, Mixed Motives and their Realization in Derived Categories. XV, 207 pages. 1995.

Vol. 1605: L. B. Wahlbin, Superconvergence in Galerkin Finite Element Methods. XI, 166 pages. 1995.

Vol. 1606: P.-D. Liu, M. Qian, Smooth Ergodic Theory of Random Dynamical Systems. XI, 221 pages. 1995.

Vol. 1607: G. Schwarz, Hodge Decomposition – A Method for Solving Boundary Value Problems. VII, 155 pages. 1995.

Vol. 1608: P. Biane, R. Durrett, Lectures on Probability Theory. Editor: P. Bernard. VII, 210 pages. 1995.

Vol. 1609: L. Arnold, C. Jones, K. Mischaikow, G. Raugel, Dynamical Systems. Montecatini Terme, 1994. Editor: R. Johnson. VIII, 329 pages. 1995.

Vol. 1610: A. S. Üstünel, An Introduction to Analysis on Wiener Space. X, 95 pages. 1995.

Vol. 1611: N. Knarr, Translation Planes. VI, 112 pages. 1995.

Vol. 1612: W. Kühnel, Tight Polyhedral Submanifolds and Tight Triangulations. VII, 122 pages. 1995.

Vol. 1613: J. Azéma, M. Emery, P. A. Meyer, M. Yor (Eds.), Séminaire de Probabilités XXIX. VI, 326 pages. 1995.

Vol. 1614: A. Koshelev, Regularity Problem for Quasilinear Elliptic and Parabolic Systems. XXI, 255 pages. 1995.

Vol. 1615: D. B. Massey, Le Cycles and Hypersurface Singularities. XI, 131 pages. 1995.

Vol. 1616: I. Moerdijk, Classifying Spaces and Classifying Topoi. VII, 94 pages. 1995.

Vol. 1617: V. Yurinsky, Sums and Gaussian Vectors. XI, 305 pages. 1995.

Vol. 1618: G. Pisier, Similarity Problems and Completely Bounded Maps. VII, 156 pages. 1996.

Vol. 1619: E. Landvogt, A Compactification of the Bruhat-Tits Building. VII, 152 pages. 1996.

Vol. 1620: R. Donagi, B. Dubrovin, E. Frenkel, E. Previato, Integrable Systems and Quantum Groups. Montecatini Terme, 1993. Editors:M. Francaviglia, S. Greco. VIII, 488 pages. 1996.

Vol. 1621: H. Bass, M. V. Otero-Espinar, D. N. Rockmore, C. P. L. Tresser, Cyclic Renormalization and Auto-morphism Groups of Rooted Trees. XXI, 136 pages. 1996.

Vol. 1622: E. D. Farjoun, Cellular Spaces, Null Spaces and Homotopy Localization. XIV, 199 pages. 1996.

Vol. 1623: H.P. Yap, Total Colourings of Graphs. VIII, 131 pages. 1996.

Vol. 1624: V. Brinzanescu, Holomorphic Vector Bundles over Compact Complex Surfaces. X, 170 pages. 1996.

Vol.1625: S. Lang, Topics in Cohomology of Groups. VII, 226 pages. 1996.

Vol. 1626: J. Azéma, M. Emery, M. Yor (Eds.), Séminaire de Probabilités XXX. VIII, 382 pages. 1996.

Vol. 1627: C. Graham, Th. G. Kurtz, S. Méléard, Ph. E. Protter, M. Pulvirenti, D. Talay, Probabilistic Models for Nonlinear Partial Differential Equations. Montecatini Terme, 1995. Editors: D. Talay, L. Tubaro. X, 301 pages. 1996.

Vol. 1628: P.-H. Zieschang, An Algebraic Approach to Association Schemes. XII, 189 pages. 1996.

Vol. 1629: J. D. Moore, Lectures on Seiberg-Witten In-variants. VII, 105 pages. 1996.

Vol. 1630: D. Neuenschwander, Probabilities on the Heisenberg Group: Limit Theorems and Brownian Motion. VIII, 139 pages. 1996.

Vol. 1631: K. Nishioka, Mahler Functions and Trans-cendence.VIII, 185 pages.1996.

Vol. 1632: A. Kushkuley, Z. Balanov, Geometric Methods in Degree Theory for Equivariant Maps. VII, 136 pages. 1996.

Vol.1633: H. Aikawa, M. Essén, Potential Theory – Selected Topics. IX, 200 pages.1996.

Vol. 1634: J. Xu, Flat Covers of Modules. IX, 161 pages. 1996.

Vol. 1635: E. Hebey, Sobolev Spaces on Riemannian Manifolds. X, 116 pages. 1996.

Vol. 1636: M. A. Marshall, Spaces of Orderings and Ab-stract Real Spectra. VI, 190 pages. 1996.

Vol. 1637: B. Hunt, The Geometry of some special Arithmetic Quotients. XIII, 332 pages. 1996.

Vol. 1638: P. Vanhaecke, Integrable Systems in the realm of Algebraic Geometry. VIII, 218 pages. 1996.

Vol. 1639: K. Dekimpe, Almost-Bieberbach Groups: Affine and Polynomial Structures. X, 259 pages. 1996.

Vol. 1640: G. Boillat, C. M. Dafermos, P. D. Lax, T. P. Liu, Recent Mathematical Methods in Nonlinear Wave Propagation. Montecatini Terme, 1994. Editor: T. Ruggeri. VII, 142 pages. 1996.

Vol. 1641: P. Abramenko, Twin Buildings and Applications to S-Arithmetic Groups. IX, 123 pages. 1996.

Vol. 1642: M. Puschnigg, Asymptotic Cyclic Cohomology. XXII, 138 pages. 1996.

Vol. 1643: J. Richter-Gebert, Realization Spaces of Polytopes. XI, 187 pages. 1996.

Vol. 1644: A. Adler, S. Ramanan, Moduli of Abelian Varieties. VI, 196 pages. 1996.

Vol. 1645: H. W. Broer, G. B. Huitema, M. B. Sevryuk, Quasi-Periodic Motions in Families of Dynamical Systems. XI, 195 pages. 1996.

Vol. 1646: J.-P. Demailly, T. Peternell, G. Tian, A. N. Tyurin, Transcendental Methods in Algebraic Geometry. Cetraro, 1994. Editors: F. Catanese, C. Ciliberto. VII, 257 pages. 1996.

Vol. 1647: D. Dias, P. Le Barz, Configuration Spaces over Hilbert Schemes and Applications. VII. 143 pages. 1996.

Vol. 1648: R. Dobrushin, P. Groeneboom, M. Ledoux, Lectures on Probability Theory and Statistics. Editor: P. Bernard. VIII, 300 pages. 1996.

Vol. 1649: S. Kumar, G. Laumon, U. Stuhler, Vector Bundles on Curves – New Directions. Cetraro, 1995. Editor: M. S. Narasimhan. VII, 193 pages. 1997.

Vol. 1650: J. Wildeshaus, Realizations of Polylogarithms. XI, 343 pages. 1997.

Vol. 1651: M. Drmota, R. F. Tichy, Sequences, Discrepancies and Applications. XIII, 503 pages. 1997.

Vol. 1652: S. Todorcevic, Topics in Topology. VIII, 153 pages. 1997.

Vol. 1653: R. Benedetti, C. Petronio, Branched Standard Spines of 3-manifolds. VIII, 132 pages. 1997.

Vol. 1654: R. W. Ghrist, P. J. Holmes, M. C. Sullivan, Knots and Links in Three-Dimensional Flows. X, 208 pages. 1997.

Vol. 1655: J. Azéma, M. Emery, M. Yor (Eds.), Séminaire de Probabilités XXXI. VIII, 329 pages. 1997.

Vol. 1656: B. Biais, T. Björk, J. Cvitanic, N. El Karoui, E. Jouini, J. C. Rochet, Financial Mathematics. Bressanone, 1996. Editor: W. J. Runggaldier. VII, 316 pages. 1997.

Vol. 1657: H. Reimann, The semi-simple zeta function of quaternionic Shimura varieties. IX, 143 pages. 1997.

Vol. 1658: A. Pumarino, J. A. Rodríguez, Coexistence and Persistence of Strange Attractors. VIII, 195 pages. 1997.

Vol. 1659: V, Kozlov, V. Maz'ya, Theory of a Higher-Order Sturm-Liouville Equation. XI, 140 pages. 1997.

Vol. 1660: M. Bardi, M. G. Crandall, L. C. Evans, H. M. Soner, P. E. Souganidis, Viscosity Solutions and Applications. Montecatini Terme, 1995. Editors: I. Capuzzo Dolcetta, P. L. Lions. IX, 259 pages. 1997.

Vol. 1661: A. Tralle, J. Oprea, Symplectic Manifolds with no Kähler Structure. VIII, 207 pages. 1997.

Vol. 1662: J. W. Rutter, Spaces of Homotopy Self-Equivalences – A Survey. IX, 170 pages. 1997.

Vol. 1663: Y. E. Karpeshina; Perturbation Theory for the Schrödinger Operator with a Periodic Potential. VII, 352 pages. 1997.

Vol. 1664: M. Väth, Ideal Spaces. V, 146 pages. 1997.

Vol. 1665: E. Giné, G. R. Grimmett, L. Saloff-Coste, Lectures on Probability Theory and Statistics 1996. Editor: P. Bernard. X, 424 pages, 1997.

Vol. 1666: M. van der Put, M. F. Singer, Galois Theory of Difference Equations. VII, 179 pages. 1997.

Vol. 1667: J. M. F. Castillo, M. González, Three-space Problems in Banach Space Theory. XII, 267 pages. 1997.

Vol. 1668: D. B. Dix, Large-Time Behavior of Solutions of Linear Dispersive Equations. XIV, 203 pages. 1997.

Vol. 1669: U. Kaiser, Link Theory in Manifolds. XIV, 167 pages. 1997.

Vol. 1670: J. W. Neuberger, Sobolev Gradients and Differential Equations. VIII, 150 pages. 1997.

Vol. 1671: S. Bouc, Green Functors and G-sets. VII, 342 pages. 1997.

Vol. 1672: S. Mandal, Projective Modules and Complete Intersections. VIII, 114 pages. 1997.

Vol. 1673: F. D. Grosshans, Algebraic Homogeneous Spaces and Invariant Theory. VI, 148 pages. 1997.

Vol. 1674: G. Klaas, C. R. Leedham-Green, W. Plesken, Linear Pro-p-Groups of Finite Width. VIII, 115 pages. 1997.

Vol. 1675: J. E. Yukich, Probability Theory of Classical Euclidean Optimization Problems. X, 152 pages. 1998.

Vol. 1676: P. Cembranos, J. Mendoza, Banach Spaces of Vector-Valued Functions. VIII, 118 pages. 1997.

Vol. 1677: N. Proskurin, Cubic Metaplectic Forms and Theta Functions. VIII, 196 pages. 1998.

Vol. 1678: O. Krupková, The Geometry of Ordinary Variational Equations. X, 251 pages. 1997.

Vol. 1679: K.-G. Grosse-Erdmann, The Blocking Technique. Weighted Mean Operators and Hardy's Inequality. IX, 114 pages. 1998.

Vol. 1680: K.-Z. Li, F. Oort, Moduli of Supersingular Abelian Varieties. V, 116 pages. 1998.

Vol. 1681: G. J. Wirsching, The Dynamical System Generated by the 3n+1 Function. VII, 158 pages. 1998.

Vol. 1682: H.-D. Alber, Materials with Memory. X, 166 pages. 1998.

Vol. 1683: A. Pomp, The Boundary-Domain Integral Method for Elliptic Systems. XVI, 163 pages. 1998.

Vol. 1684: C. A. Berenstein, P. F. Ebenfelt, S. G. Gindikin, S. Helgason, A. E. Tumanov, Integral Geometry, Radon Transforms and Complex Analysis. Firenze, 1996. Editors: E. Casadio Tarabusi, M. A. Picardello, G. Zampieri. VII, 160 pages. 1998

Vol. 1685: S. König, A. Zimmermann, Derived Equivalences for Group Rings. X, 146 pages. 1998.

Vol. 1686: J. Azéma, M. Émery, M. Ledoux, M. Yor (Eds.), Séminaire de Probabilités XXXII. VI, 440 pages. 1998.

Vol. 1687: F. Bornemann, Homogenization in Time of Singularly Perturbed Mechanical Systems. XII, 156 pages. 1998.

Vol. 1688: S. Assing, W. Schmidt, Continuous Strong Markov Processes in Dimension One. XII, 137 page. 1998.

Vol. 1689: W. Fulton, P. Pragacz, Schubert Varieties and Degeneracy Loci. XI, 148 pages. 1998.

Vol. 1690: M. T. Barlow, D. Nualart, Lectures on Probability Theory and Statistics. Editor: P. Bernard. VIII, 237 pages. 1998.

Vol. 1691: R. Bezrukavnikov, M. Finkelberg, V. Schechtman, Factorizable Sheaves and Quantum Groups. X, 282 pages. 1998.

Vol. 1692: T. M. W. Eyre, Quantum Stochastic Calculus and Representations of Lie Superalgebras. IX, 138 pages. 1998.

Vol. 1694: A. Braides, Approximation of Free-Discontinuity Problems. XI, 149 pages. 1998.

Vol. 1695: D. J. Hartfiel, Markov Set-Chains. VIII, 131 pages. 1998.

Vol. 1696: E. Bouscaren (Ed.): Model Theory and Algebraic Geometry. XV, 211 pages. 1998.

Vol. 1697: B. Cockburn, C. Johnson, C.-W. Shu, E. Tadmor, Advanced Numerical Approximation of Nonlinear Hyperbolic Equations. Cetraro, Italy, 1997. Editor: A. Quarteroni. VII, 390 pages. 1998.

Vol. 1698: M. Bhattacharjee, D. Macpherson, R. G. Möller, P. Neumann, Notes on Infinite Permutation Groups. XI, 202 pages. 1998.

Vol. 1699: A. Inoue, Tomita-Takesaki Theory in Algebras of Unbounded Operators. VIII, 241 pages. 1998.

Vol. 1700: W. A. Woyczyński, Burgers-KPZ Turbulence, XI, 318 pages. 1998.

Vol. 1701: Ti-Jun Xiao, J. Liang, The Cauchy Problem of Higher Order Abstract Differential Equations, XII, 302 pages. 1998.

Vol. 1702: J. Ma, J. Yong, Forward-Backward Stochastic Differential Equations and Their Applications. XIII, 270 pages. 1999.

Vol. 1703: R. M. Dudley, R. Norvaiša, Differentiability of Six Operators on Nonsmooth Functions and p-Variation. VIII, 272 pages. 1999.

Vol. 1704: H. Tamanoi, Elliptic Genera and Vertex Operator Super-Algebras. VI, 390 pages. 1999.

Vol. 1705: I. Nikolaev, E. Zhuzhoma, Flows in 2-dimensional Manifolds. XIX, 294 pages. 1999.

Vol. 1706: S. Yu. Pilyugin, Shadowing in Dynamical Systems. XVII, 271 pages. 1999.

Vol. 1707: R. Pytlak, Numerical Methods for Optimal Control Problems with State Constraints. XV, 215 pages. 1999.

Vol. 1708: K. Zuo, Representations of Fundamental Groups of Algebraic Varieties. VII, 139 pages. 1999.

Vol. 1709: J. Azéma, M. Émery, M. Ledoux, M. Yor (Eds), Séminaire de Probabilités XXXIII. VIII, 418 pages. 1999.

Vol. 1710: M. Koecher, The Minnesota Notes on Jordan Algebras and Their Applications. IX, 173 pages. 1999.

Vol. 1711: W. Ricker, Operator Algebras Generated by Commuting Projections: A Vector Measure Approach. XVII, 159 pages. 1999.

Vol. 1712: N. Schwartz, J. J. Madden, Semi-algebraic Function Rings and Reflectors of Partially Ordered Rings. XI, 279 pages. 1999.

Vol. 1713: F. Bethuel, G. Huisken, S. Müller, K. Steffen, Calculus of Variations and Geometric Evolution Problems. Cetraro, 1996. Editors: S. Hildebrandt, M. Struwe. VII, 293 pages. 1999.

Vol. 1714: O. Diekmann, R. Durrett, K. P. Hadeler, P. K. Maini, H. L. Smith, Mathematics Inspired by Biology. Martina Franca, 1997. Editors: V. Capasso, O. Diekmann. VII, 268 pages. 1999.

Vol. 1715: N. V. Krylov, M. Röckner, J. Zabczyk, Stochastic PDE's and Kolmogorov Equations in Infinite Dimensions. Cetraro, 1998. Editor: G. Da Prato. VIII, 239 pages. 1999.

Vol. 1716: J. Coates, R. Greenberg, K. A. Ribet, K. Rubin, Arithmetic Theory of Elliptic Curves. Cetraro, 1997. Editor: C. Viola. VIII, 260 pages. 1999.

Vol. 1717: J. Bertoin, F. Martinelli, Y. Peres, Lectures on Probability Theory and Statistics. Saint-Flour, 1997. Editor: P. Bernard. IX, 291 pages. 1999.

Vol. 1718: A. Eberle, Uniqueness and Non-Uniqueness of Semigroups Generated by Singular Diffusion Operators. VIII, 262 pages. 1999.

Vol. 1719: K. R. Meyer, Periodic Solutions of the N-Body Problem. IX, 144 pages. 1999.

Vol. 1720: D. Elworthy, Y. Le Jan, X-M. Li, On the Geometry of Diffusion Operators and Stochastic Flows. IV, 118 pages. 1999.

Vol. 1721: A. Iarrobino, V. Kanev, Power Sums, Gorenstein Algebras, and Determinantal Loci. XXVII, 345 pages. 1999.

Vol. 1722: R. McCutcheon, Elemental Methods in Ergodic Ramsey Theory. VI, 160 pages. 1999.

Vol. 1723: J. P. Croisille, C. Lebeau, Diffraction by an Immersed Elastic Wedge. VI, 134 pages. 1999.

Vol. 1724: V. N. Kolokoltsov, Semiclassical Analysis for Diffusions and Stochastic Processes. VIII, 347 pages. 2000.

Vol. 1725: D. A. Wolf-Gladrow, Lattice-Gas Cellular Automata and Lattice Boltzmann Models. IX, 308 pages. 2000.

Vol. 1726: V. Marić, Regular Variation and Differential Equations. X, 127 pages. 2000.

Vol. 1727: P. Kravanja, M. Van Barel, Computing the Zeros of Analytic Functions. VII, 111 pages. 2000.

Vol. 1728: K. Gatermann, Computer Algebra Methods for Equivariant Dynamical Systems. XV, 153 pages. 2000.

Vol. 1729: J. Azéma, M. Émery, M. Ledoux, M. Yor, Séminaire de Probabilités XXXIV. VI, 431 pages. 2000.

Vol. 1730: S. Graf, H. Luschgy, Foundations of Quantization for Probability Distributions. X, 230 pages. 2000.

Vol. 1731: T. Hsu, Quilts: Central Extensions, Braid Actions, and Finite Groups,. XII, 185 pages. 2000.

Vol. 1732: K. Keller, Invariant Factors, Julia Equivalences and the (Abstract) Mandelbrot Set. X, 206 pages. 2000.

Vol. 1733: K. Ritter, Average-Case Analysis of NumericalProblems. IX, 254 pages. 2000.

Vol. 1734: M. Espedal, A. Fasano, A. Mikelić, Filtration in Porous Media and Industrial Applications. Cetraro 1998. Editor: A. Fasano. 2000.

Vol. 1735: D. Yafaev, Scattering Theory: Some Old and New Problems. XVI, 169 pages. 2000.

Vol. 1736: B. O. Turesson, Nonlinear Potential Theory and Weighted Sobolev Spaces. XIV, 173 pages. 2000.

Vol. 1737: S. Wakabayashi, Classical Microlocal Analysis in the Space of Hyperfunctions. VIII, 367 pages. 2000.

Vol. 1738: M. Emery, A. Nemirovski, D. Voiculescu, Lectures on Probability Theory and Statistics. XI, 356 pages. 2000.

Vol. 1739: R. Burkard, P. Deuflhard, A. Jameson, J.-L. Lions, G. Strang, Computational Mathematics Driven by Industrial Problems. Martina Franca, 1999. Editors: V. Capasso, H. Engl, J. Periaux. VII, 418 pages. 2000.

Vol. 1740: B. Kawohl, O. Pironneau, L. Tartar, J.-P. Zolesio, Optimal Shape Design. Tróia, Portugal 1999. Editors: A. Cellina, A. Ornelas. IX, 388 pages. 2000.

Vol. 1741: E. Lombardi, Oscillatory Integrals and Phenomena Beyond all Algebraic Orders. XV, 413 pages. 2000.

Vol. 1742: A. Unterberger, Quantization and Non-holomorphic Modular Forms. VIII, 253 pages. 2000.

Vol. 1743: L. Habermann, Riemannian Metrics of Constant Mass and Moduli Spaces of Conformal Structures. XII, 116 pages. 2000.

Vol. 1744: M. Kunze, Non-Smooth Dynamical Systems. X, 228 pages. 2000.

Vol. 1745: V. D. Milman, G. Schechtman, Geometric Aspects of Functional Analysis. VIII, 289 pages. 2000.

Vol. 1746: A. Degtyarev, I. Itenberg, V. Kharlamov, Real Enriques Surfaces. XVI, 259 pages. 2000.

Vol. 1747: L. W. Christensen, Gorenstein Dimensions. VIII, 204 pages. 2000.

Vol. 1748: M. Růžička, Electrorheological Fluids: Modeling and Mathematical Theory. XV, 176 pages. 2001.

Vol. 1749: M. Fuchs, G. Seregin, Variational Methods for Problems from Plasticity Theory and for Generalized Newtonian Fluids. VI, 269 pages. 2001.

Vol. 1750: B. Conrad, Grothendieck Duality and Base Change. X, 296 pages. 2001.

Vol. 1751: N. J. Cutland, Loeb Measures in Practice: Recent Advances. XI, 111 pages. 2001.

4. Lecture Notes are printed by photo-offset from the master-copy delivered in camera-ready form by the authors. Springer-Verlag provides technical instructions for the preparation of manuscripts. Macro packages in T_EX, L^AT_EX2e, $L^AT_EX2.09$ are available from Springer's web-pages at

http://www.springer.de/math/authors/b-tex.html.

Careful preparation of the manuscripts will help keep production time short and ensure satisfactory appearance of the finished book.

The actual production of a Lecture Notes volume takes approximately 12 weeks.

5. Authors receive a total of 50 free copies of their volume, but no royalties. They are entitled to a discount of 33.3% on the price of Springer books purchase for their personal use, if ordering directly from Springer-Verlag.

Commitment to publish is made by letter of intent rather than by signing a formal contract. Springer-Verlag secures the copyright for each volume. Authors are free to reuse material contained in their LNM volumes in later publications: A brief written (or e-mail) request for formal permission is sufficient.

Addresses:

Professor F. Takens, Mathematisch Instituut,
Rijksuniversiteit Groningen, Postbus 800,
9700 AV Groningen, The Netherlands
E-mail: F.Takens@math.rug.nl

Professor B. Teissier
Université Paris 7
UFR de Mathématiques
Equipe Géométrie et Dynamique
Case 7012
2 place Jussieu
75251 Paris Cedex 05
E-mail: Teissier@math.jussieu.fr

Springer-Verlag, Mathematics Editorial, Tiergartenstr. 17,
D-69121 Heidelberg, Germany,
Tel.: *49 (6221) 487-701
Fax: *49 (6221) 487-355
E-mail: lnm@Springer.de